ORE MICROSCOPY
AND ORE PETROGRAPHY

ORE MICROSCOPY AND ORE PETROGRAPHY

JAMES R. CRAIG

Professor of Geology
Virginia Polytechnic Institute and State University
Blacksburg, Virginia

DAVID J. VAUGHAN

Reader in Mineralogy
University of Aston in Birmingham
Birmingham, England

A WILEY-INTERSCIENCE PUBLICATION

JOHN WILEY & SONS

New York • Chichester • Brisbane • Toronto • Singapore

Library of Congress Cataloging in Publication Data:

Craig, James R 1940
Ore microscopy.

"A Wiley-Interscience publication."
Includes bibliographies and indexes.
1. Ores. 2. Thin sections (Geology)
I. Vaughan, David J., 1946- joint author.
II. Title

QE390.C7 549'.12 80-39786
ISBN 0-471-08596-0
Printed in the United States of America
10 9 8 7 6 5 4 3 2

Preface

The study of opaque minerals or synthetic solids in polished section using the polarizing reflected light microscope is the most important technique for the identification and characterization of the opaque phases in a sample and the textural relationships between them. Since most metalliferous ores are comprised of opaque minerals, this study has been traditionally known as *ore microscopy* and has found its greatest applications in the study of mineral deposits. It may be applied, however, as a general technique in the study of igneous, sedimentary, or metamorphic rocks containing opaque minerals and even in the study of metallurgical products or other synthetic materials.

The objective of this book is to present an up-to-date introduction to ore microscopy for the student or professional scientist unfamiliar with the technique, an introduction that would accompany a course at the senior undergraduate or graduate student level or provide the professional with a first step in familiarization. No attempt has been made at a comprehensive treatment of the subject, for which the reader is referred to the excellent works of such authors as Ramdohr, Uytenbogaart and Burke, Galopin and Henry cited in relevant sections of this text. Emphasis is placed on the basic skills required for the study of opaque minerals in polished section, and information in the text and appendices on the more common ore minerals and assemblages provides examples.

The first two chapters cover the design and operation of the ore microscope and the preparation of polished (and polished thin) sections. A chapter dealing with qualitative mineral properties used in identification is followed by a discussion of reflected light optics. The quantitative measurement of reflectance, color, and microhardness are treated in Chapters 5 and 6 along with overall schemes for employing these measurements in identification. Chapters 7 and 8 deal with ore mineral textures and paragenesis and include a brief discussion of the study of fluid inclusions. Chapters 9 and 10 are concise discussions of many of the major ore mineral associations observed under the microscope. These discussions are not intended to be exhaustive, especially as regards comments made on the genesis of the ores, but are designed to aid understanding by placing the ore mineral textures and associations in a broader geological context. The final chapter deals with the applications of ore microscopy in mineral technology. The appendices contain the data necessary to identify approximately 100 of the more common ore minerals, those likely to be encountered by the student in an introductory course and those frequently encountered by the professional scientist. Key references to more detailed accounts are provided throughout the text. The dependence of the properties, textures, and associations of the minerals on their crystal chemistry, thermochemistry, and phase relations is an aspect of modern studies of ore minerals particularly emphasized.

A science such as ore microscopy relies heavily on the experience and knowledge of the teacher. We have learned much from our teachers and particularly wish to acknowledge our indebtedness to Dr. S. H. U. Bowie, Dr. N. F. M. Henry, Professor G. Kullerud, Professor A. P. Millman, Dr. E. H. Nickel, Mr. R. Phillips, Professor E. F. Stumpfl, and Professor E. A. Vincent.

Completion of this text would not have been possible without the help and encouragement of our departments and many friends and colleagues, not all of whom can be mentioned here. Dr. P. B. Barton, Jr., Dr. L. J. Cabri, Dr. R. A. Ixer, Professor S. D. Scott, and Professor F. M. Vokes critically reviewed the complete text and provided numerous suggestions for improvement. Dr. N. F. M. Henry read Chapters 4–6 and made many valuable comments. Dr. E. Roedder's criticisms of portions of Chapter 8 were most beneficial. However, any errors and imperfections that remain in the text are entirely our responsibility.

In the preparation of the typescript we wish to thank Donna Williams and Cathy Kennedy. Thanks are also due to Sharon Chiang and Martin Eiss for drafting the illustrations and to Gordon Love for help with the photomicrographs. We are much indebted to our wives, Lois and Heather, for their support and encouragement throughout the work.

Ore Microscopy and Ore Petrography is dedicated to our children, Nancie Eva and James Matthew Craig and Emlyn James Vaughan.

JAMES R. CRAIG
DAVID J. VAUGHAN

Blacksburg, Virginia
Birmingham, England
February 1981

Contents

ORE MICROSCOPY AND ORE PETROGRAPHY

1

The Ore Microscope

1.1 INTRODUCTION

The ore microscope is the basic instrument for the petrographic examination of the large and economically important group of minerals referred to collectively as "ore" or "opaque" minerals. Although neither term is strictly correct (inasmuch as pyrite is opaque but rarely, if ever, constitutes an economically viable ore and sphalerite and cassiterite are important ore minerals but are not opaque), both terms are frequently used synonymously. The ore microscope is similar to conventional petrographic microscopes in the systems of lenses, polarizer, analyzer, and various diaphragms employed, but differs in that it has an incident light source rather than a substage transmitting light source. This allows the examination of polished surfaces of opaque minerals. The increasing interest in ore-gangue relationships and the recognition that much textural information can be derived from the examination of translucent ore minerals in polished thin sections, now commonly leads to the use of microscopes equipped for both reflected- and transmitted-light study. The discussion below is concerned specifically with the design and use of the standard components of the reflected-light microscope; further details of the transmitted and of the reflected light microscopes are described by Cameron (1961), Bloss (1961), Piller (1977), and Bowie and Simpson (1977).

The variety of commercially available reflected-light microscopes tends to mask the basic similarities between them in terms of the arrangement of light source, lenses, diaphragms, reflector, objectives, and oculars. Some of this variety is evident in Figure 1.1 which shows research and student model microscopes. Each manufacturer incorporates unique design features into the ore microscopes they produce, and it is necessary for the reader to refer to the instruction manual accompanying a particular microscope for the exact placement and employment of the components described in the following and for information regarding other accessories.

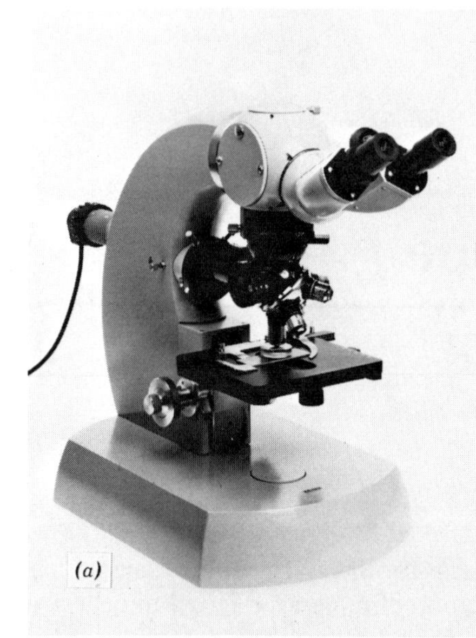

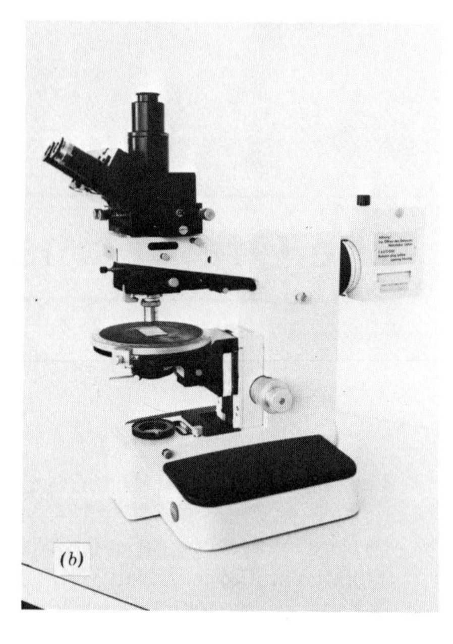

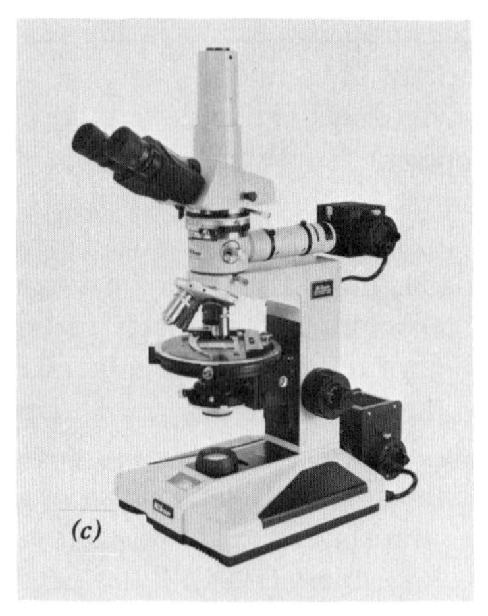

FIGURE 1.1 Representative instruments used in ore microscopy (all photographs courtesy of manufacturers): *(a)* Zeiss Universal Research Microscope; *(b)* Leitz Orthoplan-Pol Microscope; *(c)* Nikon Optiphot-Pol Microscope; *(d)* Vickers M12a Microscope with microhardness testing equipment.

1.2 COMPONENTS OF THE ORE MICROSCOPE

The components of the ore microscope and the light path from the illuminator to the observer's eye are summarized in Figures 1.2*a*–1.2*c*. Conventional orthoscopic examination may be conducted using either a glass plate reflector (Figure 1.2*a*) or with a half-field prism (Figure 1.2*b*); some microscopes are equipped with both a glass plate and a prism; others with only a prism. Initial observations are made in plane-polarized light (better described in this work as linearly

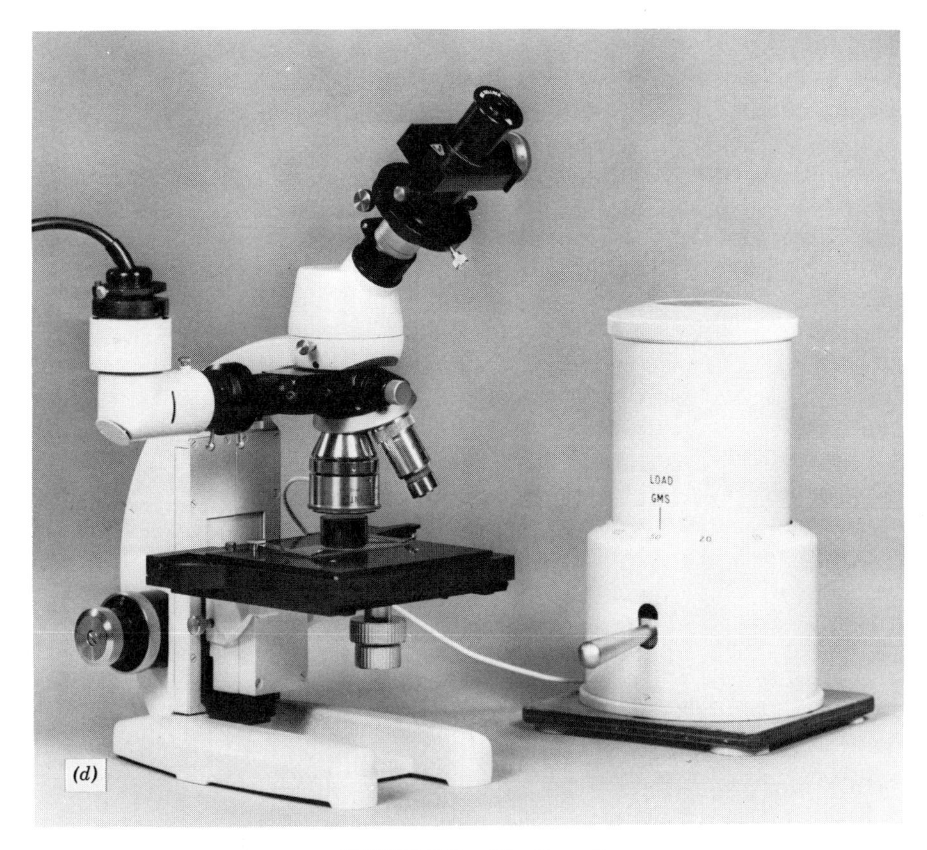

FIGURE 1.1 (*Continued*)

polarized, see Section 4.1) with only the polarizer inserted, and then under crossed polars when the analyzer is also inserted into the light path at right angles to the polarizer. Conoscopic observation may be undertaken, normally by using a glass plate reflector (Figure 1.2*c*) and inserting the analyzer and the Bertrand lens. If the microscope is not equipped with a Bertrand lens, the same effects may be observed by substituting a pinhole eyepiece for the ocular.

1.2.1 Rotatable Stage

The microscope stage, on which the polished sections are placed, should rotate freely, be perpendicular to the axis of light transmission through the microscope, and be centered relative to the objectives. Angular measurements can be made by means of the degree markings at the edge of the stage and the verniers provided. Most microscopes accept a mechanical stage equipped with X and Y movement for systematic examination or point counting of grains in specimens.

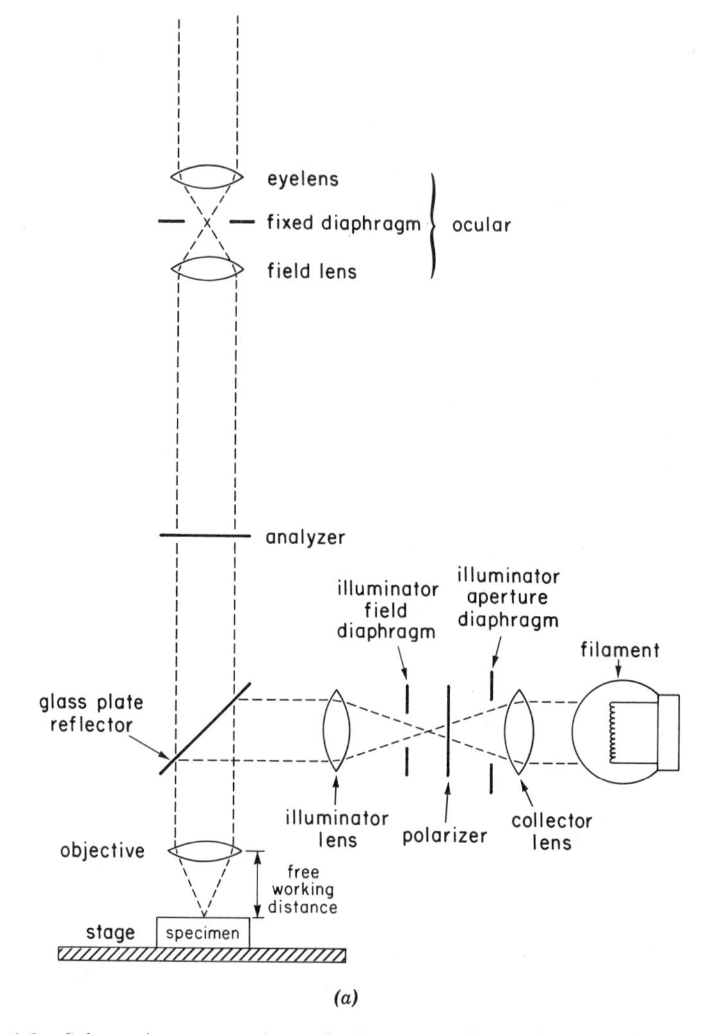

(a)

FIGURE 1.2 Schematic cross sections of microscopes illustrating essential components and the path of light through the systems involving: *(a)* whole-field glass plate reflector; *(b)* half-field prism; *(c)* whole-field glass plate reflector for conoscopic viewing.

1.2.2 Objective Lenses

Microscope objectives may be classified in terms of lens type (achromat, apochromat, or fluorite), their magnification and numerical aperture, and whether they are for oil immersion or air usage. Occasionally, the focal length or working distance is also considered.

The *achromat* is the most common and least expensive lens found on most

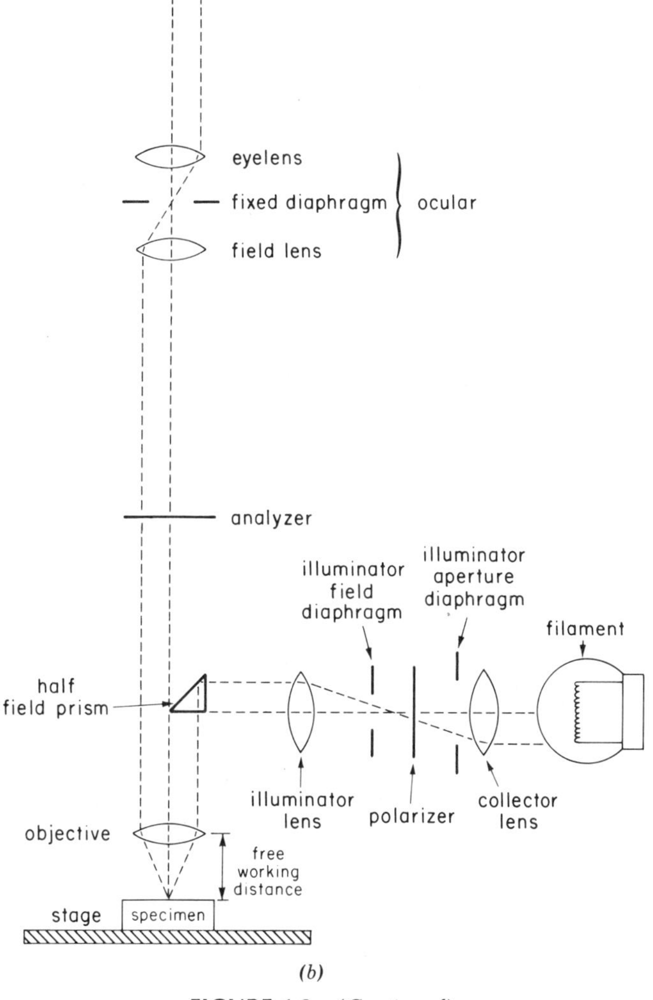

(b)

FIGURE 1.2 (*Continued*)

microscopes. It is corrected for spherical aberration for only one color, usually yellow-green, and for chromatic aberration for two colors. Thus, when used with white light, color fringes may appear at the outer margins of the image; when black and white film is used in photomicroscopy the fringes may contribute to some fuzziness. However, if monochromatic light (especially green light) is used, the image either to the eye or on a black and white film, is sharper.

The *apochromat* is a better and more expensive microscope objective. It is corrected for spherical aberration for two colors (blue and green) and for chromatic aberration for the primary spectral colors of red, green, and blue. Thus the

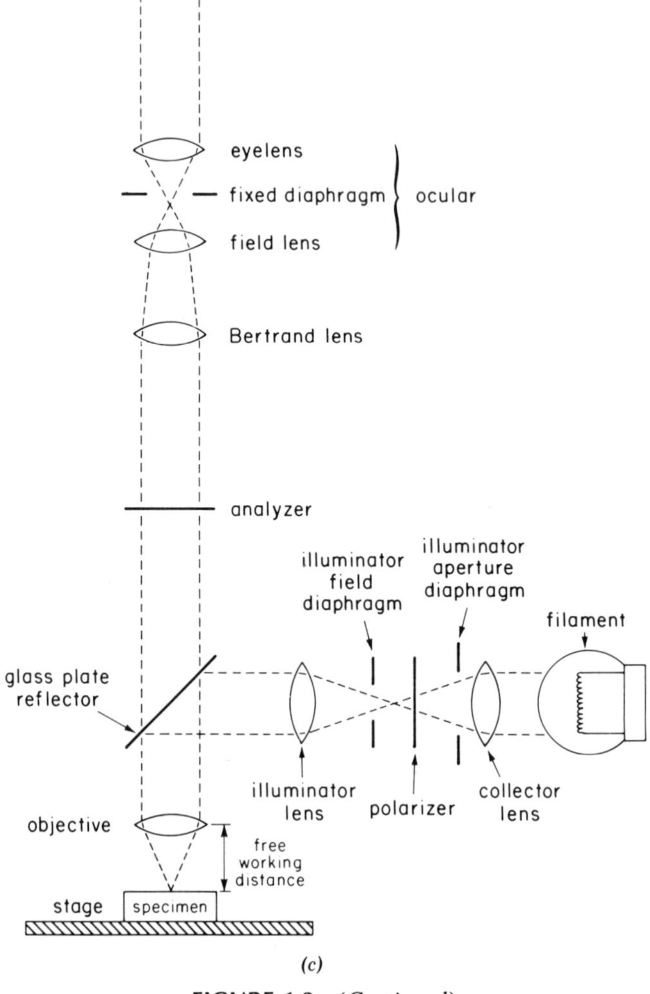

(c)

FIGURE 1.2 (*Continued*)

apochromat presents a sharper image and is better for color photomicroscopy than any other lens. To achieve the finest performance from apochromats it is necessary to use them in conjunction with "compensating" eyepieces.

Fluorite lenses (also known as "semiapochromats") are a compromise in terms of price and quality between the other two types of lenses. Fluorite objectives must be used with compensating eyepieces for best performance.

Most objectives give a primary image in a curved plane; however, with additional corrections these may be made to give a flat primary image. Such lenses are indicated by the prefix "flat-field" or "plan" and are especially useful for large fields of view and for photomicroscopy.

The *magnification* of an objective is the degree by which the image is enlarged as light passes through the objective. The magnification is classified as 5×, 10×, 20× ... up to about 125×. The projection of the primary image takes place within the body tube of the microscope and the distance from the back focal plane of the objective to the primary image is the "optical tube length."

The *numerical aperture* (N.A.) is a measure of ability to distinguish fine structural details in a specimen and determines the depth of focus and the useful range of magnification. Mathematically this lies in the range 0.04 to 1.3 and is equivalent to the product of the refractive index (n) for the medium in which the lens operates and the sine of the angle (μ) equal to one-half the angular aperture of the lens:

$$\text{N.A.} = n \sin \mu$$

The value of the N.A. is imprinted on the side of each objective. A 10× achromatic objective usually has an N.A. of 0.20 and a 20× objective of 0.40. Apo-

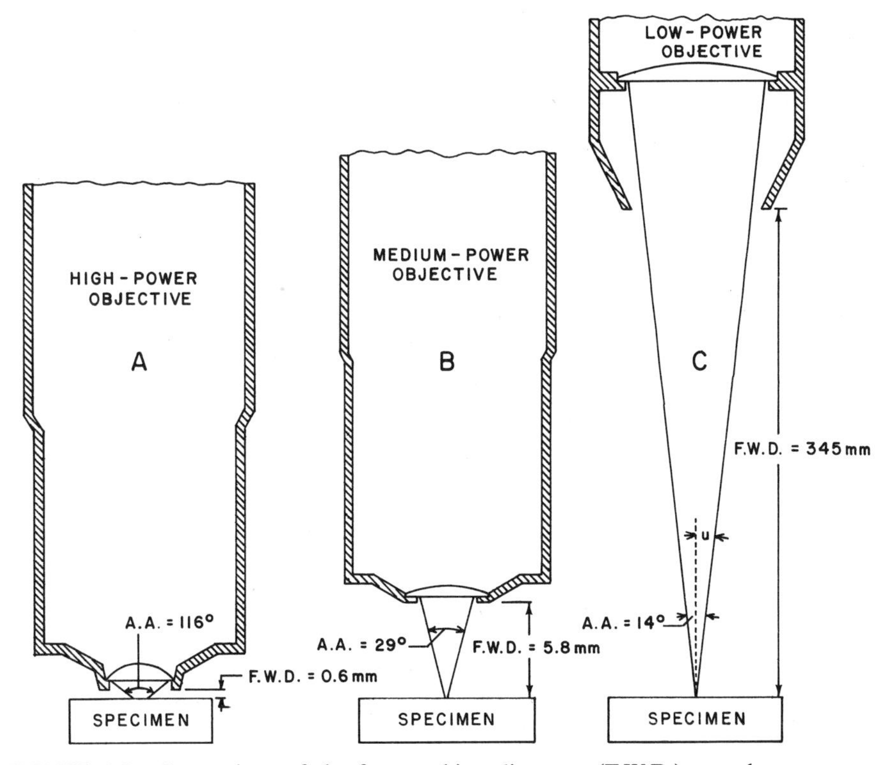

FIGURE 1.3 Comparison of the free working distances (F.W.D.), angular apertures (A.A.), and one-half angular aperture *(u)* for typical objectives used in ore microscopy. (Reproduced from *An Introduction to the Methods of Optical Crystallography* by F.D. Bloss 1961; Holt, Rinehart and Winston Inc., with the publisher's permission).

TABLE 1.1 Properties of Five Objectives* Commonly Used in Ore Microscopy

Initial Magnification	Angular Aperture (A.A.)	Numerical Aperture (N.A.)	Free Working Distance (F.W.D.) in mm	Focal Length in mm
5× (oil)	7°	0.09	0.35	50
5× (air)	10°	0.09	12	50
10× (air)	23°	0.20	14	25
20× (oil)	31°	0.40	0.23	12.5
50× (oil)	69°	0.85	0.28	2.0

*Ordinary objectives, separated by an air space from the object being viewed, are called "dry" objectives. This is in contrast to the more powerfully magnifying oil-immersion objectives in which an immersion oil fills the gap between objective and object being viewed.

chromatic objectives have higher N.A. values than achromats. The upper limit of the N.A. of a dry objective is 0.95, which corresponds to an angle of 70° as the maximum angle of incidence on the stage object (= 140° angular aperture). The maximum N.A. of immersion objectives that use an oil with a refractive index of 1.5 is about 1.4. Examples of the relationship between the angular aperture, magnification, and free working distance of objectives are shown in Figure 1.3 and are summarized in Table 1.1.

The most commonly used low and medium magnification objectives for ore microscopic work are *"dry"* or *air* lenses, which are designed to have only air between the objective and the sample. It should be noted that objectives intended for transmitted light observation are corrected for use with specimens covered by a 0.17 or 0.18 mm "cover glass." These lenses may yield poor or distorted images if used for reflected light microscopy.

Immersion objectives are commonly used for reflected light microscopy, especially when high magnification and high resolution are required. Such lenses require a drop of immersion oil (usually with a refractive index of 1.515) between the sample and the objective; some objectives are designed for use with water instead of oil. The presence of the immersion medium (oil or water) reduces the reflectance of the minerals (see Section 4.1.2), but enhances color differences, reduces diffuse light scattering, and generally permits the observation of weak anisotropism and bireflectance (see Sections 3.2.3 and 3.2.4), which are not visible with dry objectives. The slight inconvenience of cleaning objectives and samples when using immersion objectives is more than compensated by the increased information derived. Ramdohr (1969, p. 297) has summed up the value of immersion lenses in the statement that "It has to be emphasized over and over again that whoever shuns the use of oil immersion misses an important diagnostic tool and will never see hundreds of details described in this book." All immersion objectives have very short working distances; hence great care must be taken in focusing not to damage the objective through collision with the sample. The immersion oils should be removed from lenses with solvents to prevent their gathering dust. Water immersion objectives, which are so labeled, are especially convenient because of their ease in cleaning.

1.2.3 Ocular Lenses

The ocular or eyepiece system of the microscope serves to enlarge the primary image formed by the objective and to render it visible to the eye. Most microscopes are equipped with "Huygenian" oculars of between 5× and 12× magnification that consist of two lenses and an intermediate fixed diaphragm. The diaphragm commonly contains perpendicular crosshairs but may instead be equipped with a micrometer disk or a grid with a fixed rectangular pattern useful for particle size measurement or estimation. Oculars designed for photography do not contain crosshairs and are often designed to be of the "compensating" type to correct for chromatic aberration. Large research microscopes may have "wide field" oculars, which are designed to give a larger clear field of view.

1.2.4 Illuminating Systems

Two main types of lamp are commonly used in ore microscopes: the incandescent filament lamp and the gas discharge lamp. For most routine work, especially on student microscopes, the incandescent tungsten filament lamp is adequate. These lamps range from 6 to 12 V and 15 to 100 W with minimum bulb lifetimes of 100–300 hours and are generally operated by a variable rheostat. If the lamp is operated at too low a wattage, if the bulb is old, or if the microscope is misaligned, filament images and variably colored filament-shaped zones may be visible. Insertion of a frosted glass screen helps to eliminate the image but microscope adjustment or even servicing may be required. The lamp should provide adequate light, evenly distributed throughout the field of view, without being uncomfortable to the eyes. The color temperature of tungsten filament lamps varies from about 2850°K for 6 V 15 W bulbs to about 3300°K for halogen gas filled 12 V 100 W bulbs. These temperatures are far below the approximately 6100°K color temperature of xenon discharge lamps and if used without filters tend to bias the colors observed under the microscope toward the yellows and reds. Accordingly, most workers insert a pale blue filter between the lamp and the remainder of the illuminating system to provide a more daylight-like color temperature. The minor variations in the colors of the same minerals when viewed through different microscopes is due in large part to small differences in the effective color temperature of the light source. Knowledge of the actual color temperature of the lamp is not important in routine polished section examination but it is important in photomicroscopy because of the specific requirements of different types of film. It is also important when measuring the color of minerals quantitatively, since the color observed is partly a function of the light source. (see Chapter 5).

The standard illuminating system (Figures 1.2*a*–1.2*c*) contains two lenses, two or three diaphragms, and a polarizer in addition to the light source. The illuminator aperture diaphragm is used to restrict the field of view and to reduce stray scattered light. The illuminator field diaphragm controls the angle of the cone of light incident on the specimen and should be set to just enclose the field of view; this restricts the light to the most parallel rays, minimizes elliptical polarization (see Chapter 4), and maximizes the contrast. In many microscopes a third diaphragm helps sharpen the image.

Reflectance measurement, although sometimes carried out using standard low wattage incandescent filament lamps, usually requires either high intensity halogen filament incandescent lamps or xenon discharge lamps to provide sufficient light intensity through monochromators in the range 400–700 μm (see Chapter 5).

1.2.5 The Reflector

The reflector is a critical component of the ore microscope, being the means by which light is brought vertically onto the polished specimen surface. Reflectors are of two types: the glass plate reflector (Figure 1.2*a*) and the half-field prism (Figure 1.2*b*). Many microscopes have both types mounted on a horizontal slide such that either may be employed.

When the glass plate reflector is employed, part of the light from the illuminator is reflected downward through the objective onto the sample and part of the light passes through the reflector and is lost. The light that passes downward is then reflected back up through the objective until it reaches the glass plate again. At this point, some of the light passes through the glass plate up the microscope tube to the ocular and some is reflected back towards the illuminator and is lost. Ideally, the mirrored glass plate should reflect all the light from the illuminator down onto the specimen, but then let all the light reflected from the specimen pass through on its way up to the ocular. In fact, coated glass plate reflectors of maximum efficiency result in only about 25% of the illuminator light that first reached the reflector ultimately reaching the ocular. This efficiency is sufficient for most light sources; moreover, only with this type of reflector is there truly vertically incident light and illumination over the full aperture of the objective.

The alternative to the glass plate reflector is a totally reflecting prism or mirror system in which light is reflected downward through one-half of the aperture of the objective, strikes the specimen, is reflected back upwards through the other half of the objective, and passes behind the prism on its path to the ocular. In this situation, light is obliquely incident on the specimen and mirror aberrations arise. In conoscopic observation only half of the polarization figure is visible because half of the optical path is occupied by the prism. The advantage of the prism or totally reflecting mirror is that it permits a greater proportion of light (up to ~50%) to reach the ocular. Modern intense light sources now make the use of the prism less important. In addition, the early models of totally reflecting prisms and mirrors introduced some elliptical polarization but these problems have been overcome by the development of multiply reflecting prisms and the introduction of polaroid plates on the lower face of the prism. Most workers find the glass plate reflector adequate or superior for routine studies.

1.2.6 Polarizer and Analyzer

The polarizer in a standard ore microscope is usually positioned within the illuminating system between the lamp and the collector lens but may be located between the diaphragms. It is either a calcite prism or, more commonly, a polaroid plate that permits only the passage of light which is plane (or "linearly," see Chapter 4) polarized, usually in a north-south orientation. In standard transmitted-light thin section or grain mount petrography, the polarizer and analyzer are perpendicular to one another. However, many ore microscopists find that polarization effects are more readily observed if the polars are a few degrees from the true 90° position. This is especially true of very weakly anisotropic minerals and even some moderately anisotropic grains if they occur in a matrix of more strongly anisotropic minerals. The slight uncrossing may be accomplished either by having a rotatable analyzer or by slightly adjusting (by 3–5°) the polarizer from the crossed position. Rotation of the microscope stage to observe anisotropism and extinction is not always unambiguous because of the combination of movement and the variable anisotropism of adjacent grains. Alternatively, the stage may be left stationary and the analyzer or the polarizer rotated back and forth through the extinction position. This eliminates the distraction of movement of the specimen and may allow an unequivocal determination of the presence or absence of anisotropism (see Chapter 3).

1.3 ACCESSORIES

1.3.1 Monochromators

Because the optical properties of minerals vary as a function of wavelength, it is frequently necessary to provide incident light of specified wavelength. The operable range of most microscopes extends several hundred nanometers above and below the visible light range of approximately 400 to 700 nm wavelength. The two most commonly employed means of providing light of specified wavelength through this region are fixed monochromatic interference filters and continuous-spectrum monochromators (see Figure 1.4). Fixed interference filters consist of a glass substrate on which alternating layers of low reflecting trans-

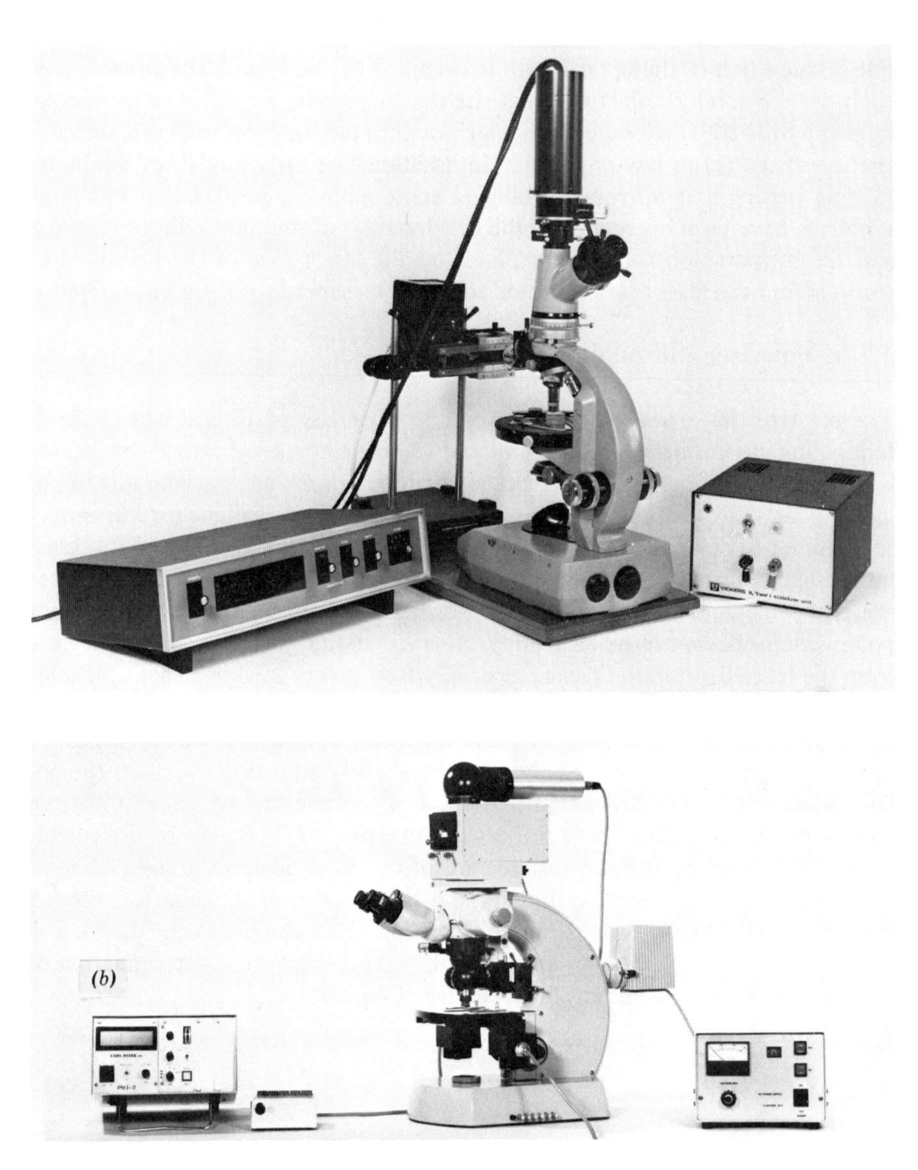

FIGURE 1.4 *(a)* Microphotometer system mounted on Vickers M74c microscope; also shown is continuous spectrum monochromator mounted in front of the light source. *(b)* Zeiss microscope photometer 03 mounted on the Universal research microscope. (Photographs courtesy of manufacturers)

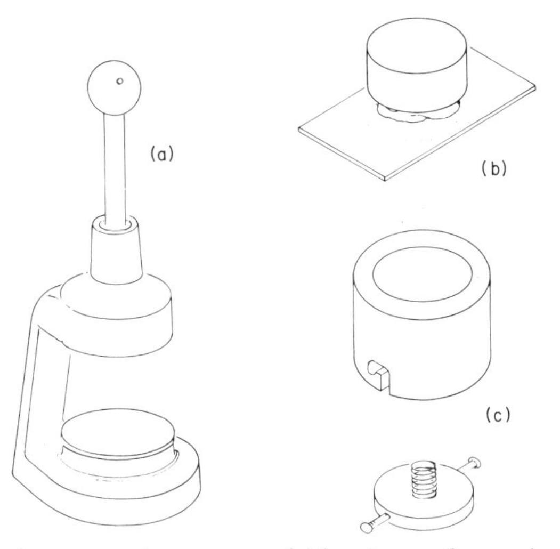

FIGURE 1.5 Specimen mounting systems: *(a)* hand press for specimen mounting; *(b)* clay on glass plate system; *(c)* spring-loaded holder.

parent dielectric substances and higher reflecting semitransparent metal films or dielectrics of high refractive index have been deposited. The light that passes through such filters is not truly monochromatic, but lies within a specified band width, usually < 15 nm if "narrow" band type and 15–50 nm if "broad" band type. The difficulty with such interference filters is that a separate filter is needed for each wavelength to be investigated.

The continuous-spectrum monochromator is an interference filter for which the wavelength of light transmitted varies continuously along the filter. A window, the width of which may be varied to control the passband width, may be slid along the monochromator to whatever wavelength is desired. In this way, a single device may be used to provide monochromatic light over the entire visible range and even beyond. Some commercially available units for reflectance measurement are designed to fit directly onto the microscope and have built-in adjustable monochromators. Otherwise the monochromator must be installed in the light path, usually immediately after the illumination source or immediately before a photometer that attaches to, or replaces, the ocular (see Chapter 5).

1.3.2 Photometers

Photometers, either built into large research microscopes or available as attachments to standard microscopes (Fig. 1.4), are used to measure the reflectance of mineral grains. Most photometers consist of a photomultiplier tube that has high sensitivity throughout the visible spectrum. To achieve meaningful results, photometers must be used in conjunction with stabilized light sources, high quality monochromators, and reflectance standards. The use of photometers in quantitative reflectance measurement is treated in detail in Chapter 5.

1.3.3 Stage Micrometers

All textural studies of ore minerals, mill products, and industrial materials require the accurate measurement of grain sizes. The stage micrometer, usually a 1 mm scale subdivided into hundredths, is invaluable in estimating grain sizes and in the calibration of a scale or grid set within an ocular. Stage micrometers are commercially available as small mounted metal disks on which the scale has been inscribed; they are positioned and observed in the same way as the polished section.

1.3.4 Sample Holder

Observation under the ore microscope requires that the sample surface is perpendicular to the incident light beam. This can be accomplished by carefully machining samples so that the top and bottom surfaces are flat and parallel or by using simple mechanical leveling devices (Fig. 1.5*a*) that press the sample down on a lump of molding clay on which it then is held level (Fig. 1.5*b*). More sophisticated devices include spring loaded cylinders (Figure 1.5*c*), in which the specimen is held against a lip that is machined parallel to the microscope stage, and more elaborate rapid specimen changers in which specimens are spring loaded into holders that are held by leveling screws. The means of securely holding a specimen with its polished surface normal to the incident light beam is a matter of personal convenience and equipment availability.

BIBLIOGRAPHY

Bloss, F. D. (1961) *An Introduction to the Methods of Optical Crystallography.* Holt, Rinehart & Winston, New York.

Bowie, S. H. U. and Simpson, P. R. (1977) Microscopy: Reflected Light. In J. Zussman, ed., *Physical Methods in Determinative Mineralogy*, 2nd ed. Academic, London, pp 109–166.

Cameron, E. N. (1961) *Ore Microscopy.* Wiley, New York.

Eastman Kodak Company (1970) *Photography Through the Microscope.* Rochester, New York.

Galopin, R. and Henry, N. F. M. (1972) *Microscopic Study of Opaque Minerals.* W. Heffer and Sons, Ltd, Cambridge, England.

Piller, H. (1977) *Microscope Photometry.* Springer-Verlag, Berlin.

2

The Preparation of Samples for Ore Microscopy

2.1 INTRODUCTION

The preparation of polished surfaces free from scratches, from thermal and mechanical modification of the sample surface, and from relief* is essential for the examination, identification, and textural interpretation of ore minerals using the reflected-light microscope. Adequate polished surfaces can be prepared on many types of materials with relatively little effort using a wide variety of mechanical and manual procedures. However, ore samples often present problems because they may consist of soft, malleable sulfides or even native metals intimately intergrown with hard, and sometimes brittle, silicates, carbonates, oxides, or other sulfides. Weathering may complicate the problem by removing cements and interstitial minerals and leaving samples friable or porous. Delicate vug fillings also cause problems with their open void spaces and poorly supported crystals. Alloys and beneficiation products present their own difficulties because of the presence of admixed phases of variable properties and fine grain sizes. In this chapter, the general procedures of sample selection and trimming, casting, grinding, and polishing (and in special cases, etching) required to prepare solid or particulate samples for examination with the ore microscope are discussed. The preparation of these *polished sections* and also polished and doubly polished *thin sections*, which are useful in the study of translucent or coexisting opaque and translucent specimens, is described.

2.2 GRINDING AND POLISHING EQUIPMENT

Several types of equipment are commonly used in the grinding and polishing of specimens for ore microscopy. Most, if properly operated, are capable of pre-

**Relief* is the uneven surface of a section resulting from hard phases being worn away less than soft phases during polishing.

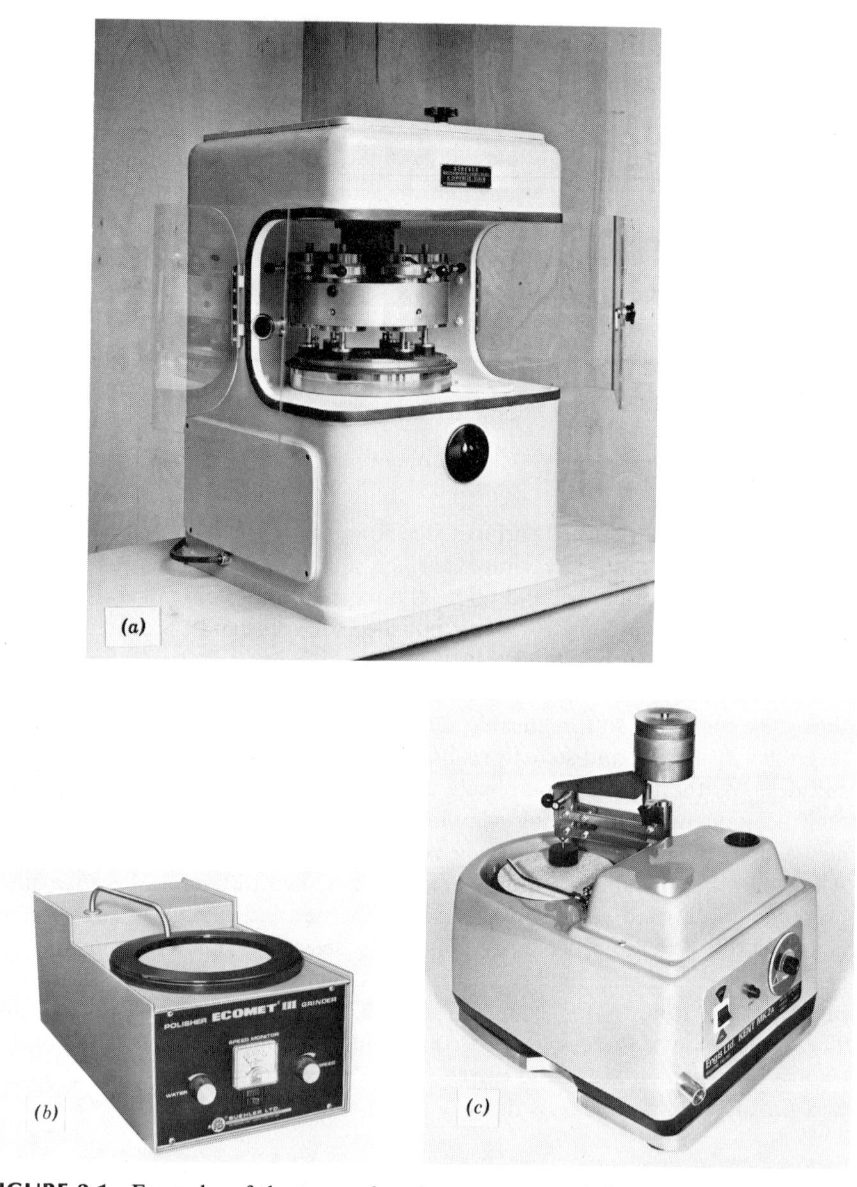

FIGURE 2.1 Examples of the types of equipment commercially available for the preparation of polished sections: *(a)* large automated Dürener grinding and polishing machine capable of preparing several samples simultaneously; *(b)* variable speed manual grinding and polishing machine, Buehler No. 49-1650 Ecomet III; *(c)* small automated grinding and polishing machine capable of handling one to three samples, Engis MK2a; *(d)* Syntron Electro-Magnetic Vibrating polishing machine, FMC Corporation. (All photographs courtesy of the manufacturers.)

FIGURE 2.1 *(Continued)*

paring adequate surfaces, free from scratches and with a minimum of relief. A few of the more popular varieties of equipment (as shown in Figure 2.1) include the following:

1. *Multiple specimen automated grinding-polishing machines* (Figure 2.1*a*). These units hold six or more samples, which are mounted on spindles that rotate clockwise as they move on a lap that rotates counterclockwise. Polishing is accomplished on grooved metal (usually lead) laps coated with a thin layer of abrasive mixed with oil.
2. *Variable speed manual grinding and polishing machines* (Figure 2.1*b*). These are popular units because of their rugged design and easy adaptability to perform a wide variety of tasks. Samples are manually (or sometimes mechanically) held as they are ground and polished on interchangeable fixed abrasive paper- or fabric-covered laps. Most units have variable speed motors although less expensive models may possess only one or two fixed speeds. Most polishing is done on napless cloths in which the polishing compounds become embedded. Final polishing or buffing of surfaces commonly employs a napped cloth.
3. *Small automated polishing machines* (Figure 2.1*c*). These units have become very popular because of their low price, compact size, and ease of operation. Samples are held on an arm that moves them in an eccentric motion on paper or fabric lap surfaces that adhere to glass plates. The glass plates are contained in individual bowls that are conveniently interchanged between steps.
4. *Vibropolishing equipment* (Figure 2.1*d*). Vibropolishing equipment is convenient in the preparation of large numbers of sections with the minimum of operator attention. Polished sections are held in cylindrical brass or stainless steel weights and ride about the circular lap as a result of the vibratory

motion. The lap surface is covered with a silk, nylon, or other hard cloth that is stretched taut and covered with an abrasive slurry or water; a small amount of MgO or Al_2O_3 is dispersed on the lap. Samples polish slowly but effectively and require virtually no attention except watching the water level and the wear of the polishing cloth.

2.3 PREPARATION AND CASTING OF SAMPLES

The procedures employed in sample preparation depend on the nature of the material under study and the objectives of the study. Coherent materials are easily cut to size with a diamond saw. Lubrication of the saw blade by water, kerosene, or a cutting oil to prevent heating of the specimen during cutting is important. If the sample is suspected of containing phases that are either soluble (e.g., metal sulfates or chlorides) or reactive with water (e.g., some rare sulfides in meteorites), it will be necessary to conduct all cutting and subsequent operations using appropriate nonreactive fluids (oils, alcohols, kerosene, etc.).

The size and shape of the sample to be mounted is somewhat arbitrary and must be dictated by the nature of the material and the intended study; most workers find that circular polished sections from 2.5 to 5 cm in diameter are adequate. The casting of samples may be in one- or two-piece cylindrical plastic, polyethylene, or metal molds (Figure 2.2) or in cylindrical plastic rings that remain a permanent part of the sample. When using molds, a thin coating of wax or other nonreactive lubricating agent (e.g., silicon stopcock grease, or vaseline) is extremely helpful in the removal of specimens once the casting resin has hardened. Sample thickness is also arbitrary but the total thickness of the polished section generally need not exceed 1 to 2 cm. The maximum thickness is often governed by the working height between the microscope stage and the lens.

If the specimen is coherent and has a low porosity, it may be cut to the desired size and polished directly. Commonly, however, specimens are cast in a mounting resin to facilitate handling and to minimize the problems of crumbling and plucking. The preparation of a flat surface on the specimen by cutting and grinding prior to casting is a very useful step because it allows the sample to lie flat

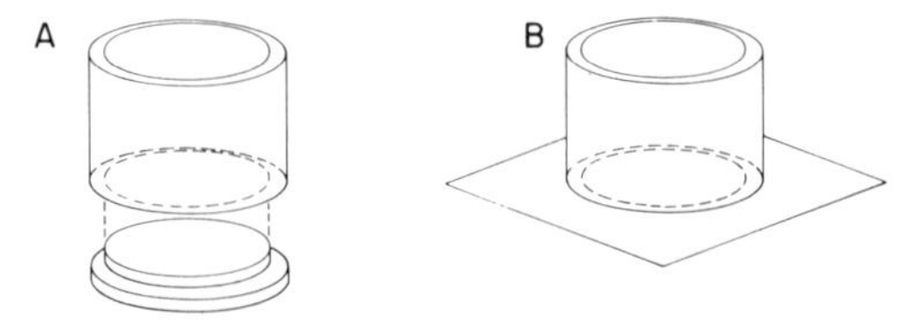

FIGURE 2.2 Sample cups used for the casting of cylindrical polished sections: *(A)* two-piece cup; *(B)* plastic ring that becomes a portion of the polished section after emplacement of the sample and the mounting resin.

on the bottom of the mold; irregular specimens may trap air bubbles and require extensive grinding of the polished section to expose sufficient surface area. It is useful to level the edges of the surface to be polished; this helps to prevent material being dislodged from, or trapped between, the sample and mounting medium. Also, it may be desirable to cut and mount a sample in specific orientations because *plucking* (the tearing of material from the surface during polishing) may occur when cleavage or fracture directions are appropriately oriented. If the specimen is friable, fractured, porous, or consists of loose grains or powders, impregnation is a useful procedure to prevent grain plucking and trapping of abrasive during grinding and polishing. Impregnation is readily accomplished by placing the mold containing the specimen and mounting resin under vacuum for several minutes; this draws air from the voids and the subsequent release of the vacuum tends to drive the mounting resin into the voids. Sometimes, repeated brief vacuum treatments are needed to drive the resin into deep cracks and voids.

It is important that the procedures involved in specimen preparation preserve the precise mineralogical and textural character of the samples. The recognition in recent years that certain common sulfide minerals (e.g., anilite, Cu_7S_4, and djurleite, $Cu_{1.96}S$) break down below 100°C and that many other ore minerals have unknown thermal stabilities, points to the necessity of using cold-setting mounting media, requiring neither heat nor pressure for preparation. Older procedures that require elevated temperatures and high pressures (e.g., the use of bakelite as a mounting medium) and that are still commonly used in metallurgical laboratories, should be avoided. Such procedures may not alter refractory ore mineral assemblages (e.g., chromite, magnetite, pyrite) but may profoundly modify low temperature or hydrous mineral assemblages. Even heating for brief periods at temperatures as low as 100°C will cause decomposition of some phases and induce twinning or exsolution in others. Accordingly, virtually all mineralogical laboratories now use cold-setting epoxy resins or plastics; these are readily available and easily handled. Most of the epoxy resins are translucent and nearly colorless, which facilitates labeling because paper labels can be set in the resin with the specimen. Lower viscosity mounting media generally penetrate cracks and voids in samples better than higher viscosity ones, an important point when dealing with friable or porous samples. If samples are also to be used for electron microprobe analysis (see Appendix 3), one must be careful to choose a resin that does not create problems in the instrument sample chamber by degassing or volatilizing.

2.4 GRINDING AND POLISHING OF SPECIMENS

Once a sample has been cut to an appropriate size and cast in a mounting medium, it is ready for grinding and polishing. Large research or industrial laboratories commonly employ automated grinding and polishing machines capable of handling large numbers of specimens. Smaller laboratories are usually equipped with either fixed or variable speed rotary lap equipment designed to prepare one sample at a time. The procedures described below are general and apply to any laboratory; experienced workers invariably develop favorite "tricks" and minor modifications but follow the same basic steps outlined below.

The purpose of grinding is to remove surface irregularities, remove casting resin that covers the sample, reduce thickness, prepare a smooth surface for further work, and remove any zone of major deformation resulting from initial sample cutting (Figure 2.3). Some workers have found that fixed abrasives (such as adhesive backed emery paper or diamond embedded in metal or epoxy) are superior to loose abrasives because the latter tend to roll and leave irregular depth scratches rather than planing off a uniform surface. Grinding and polishing compounds may be designated in terms of *grit, mesh,* or *micron* grain sizes; the equivalence of terms is shown in Figure 2.4 and in Table 2.1. Successive grinding steps using 400- and 600-mesh silicon carbide have proven adequate in preparing most surfaces, although the 400-mesh abrasive is really needed only for harder materials and many sections made up of soft sulfides can be started on a 600-mesh abrasive. It is important at each step to prevent any heating of the specimen or contamination by the carry over of abrasive from one step to the next. Samples should be thoroughly washed and cleaned (with an ultrasonic cleaner, if available) between steps and it is convenient to process samples in batches. This requirement for cleanliness also applies to an operator's hands, because abrasives and polishing compounds are readily embedded in the ridges of one's fingertips. Grinding and polishing laps should be covered when not in use to prevent contamination by stray particles and airborne dust. If samples begin to pluck, one is well advised to stop the polishing and coat or impregnate with a mounting medium and then begin again; otherwise the plucked grains probably will contaminate all grinding and polishing laps and leave the plucked sample surface and all subsequent samples highly scratched. Careful attention to the grinding steps is important in the preparation of good polished surfaces because these steps should provide a completely flat surface and progressively remove surface layers deformed in previous steps. Even in compact specimens, the fracturing from a high speed diamond saw may extend a millimeter or more from the cut surface.

Grinding may be completed with a 600-mesh abrasive compound or with

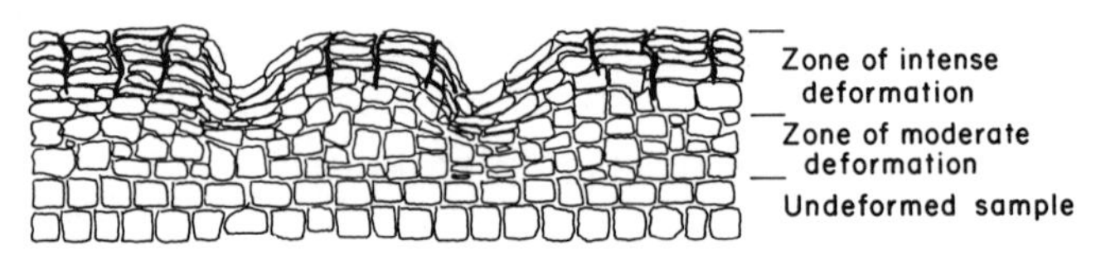

FIGURE 2.3 Schematic cross section of a polycrystalline sample illustrating the nature of deformation from the cut or coarsely ground surface down to the undeformed sample. The nature and depth of deformation varies with the nature of the sample.

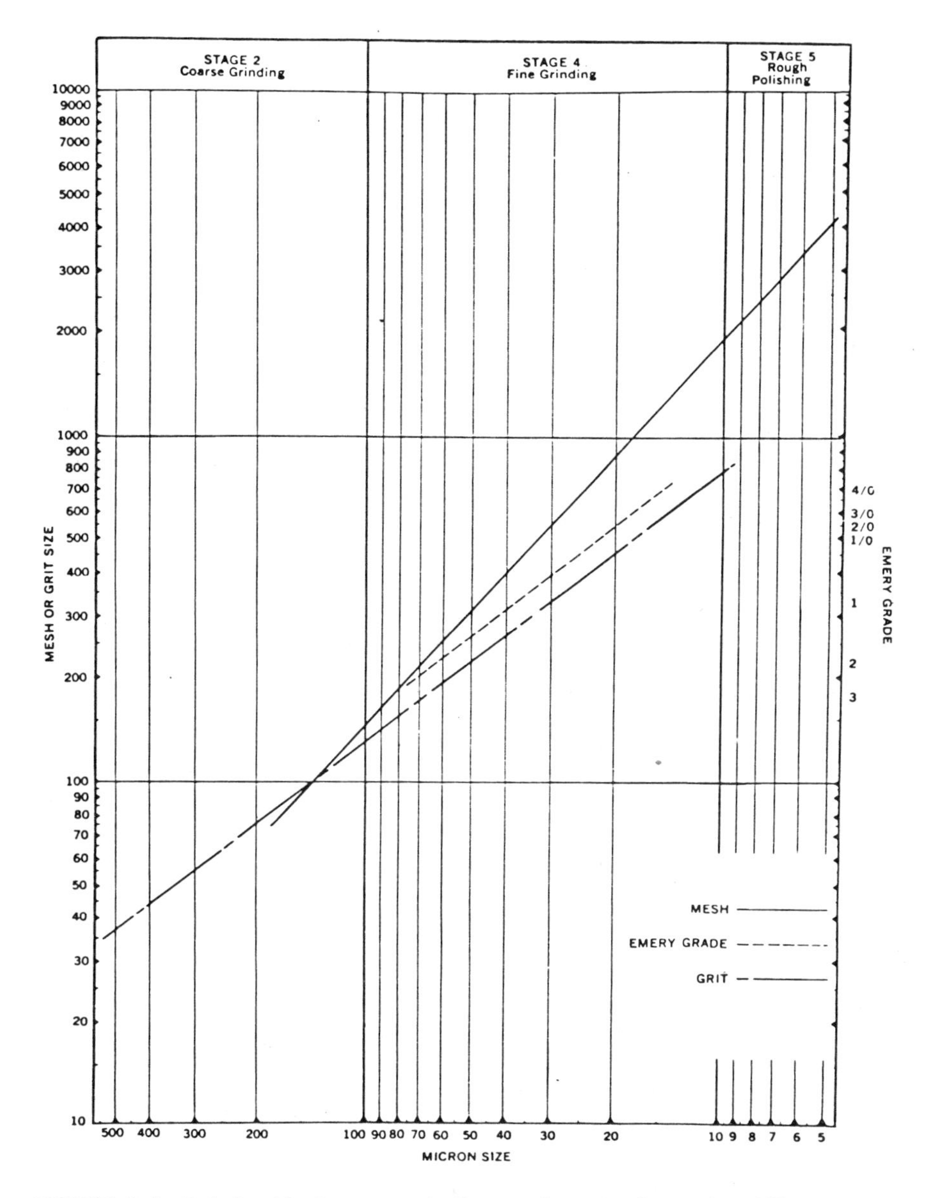

FIGURE 2.4 Relationship between grit size, mesh, and micron size. Silicon carbide abrasive powders are generally classified in mesh or grit size. These are plotted on the left of the graph. The right-hand ordinate is scaled for emery grit designation. For conversion to micron size, locate the size range on the proper ordinate and transpose this point to the correct graph line while locating this point on the abscissa. Reverse this procedure to transpose micron to grit size. (Diagram courtesy of Buehler Ltd.)

TABLE 2.1 Dimensions of Mesh or Grit Size Abrasives

Mesh or Grit Size	Microns	Inches
100	150	0.0059
200	75	0.0030
260	60	0.0024
325	45	0.0018
400	38	0.0015
600	30	0.0012
1,200	15	0.00059
1,800	9	0.00035
3,000	6	0.00024
8,000	3	0.00012
14,000	1	0.00004
60,000	½	0.00002
120,000	¼	0.00001

1200-mesh abrasive used manually on a glass plate; the matte-like surface of a typical sample at this stage is shown in Figure 2.5*a*. Rough polishing, involving the use of a fabric loaded with 15 and 6 μm abrasives, removes most or all of the remaining zone of surface deformation and the deeper scratches and prepares the sample for final polishing; the appearance of the sample after the 15 and 6 μm steps is shown in Figures 2.5*b* and 2.5*c*. The first polishing is best accomplished using diamond abrasives embedded in a napless cloth. The embedding permits the diamond grains to plane the surface without rolling and causing irregular scratching or gouging. Use of a hard napless fabric without excessive weight pressing down on the sample minimizes the development of surface relief. The distributors of diamond abrasives recommend that a 1–2 cm strip of diamond paste be placed on the polishing lap at right angles to the direction of movement and that subsequent dispersal on the surface of the lap should be by dabbing the diamond about with a clean fingertip. Polishing with 6 μm abrasives is the most important stage and should be carried out fairly slowly with the surface just sufficiently lubricated for a smooth action and without too much weight applied to the section. Polishing is continued until no deep scratches are seen in even the hardest phases. It may be useful to study the section at this stage since information lost later in polishing (e.g., grain boundaries of isotropic minerals) may be obtained. Again, care must be taken in this and subsequent steps to avoid excessive heating of the sample surface during the polishing.

Final polishing, using abrasives of less than 6 μm, removes only a very small amount of the specimen surface and should produce a relatively scratch-free surface (Figures 2.5*d*–2.5*f*). This step may be accomplished using 1 μm diamond embedded in a napless cloth, perhaps followed by 0.25 μm diamond cloth, or by using α–Al_2O_3 (1–0.3 μm) or γ–Al_2O_3 (0.05 μm) suspended in water on a

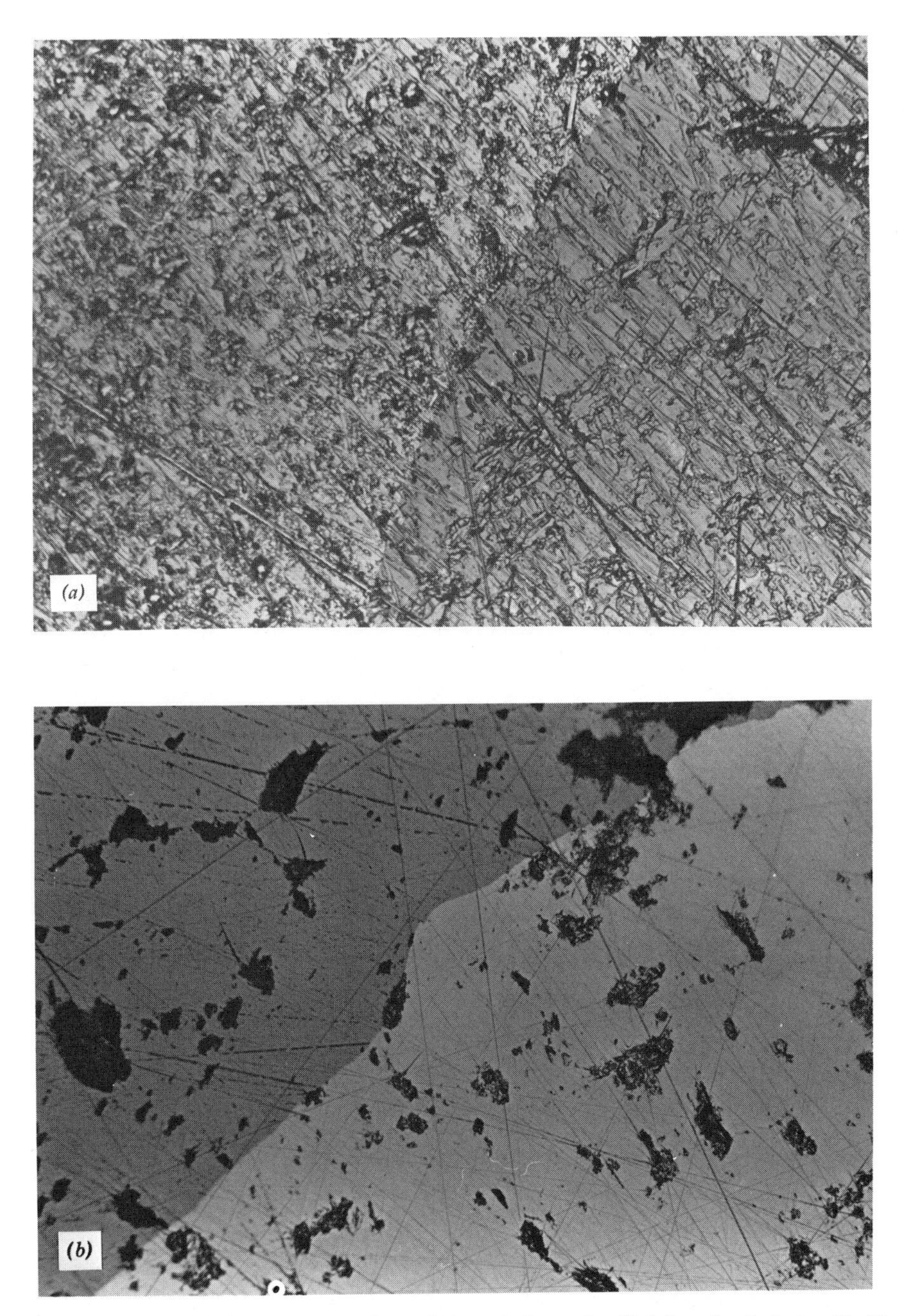

FIGURE 2.5 Polished surfaces of coexisting chalcopyrite (light) and sphalerite (dark) after abrasion by *(a)* 600 mesh silicon carbide; *(b)* 15 μ diamond; *(c)* 6 μ diamond; *(d)* 1 μ diamond; *(e)* 0.05 μ γ-AL_2O_3 on microcloth; *(f)* γ-AL_2O_3 on silk in a vibropolisher but without the intermediate steps of 15, 6, and 1 μ diamond. (Width of field = 520μm.)

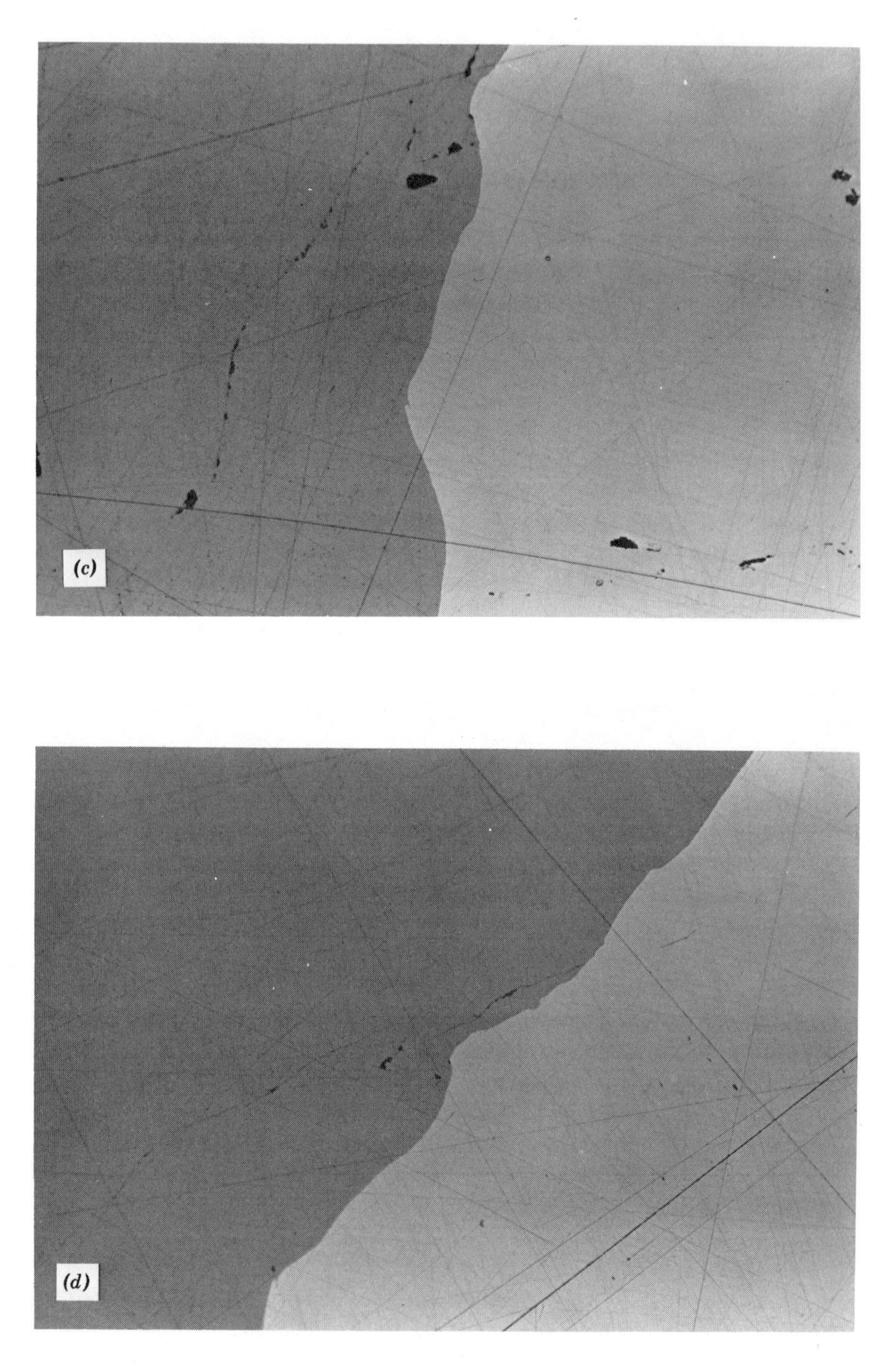

FIGURE 2.5 *(Continued)*

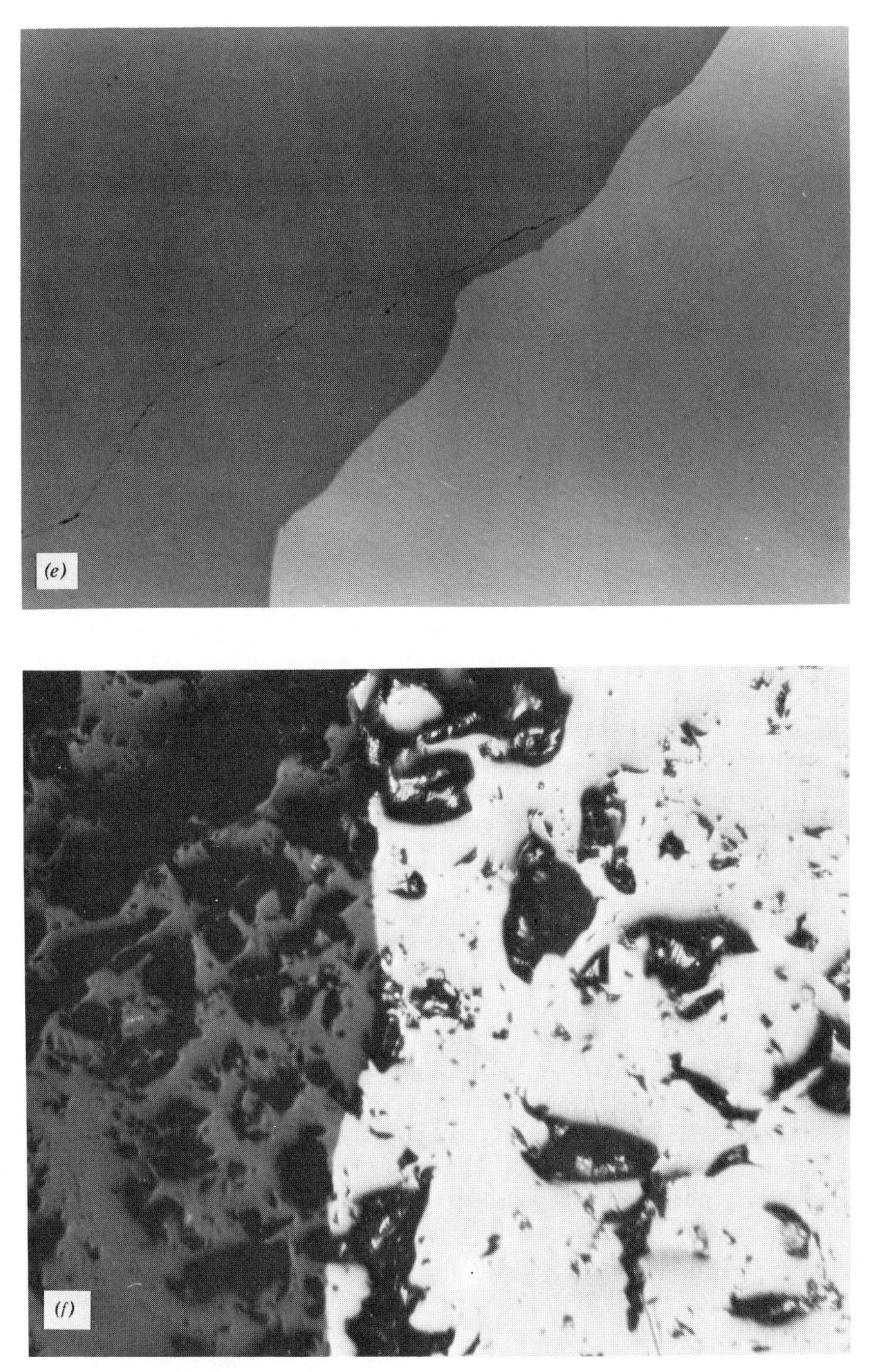

FIGURE 2.5 *(Continued)*

napped cloth. Some laboratories have also found chromic oxide, magnesia, and cerium oxide to be useful in final polishing, especially of metallurgical samples. Napped cloths are very useful in the final polishing and when a quick buffing of a slightly oxidized surface is required, but such cloths very rapidly produce relief on samples containing several minerals of differing hardness.

Some laboratories prepare sections using machines capable of polishing six or more specimens simultaneously on grooved cast-iron, copper, or lead laps. Grooved laps require constant resurfacing but if properly maintained allow preparation of excellent relief-free sections. Specific instructions accompany each type of machine but the polishing steps broadly parallel those outlined above.

Vibropolishing equipment is efficient and effective in the final polishing of many types of specimens. The quality of the surface and time necessary to achieve a satisfactory polish depends on the prior grinding steps, the hardness, and the homogeneity of the specimen. The lap of the vibropolisher is usually covered with a taut piece of fine-weave silk, nylon, or other hard fabric; the unit is filled with water to a depth of 2 to 3 cm and a small amount of 0.3 μm α-alumina or 0.05 μm γ-alumina. Because these very fine grained compounds remove surface layers only slowly, careful prior grinding down to the 600-mesh abrasive size, or finer, is necessary to ensure flatness and uniformity of sample surface. Otherwise polishing, which for the common sulfides and oxides takes only a few hours, will require several days and even then produces only a mediocre polish.* A drawback of vibropolishing units is the tendency for polished surfaces to develop considerable relief (Figure 2.6), especially when soft and hard minerals are intergrown. Such relief may obscure contact relationships between minerals, make the identification of soft rimming phases and inclusions difficult, and handicap attempts at photomicroscopy and electron microprobe analysis. Nevertheless, the simplicity of operation, easy maintenance, and capability of handling many samples simultaneously make vibropolishers useful.

2.5 PREPARATION OF POLISHED THIN SECTIONS

The thin section, a ~0.03 mm thick slice of rock, has for many years been the standard type of sample employed in petrological microscopy. The need to examine the opaque minerals in rocks and to be able to analyze all of the minerals in thin sections using the electron microprobe has led to the preparation of polished thin sections by simply polishing the upper uncovered surface of a thicker-than-normal thin rock slice.

Although the study of most ore samples still involves conventional polished sections, there has been an increased awareness that important textural informa-

*Samples are generally held in stainless steel or brass weights in order to increase the polishing speed and to insure that samples do not merely slide about without receiving any polish.

FIGURE 2.6 Comparison of the relief developed in the same sample, which contains a hard mineral (pyrite as cubes) and a soft mineral (pyrrhotite) when polished: *(a)* by polishing with 15, 6, 1 μ diamond followed by 0.05 μ γ-AL_2O_3 on microcloth (briefly) and *(b)* by using a vibropolishing machine charged with 0.05 γ-AL_2O_3. The relief is evident in the photomicrograph as the dark shadow around the edges of the pyrite. (Width of field = 100 μm.)

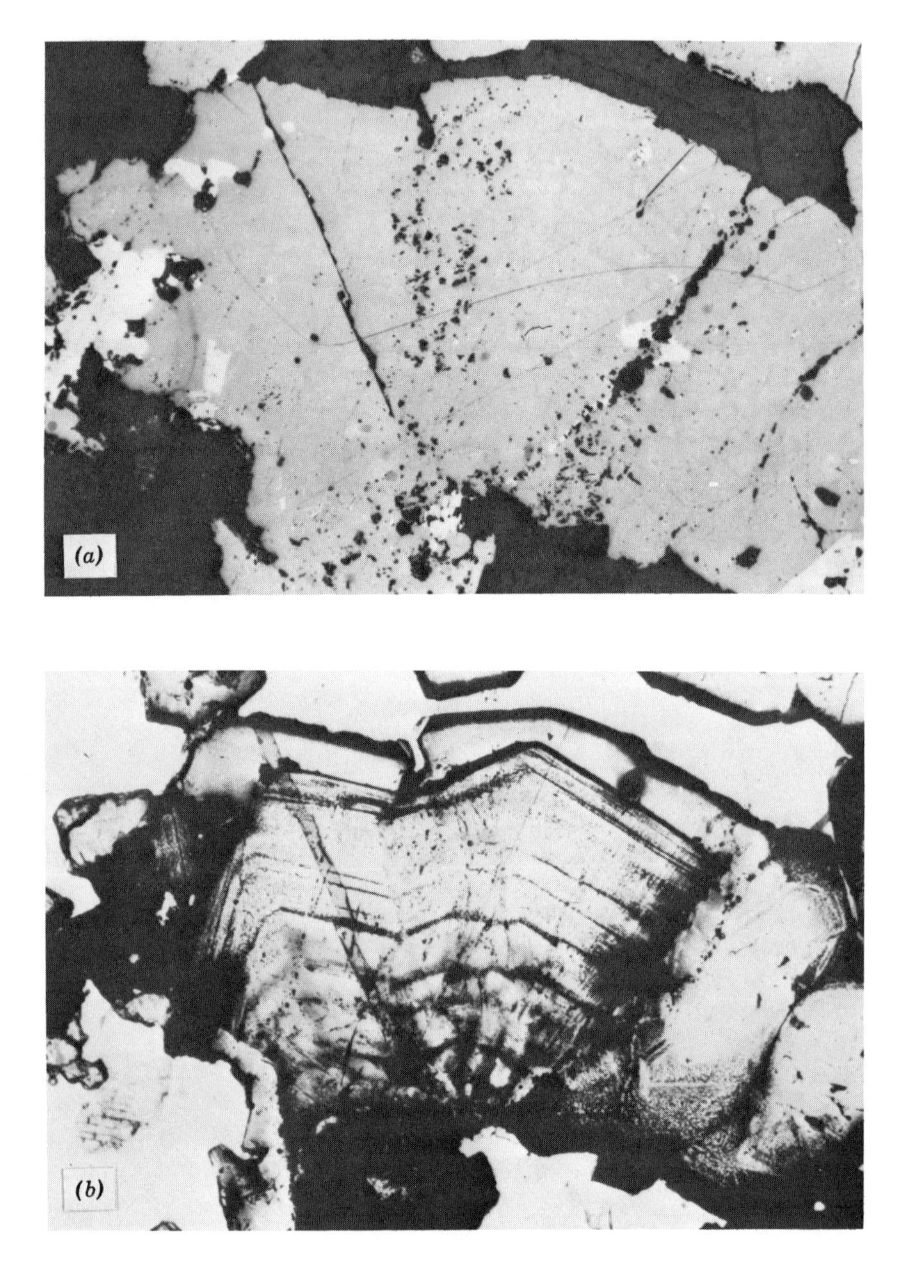

FIGURE 2.7 Comparison of the structure visible in the same sample of sphalerite when viewed: *(a)* in normal reflected light and *(b)* in transmitted light in a doubly polished thin section. (Figures courtesy of P. B. Barton reproduced from *Min. Geo.* **28,** 296, 1978, with the publisher's permission.)

tion in some ore minerals is not readily visible in conventional polished sections, thin sections, or even polished thin-sections. In polished sections or polished thin sections it is evident that most sphalerite, cassiterite, cinnabar, tetrahedrite, tennantite, ruby silvers, rutile and copper oxides transmit light, but the diffuse light scattering from the rough underside of the thin section makes recognition of some internal features difficult. The difficulty is readily overcome if doubly polished thin sections are employed. These are prepared either by cutting standard thin-section blanks, which are polished and then glued to a glass slide, or by gluing conventional polished sections to a glass slide. The excess blank or the polished section is then cut away from the slide by a *wafering* saw (a slow speed saw is especially useful because it minimizes the amount of fracturing and distortion within the sample). The thin slice on the glass slide is then gently ground until it transmits sufficient light to suit the worker's needs, and it is given a final polish in the normal manner. The result is a section that combines certain advantages of both the polished section and the thin section by allowing simultaneous study of ore and gangue minerals and a section that may reveal internal structure otherwise missed (see Figure 2.7 and Chapter 7).

The preparation of samples for fluid inclusion studies is very similar and is described in detail in Chapter 8.

2.6 PREPARATION OF GRAIN MOUNTS

Loose mineral grains from placer deposits or extracted from larger specimens may be cast into polished sections in several ways. The most direct technique is that of pouring embedding resin into the mold and then dispersing the mineral grains over the surface and allowing them to sink to the bottom. Two potential problems faced in this technique are the trapping of air bubbles and the stratification of mineral grains due to differential rates of settling through the resin. Air bubbles can usually be removed by using the vacuum impregnation techniques described in Section 2.3. Stratification can sometimes be avoided by merely dispersing grains such that all, or at least representatives of all, types of grains rest on the bottom of the mold. Alternatively, one can cast the grains and then section the grain mount vertically to expose any layering present. This vertical section through the grain mount may be polished directly or recast in another section such that the layering of grains is visible.

If one is dealing with only one or a few selected grains (e.g., in placer gold samples), the most efficient technique is often to place the grains either on a glass slide or on a polished section "blank" and to cover them carefully with a few drops of resin. In this manner, the grains may be placed in a pattern for easy recognition and their relative heights controlled so that all are exposed simultaneously during polishing. Grains may even be specifically oriented, if desired, by the use of small supporting devices or by cutting holes into the underlying polished section blank.

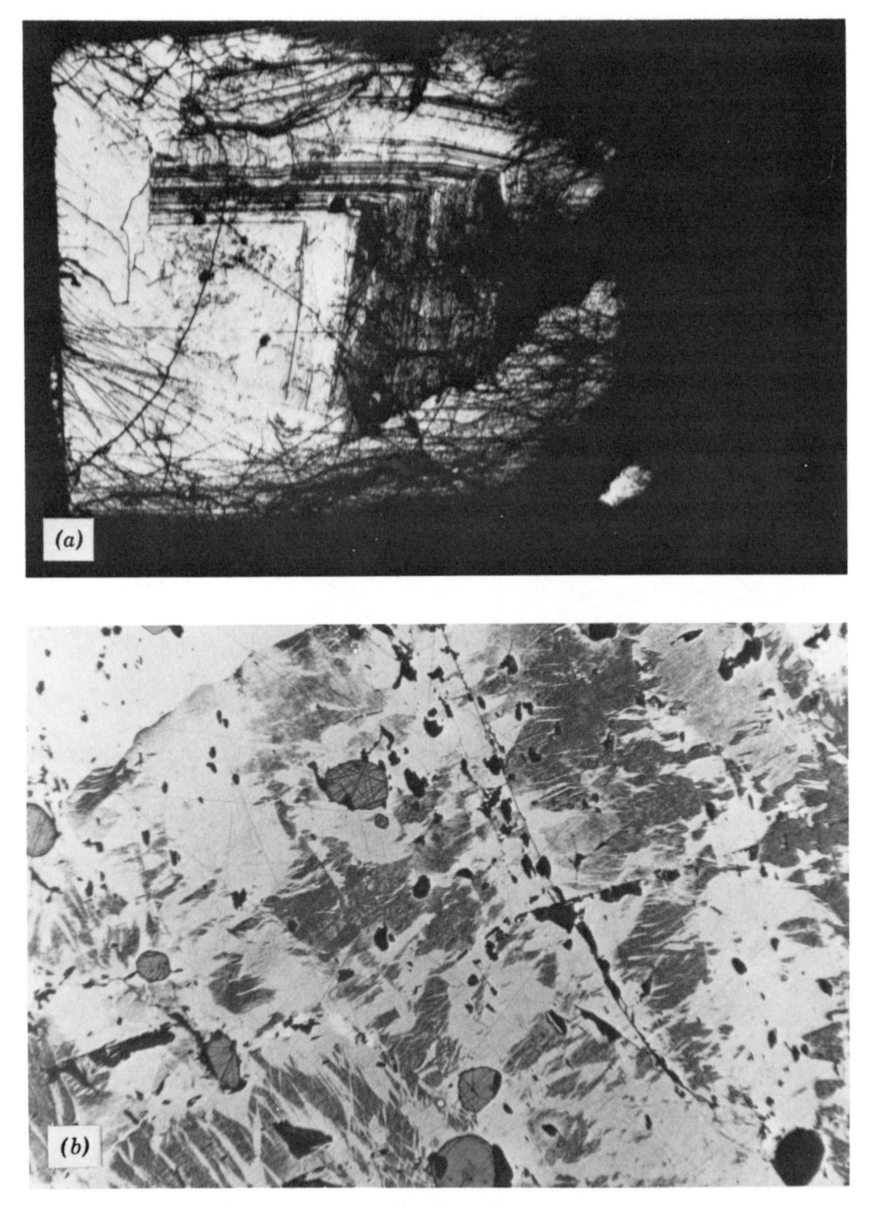

FIGURE 2.8 The use of etchants to enhance the internal structure of a mineral and to distinguish between two otherwise very similar minerals: *(a)* pyrite after etching to bring out the growth zoning (width of field = 2000 μm) (courtesy of L. Cox); *(b)* a mixture of hexagonal and monoclinic pyrrhotite after etching to enhance the difference (in the portion of the section that has been etched, the hexagonal pyrrhotite appears much darker). (Width of field = 1300 μm.) (Reproduced from *Econ. Geol.* **66,** 1136, 1971, with the publisher's permission.)

2.7 ELECTROLYTE POLISHING AND ETCHING TECHNIQUES

Metallurgists, and to a lesser extent mineralogists, have supplemented conventional polishing methods with electrolytic and chemical-mechanical techniques. In the former, electrically conductive samples, ground only as far as 600-mesh abrasive, are polished with 0.05 μm Al_2O_3 suspended in an electrolyte fluid. The sample is attached as anode and the polishing lap as cathode in a 25 V DC circuit. Polishing occurs by the removal of material from the sample by both me-

TABLE 2.2 Etchants Useful in the Enhancement of Mineral Textures

Some general etchants
- HNO_3—concentrated; 1 part HNO_3 to 1 part H_2O; 1 part HNO_3 to 7 parts H_2O
- HCl—concentrated; 1 part HCl to 1 part H_2O; 1 part HCl to 5 parts H_2O
- H_2SO_4—concentrated; 1 part H_2SO_4 to 1 part H_2O; 1 part H_2SO_4 to 4 parts H_2O
- HI—concentrated; 1 part HI to 1 part H_2O
- HBr—concentrated;
- Aqua Regia—1 part conc. HNO_3 to 3 parts conc. HCl
- HCl—1 part conc. HCl to 3 parts dilute (100 g/1) thiourea soln.
- Picric acid—4 g in 100 cm^3 ethyl alcohol
- $KMnO_4$—2.5 g in 100 cm^3 H_2O; 1 part to 1 part conc. HNO_3; 1 part to 1 part conc. H_2SO_4; 1 part to 1 part KOH
- $(NH_4)_2Cr_2O_7$—400 mg in 25 cm^3 of 15% HCl
- KCN—20 g in 100 cm^3 H_2O
- $FeCl_3$—20 g in 100 cm^3 H_2O; 50 g in 100 cm^3 H_2O
- $SnCl_2$—1 part saturated soln. to 1 part conc. HCl
- KOH—saturated soln.
- H_2O_2—16%
- NH_4OH—1 part conc. NH_4OH to 1 part 16% H_2O_2; 5 parts conc. NH_4OH to 1 part 3% H_2O_2

For pyrrhotites and iron-rich sphalerite:
1. 400 mg $(NH_4)_2Cr_2O_7$ dissolved in 25 cm^3 of 15% HCl soln.*
2. 50% HI

For pyrite:
conc. HNO_3 followed by brief conc. HCl

For monoclinic pyrrhotite: Magnetic colloid
1. Dissolve 2 g $FeCl_2 \cdot 4\,H_2O$ and 5.4 g $FeCl_3 \cdot 6\,H_2O$ in 300 cm^3 of distilled water at 70 °C
2. Dissolve 5 g NaOH in 50 cm^3 of distilled H_2O
3. Mix the two liquids; filter black precipitate and rinse with distilled water and 0.01N HCl. Place black precipitate in 500 cm^3 0.5% sodium oleate solution and boil for a few minutes; the resulting colloid should be stable for several months.

*Although this etchant has commonly been used to distinguish between hexagonal and monoclinic pyrrhotite, which mineral has received the darker stain has varied from one author to another.

chanical and electrical stripping. Chemical-mechanical polishing employs a fine polishing compound in a chemically reactive fluid (commonly an etchant). The surface is prepared by the combined mechanical polishing and the chemical etching. Care must be taken when using chemical polishing techniques because they may selectively remove some phases (e.g., chromic acid/rouge may result in loss of native silver or bismuth).

Chemical etchants were widely used in mineral identification prior to the development of the electron microprobe (see Appendix 3). Although now less widely used for identification, etchants are still useful to enhance mineral textures. Reaction of the specimen with various chemicals in liquid or vapor form will often reveal mineral zoning, twinning, and grain boundaries in apparently homogeneous material and will accentuate differences in phases that are nearly identical in optical properties. A typical example of etching to enhance growth zoning is illustrated in Figure 2.8*a* in which pyrite is shown after etching with concentrated HNO_3. In Figure 2.8*b* the distinction between monoclinic and hexagonal pyrrhotite is made by etching with ammonium dichromate–HCl solution. Monoclinic pyrrhotite is also distinguished by use of a magnetic colloid, which will coat grains of this mineral while leaving adjacent grains of hexagonal pyrrhotite uncoated. Metallurgical laboratories have for many years found that a wide variety of etchants bring out alloy microstructures that are otherwise invisible. It is not possible to record all of the numerous etch reagents described in the literature, but Table 2.2 includes many of the most useful etchants used in mineralogical and metallurgical studies.

BIBLIOGRAPHY

Buehler (1973) Metallographic sample preparation. *Metal Digest* **11**, No. 2/3.

McCall, J. L. and Mueller, W. M. (1974) *Metallographic Specimen Preparation: Optical and Electron Microscopy*. Plenum, New York.

Muller, W. D. (1972) Discussion on preparation and polishing of specimens. *Mineral. Mater. News Bull. Quant. Microsc. Methods*. No. 4, pp. 13–14.

Taggart, J. E. (1977) Polishing technique for geologic samples. *Am. Mineral.* **62**, 824–827.

van de Pipjekamp, B. (1971–1972) Discussion on preparation and polishing of specimens. *Mineral. Mater. News Bull. Quant. Microsc. Methods* 1971, No. 2, p. 7; 1971, No. 3, pp. 10–11; 1971, No. 4, pp. 15–16; 1972, No. 2, pp. 6, 8.

Ramdohr, P. and Rehwald, G. (1966) The selection of ore specimens and the preparation of polished sections. In H. Freund, ed., *Applied Ore Microscopy*. Macmillan, New York, pp. 319–382.

3

Mineral Identification—Qualitative Methods

3.1 INTRODUCTION

When a polished section or polished thin section is placed on the stage of a standard reflected-light microscope, the first objective of any examination of the section is usually the identification of the minerals present. A variety of properties exhibited by each phase can be studied using the microscope without any modification of the instrument or ancillary apparatus. However, in contrast with the study of thin sections in transmitted light, these properties are qualitative* and may not be sufficient for unequivocal identification. The extent to which individual phases may be identified using these qualitative methods depends considerably on the knowledge and experience of the microscopist. To the beginner, very few minerals are immediately identifiable, and although the experienced microscopist can identify many more minerals, there will still be certain phases (or groups of phases) that cause confusion. The tendency to rely on experience has led to the view (an unfortunate misconception in the opinion of the present authors) that ore microscopy is a "difficult art." The developments in quantitative methods of reflectance and microhardness measurement discussed in Chapters 5 and 6 and the more widespread availability of such ancillary methods as electron microprobe analysis and X-ray diffraction have done much to dispel this view. There is still much to be gained by a careful qualitative examination of minerals in polished section however; given a moderate amount of experience, an appreciable number of phases can be identified positively without the use of elaborate ancillary equipment.

In this chapter, the qualitative properties of value in mineral identification are discussed. There are three major categories: (1) optical properties; (2) properties dependent on hardness; and (3) properties dependent on the structure and morphology of phases. As noted at the end of this chapter, information can also be derived from the associated minerals. The importance of spending as much time

*The study of rotation properties further discussed in Chapter 4 are an exception but the techniques have not proved satisfactory for routine work in most laboratories.

as possible looking at different minerals in various associations cannot be overemphasized for the beginning student. Appendix 1 incorporates data on qualitative properties for the more common minerals; additional detail or information on uncommon minerals may be found in reference texts such as Ramdohr's *The Ore Minerals and Their Intergrowths* and Uytenbogaardt and Burke's *Tables for the Microscopic Identification of Ore Minerals.*

3.2 QUALITATIVE OPTICAL PROPERTIES

Observations with the ore microscope are usually made either with (1) only the polarizer inserted (i.e., using linearly or "plane" polarized light) or (2) with both polarizer and analyzer inserted (i.e., under "crossed polars," the analyzer being at 90° to the polarizer). Observations are also made using air and oil immersion objectives. Color, reflectance, bireflectance, and reflection pleochroism are observed using linearly ("plane") polarized light; anisotropism and internal reflections are observed under crossed polars.

3.2.1 Color

A very small number of ore minerals are strongly and distinctively colored (e.g., covellite, bornite, gold) but most are only weakly colored and may appear to the beginner as white through various shades of gray. However, with some practice, many of the subtle color differences become apparent. Although the eye is generally good at distinguishing minor color differences between associated phases, it has a poor "memory" for colors. A further problem is that the apparent color of a mineral depends on its surroundings (e.g., the mineral chalcopyrite appears distinctly yellow against a white or gray phase, but a greenish-yellow when seen next to native gold). This phenonenon of "mutual color interference" means that it is important to see a mineral in a range of associations. Accordingly, the colors of minerals are best described in comparison with other common minerals with which they are often associated. It is important for individual observers to make their own descriptions of colors because of differences in the perception of color by different microscopists. It is also difficult to rely only on other observers' descriptions of colors; thus the information given in Appendix 1 should be regarded as only a rough guide. Another reason for caution in using qualitative color descriptions is the fact that colors are also dependent on the illumination employed (see Section 1.2.4) and will show subtle differences when different microscopes are used. When beginning work with a new instrument it is necessary to "get your eye in" by examining a few common and easily identifiable minerals, some of which may even be recognized before placing the section under the microscope.

In Section 5.5, the problem of quantitatively representing color differences is discussed and this both highlights the limitations of the qualitative approach and shows a direction in which future discussions of colors of minerals in polished section will probably progress.

3.2.2 Reflectance

The amount of light incident on a polished surface of a particular mineral that is reflected to the observer depends on an important property of that mineral, its reflectance. The reflectance of a phase (what could colloquially be termed its "brightness") is a fundamental property that will be discussed in much greater detail in Chapters 4 and 5; it is a property that can be accurately measured using equipment added to the standard reflecting microscope and is defined on a percentage scale as

$$\text{reflectance } (R\%) = \frac{\text{intensity of reflected light}}{\text{intensity of incident light}} \ (\times 100)$$

Two other cautionary notes regarding color concern the effect on color of tarnishing (e.g., bornite may appear purple rather than brown after a section has been left exposed to air) and of incorrect polishing procedures, particularly "overpolishing" (e.g., chalcopyrite, if overpolished, may appear white although it will still exhibit its characteristic yellow color at grain margins and along fractures). Both these sources of error are best eliminated by repetition of the final stages of polishing. The reflectance of a phase may vary with its orientation (see the following), with the wavelength of light being reflected (i.e., the mineral may preferentially reflect certain wavelengths and hence be colored), and with the angle of incidence of the light (although in ore microscopy, illumination is always at effectively normal incidence).

Although the eye cannot "measure" reflectance, it is easy to arrange the minerals in a section in order of increasing reflectance by visual inspection. Given one or two readily identifiable phases, the reflectances of which can be easily checked (e.g., magnetite $R\%$ ~20; pyrite $R\%$ ~55; galena $R\%$ ~43) it is possible with a little experience to estimate the reflectances of "unknown" phases by comparison with these visual "standards." It is useful also to note that quartz gangue and many mounting plastics have reflectance values of ~5%. The reflectance values of the common ore minerals are given in Appendix 1.

A number of factors can cause confusion in estimating reflectance. Minerals that take a poor polish will appear of lower reflectance than those that polish well, even though reflectance values may be similar. Color, not surprisingly, may cause difficulties since it arises as a result of the sampling by the eye of a range of light wavelengths of differing reflectances (see Chapter 4). One solution to this problem is to insert a filter to limit the light illuminating the section to a narrow range of wavelengths (say of "green" or "yellow" light). However, provided these problems are kept in mind, reflectance can be estimated by eye with sufficient accuracy to considerably aid identification.

3.2.3 Bireflectance and Reflection Pleochroism

Cubic minerals remain unchanged in reflectance and color on rotation of the stage whatever the orientation of the grains. Most minerals of other crystal symmetries show changes in reflectance or color (or both) when sections of certain orientations are rotated. The change of reflectance is a property termed *bireflectance*, and the change of color (or tint) is the property called *reflection pleochroism.* However, the isometric (basal) sections of hexagonal and tetragonal crystals do not exhibit either of these properties and appear the same as cubic minerals. In addition to noting that either of these (or both) properties are shown by a mineral, it is usual to note the intensity with which the property is exhibited (very weak, weak, moderate, strong, very strong) and, in the case of reflection pleochroism, to note the colors observed in different orientations. Examples of minerals that show reflection pleochroism are listed in Table 3.1 and illustrated in Figure 3.1.

The difference in maximum and minimum values of the percent reflectance is, of course, a measure of the bireflectance. Examples of minerals exhibiting strong bireflectance are graphite, molybdenite, covellite, stibnite, valleriite; moderate bireflectance is shown by marcasite, hematite, niccolite, cubanite, pyrrhotite; weak bireflectance by ilmenite, enargite, arsenopyrite (see Appendix 1 and Table 3.1).

It is important to remember that both properties are a function of the orientation of the crystal relative to the polarized light beam with various orientations showing anything from no effect to the maximum. In searching for weak to

TABLE 3.1 Examples of Minerals That Exhibit Reflection Pleochroism (in Air) and Bireflectance

Mineral	Color Range (Darker–Lighter)	Bireflectance Range (approximate $R\%$ at 546 nm)
Covellite	Deep blue–bluish white	7–24
Molybdenite	Whitish grey–white	19–39
Bismuthinite	Whitish grey–yellowish white	38–45
Pyrrhotite	Pinkish brown–brownish yellow	34–40
Niccolite	Pinkish brown–bluish white	47–52
Cubanite	Pinkish brown–clear yellow	35–40
Valleriite	Brownish grey–cream yellow	14–22
Millerite	Yellow–light yellow	50–56
Graphite	Brownish grey–greyish black	7–18

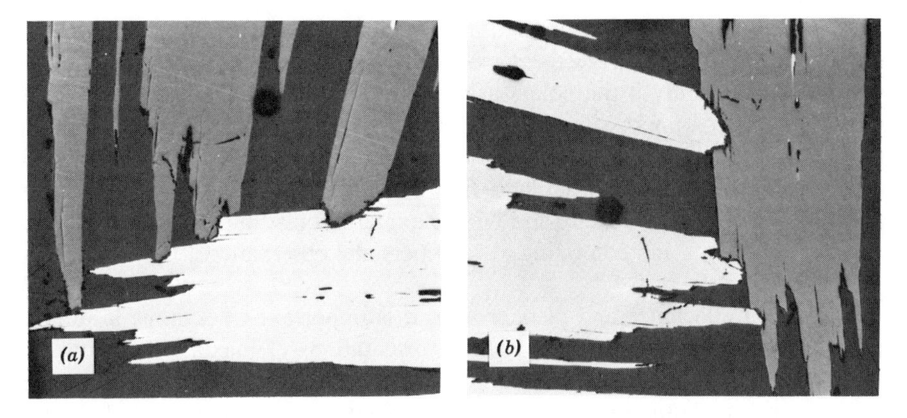

FIGURE 3.1 The properties of bireflectance and reflection pleochroism illustrated by the mineral covellite photographed in two orientations at 90° to one another in plane (linearly) polarized light. The darker gray blades in the black and white photograph are actually deep blue; the light gray blades are pale bluish white. (Width of field = 400 μm.)

moderate bireflectance or pleochroism, it is useful to examine closely adjacent grains where the eye can distinguish small differences.

3.2.4 Anisotropism

When a polished surface of a cubic mineral is examined under crossed polars, it is found to remain dark (in *extinction*) in all positions of the stage whatever the crystallographic orientation of the polished surface (in some cases the section may not be completely dark but will nevertheless remain unchanged on rotation). Such minerals are termed *isotropic.* Minerals that form crystals of lower-than-cubic symmetry will not remain unchanged as the stage is rotated for polished surfaces of most orientations; that is, they are *anisotropic.* Sections of certain special orientations of anisotropic phases may be isotropic (e.g., the basal sections of hexagonal and tetragonal crystals) but most will show variations in brightness or color, or in both, as the stage is rotated through 360°. As with bireflectance and pleochroism, the anisotropy can range from a maximum to zero (i.e., isotropic) depending on which section through the crystal has been polished. In a section showing anisotropy, there will be four positions, 90° apart, in one 360° rotation where the section is dark (in extinction) or at least shows minimum brightness. Between these (at ~45° to extinction positions) lie the positions of maximum brightness. Having noted that a mineral exhibits anisotropism, the first observation of the microscopist is of its intensity; again, the terms very weak, weak, moderate, strong, very strong are commonly used.

The colors exhibited by an anisotropic mineral on rotating the stage (*aniso-*

tropic colors) may be of value in identification when used with caution, and some are quite distinctive (e.g., the deep green colors of marcasite). However, the colors are constant only if the polars are exactly crossed and change in a characteristic manner on uncrossing the polars for a particular mineral. Aside from the problems of qualitative color description discussed in Section 3.2.1., anisotropic colors may be sensitive to differences between microscopes and illumination systems. However, used carefully and with the microscopist maintaining consistent viewing conditions and compiling a set of personal observations, they are of value in identification.

As with bireflectance and pleochroism, it is important to examine a number of grains in a polished section to obtain orientations of maximum anisotropy; when the rock itself has an oriented fabric it may be necessary to cut and polish further sections. Sometimes the combined effects of stage rotation and the anisotropism of adjacent grains make it difficult to determine whether a given grain is isotropic or weakly anisotropic. Such a determination may be more easily accomplished if the specimen is left stationary and the analyzer slowly rotated 5–10° back and forth through its "crossed" position at 90° from the polarizer; this eliminates the distraction of movements while trying to observe minor variations in color or reflectance. A well-centered objective and the use of the field diaphram to eliminate extraneous grains from the field of view also helps.

Note also that fine, parallel scratches left from incomplete polishing or careless buffing can produce effects similar to anisotropism and bireflectance.

3.2.5 Internal Reflections

Some minerals examined in polished section are transparent and others completely opaque, with some being intermediate in their "opacity." Transparent phases are, of course, best studied in transmitted light but in the intermediate category are many that are best studied in reflected light. Such phases may be sufficiently transparent to allow light to penetrate deep below the surface and be reflected back to the observer from cracks or flaws within the crystal. Such light will appear as diffuse areas or patches known as *internal reflections*. Both the occurrence of internal reflections and their colors are of diagnostic value, the latter because certain wavelengths of the incident white light (see Section 4.1.1) are preferentially absorbed by the crystal that exhibits a characteristic "body color." For example, cassiterite shows yellow or yellow-brown internal reflections, and proustite shows ruby-red internal reflections. The nonopaque phases in a polished section may, under certain illumination conditions, be a mass of internal reflections.

Internal reflections are best seen under crossed polars with intense illumination; they may also be visible in linearly ("plane") polarized light. It is important to realize that many grains of a phase that could show internal reflections may not exhibit them and a careful search must be made over the whole section. The

TABLE 3.2 Examples of Minerals That Exhibit Internal Reflections

Mineral	Color of Internal Reflections
	Often seen in air, strong in oil
Sphalerite	Yellow to brown (more rarely to green to red)
Cinnabar	Blood red
Proustite–pyrargyrite	Ruby red
Rutile	Clear yellow to deep red-brown
Anatase	Blue
Azurite	Blue
Malachite	Green
Cassiterite	Yellow brown to yellow
	Sometimes seen in air, often in oil
Hematite	Blood red
Wolframite	Deep brown
Chromite	Very deep brown

visibility of internal reflections is also enhanced by using oil immersion objectives and by using high-power magnification; they are often best seen at the edges of grains or in small grains. Some examples of common ore minerals exhibiting internal reflections are given in Table 3.2, and the internal reflections exhibited by sphalerite are shown in Figure 3.2.

3.3 QUALITATIVE EXAMINATION OF HARDNESS

A detailed discussion of hardness and its quantitative determination is the subject of Chapter 6. As will be explained further, hardness is a complex property of a mineral and the term hardness as used in ore mineralogy may refer to a number of phenomena. Three types of hardness are particularly important:

1 Polishing hardness.

2 Scratch hardness.

3 Microindentation hardness.

The third of these will be dealt with in Chapter 6 and forms the basis of quantitative hardness determination; the first two can be examined with the standard ore microscope by comparing the relative hardness of adjacent phases and can be very helpful in mineral identification. It is important to realize that these three forms of hardness are not entirely equivalent, being the response of the material to different kinds of deformation or abrasion.

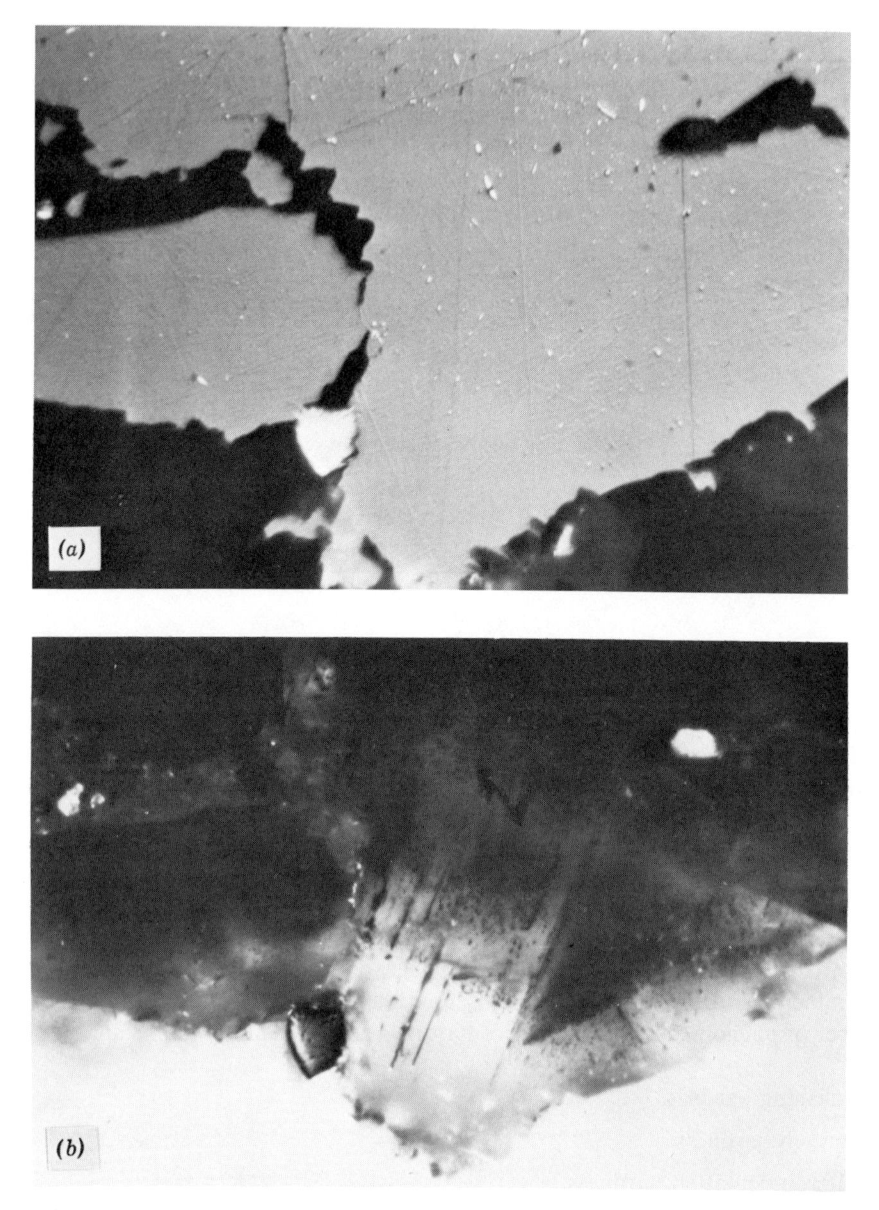

FIGURE 3.2 Internal reflections in sphalerite: *(a)* the sphalerite, which contains inclusions of chalcopyrite, photographed in plane (linearly) polarized light; *(b)* the same grain photographed under crossed polars. The reflection from the surface, and hence the isotropic nature of the phase, is masked by the intense internal reflections. (Width of field = 520 μm.)

3.3.1 Polishing Hardness

This is the resistance of a particular mineral to abrasion during the polishing process. The fact that hard minerals are worn away more slowly than soft minerals means that they may stand slightly above the surfaces of softer grains in the section—the effect known as *polishing relief.* Although in the preparation of a polished section (see Section 2.4) every attempt is made to minimize the amount of relief, the presence of some relief enables relative hardness to be rapidly estimated. This determination involves a simple test using the *Kalb light line*, which is a phenomenon analogous to the Becke line used in transmitted light work, although it is of *wholly different origin*. The procedure is as follows:

1 Focus on a clear boundary line between two mineral grains.

2 Lower the stage (or raise the microscope tube) so that the sample begins to go out of focus as the distance between the specimen and the objective increases.

3 Observe a "line" of light (cf. the Becke line) which will move toward the *softer* mineral provided there is significant relief.

The origin of the Kalb light line can be considered using Figure 3.3. At the junction between the soft and hard minerals light can be reflected at non-normal incidence. This light is not apparent when the boundary area is in focus (focal plane F) but during defocusing to F_2 is seen as a line of light moving towards the softer mineral. It is important to realize that this effect will only be seen when there is appreciable relief and this depends on the relative hardness of the adjacent grains and the method of polishing. The ore minerals can be arranged in a sequence of increasing polishing hardness although factors such as polishing method and state of aggregation of the mineral may slightly influence its position in such a table. Commonly, the polishing hardness of a mineral is compared to one or several relatively abundant minerals (e.g., a sequence where minerals are described as "less hard," "as hard," or "harder than" galena, chalcopyrite, or pyrite is sometimes used).

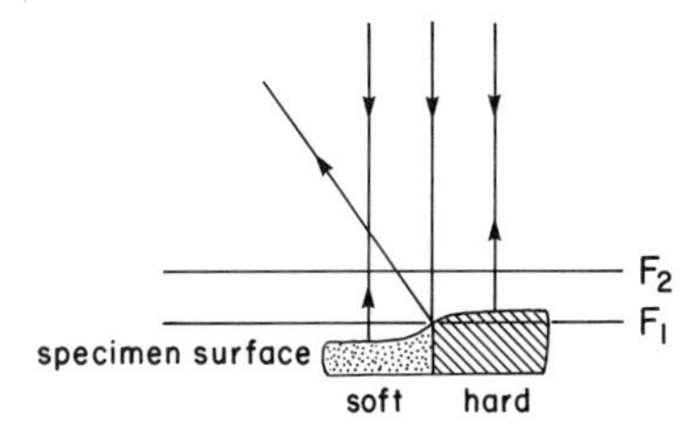

FIGURE 3.3 Cross section (schematic) of polished section surface showing origin of Kalb line at the boundary of two minerals of different hardness.

In favorable circumstances, polishing hardness can enable a fairly accurate estimate of the hardness of an unknown phase to be made.

3.3.2 Scratch Hardness

Although the perfect polished section is completely scratch-free, in practice the surface of a section always has some scratches. Again, the relative amount of surface scratching and the depth of scratches that cross grain boundaries may be used in favorable circumstances to estimate relative hardness. Although some soft minerals can acquire a smooth brilliant polish (covellite, bismuthinite), others nearly always retain a scratched appearance (graphite, molybdenite, gold, silver). Some hard minerals readily acquire a smooth polish (arsenopyrite, ilmenite, niccolite) while others require much longer polishing (magnetite, wolframite), although this also depends on the crystallographic orientation of the surface being polished. A scratch extending across the boundary between two minerals may indicate relative hardness by being more deeply incised in the softer mineral (Figure 3.4*a*). However, this test must be applied with caution because deep scratches remaining from an early stage of polishing may occur in a hard phase (pyrite, marcasite—Figure 3.4*b*), having been removed by later stages in an adjacent soft phase.

3.4 STRUCTURAL AND MORPHOLOGICAL PROPERTIES

These are the properties of minerals that depend chiefly on their crystal structures and comprise: (1) crystal form and habit; (2) cleavage, and (3) twinning. They are, of course, an essential aspect of the textures of ore minerals and mineral assemblages and are further discussed in Chapter 7. Their importance in mineral identification will be briefly considered here.

3.4.1 Crystal Form and Habit

The full range of crystal forms and habits encountered in the study of minerals in hand specimen and thin section can also be seen in polished section and the same terminology can be applied. Some minerals commonly develop well-formed crystals or *euhedra* (e.g., pyrite, arsenopyrite, magnetite, hematite, wolframite), whereas others are characteristically *anhedral* (e.g., chalcopyrite, bornite, tetrahedrite). In the average ore, the majority of minerals are not bounded by crystal outlines. It is important to remember that a polished surface only gives a two-dimensional view of a three-dimensional object; thus, for example, a cube may appear on a polished surface as a square, rectangle, or triangle.

All the standard mineralogical terms to describe crystal habit can be employed in ore microscopy (e.g., cubic, octahedral, tabular, acicular, columnar,

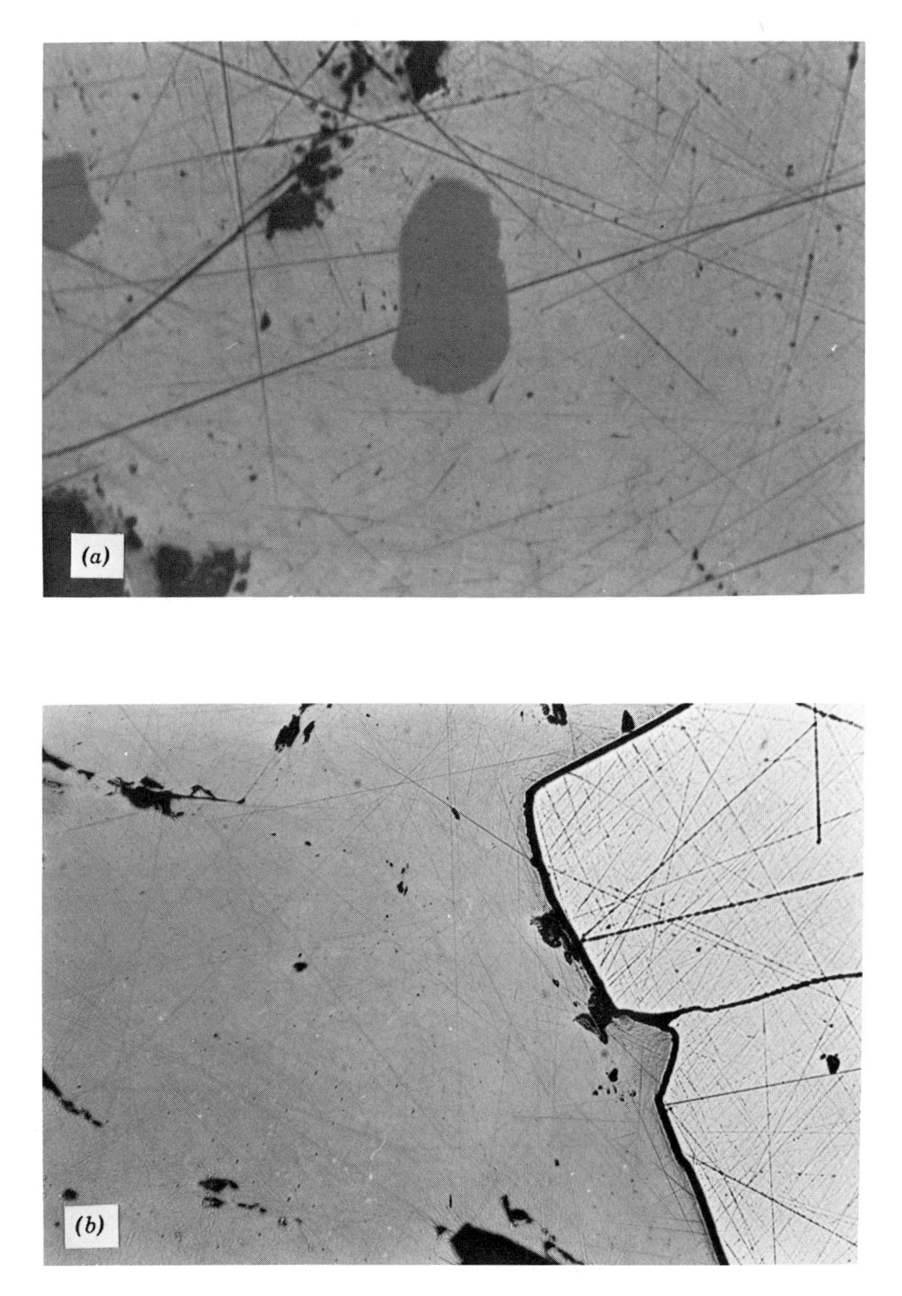

FIGURE 3.4 *(a)* Example of scratch hardness as a scratch extends from a softer mineral (galena) across a harder one (tetrahedrite) and back to the softer one. (Width of field = 1000 μm.) *(b)* Residual scratches in pyrite (white) remaining after the scratches in the adjacent sphalerite have been removed in the last polishing steps. (Width of field = 2000 μm.)

bladed, fibrous, colloform, micaceous, prismatic; see Figures 3.1, 7.4, 7.8, etc.) and the observation of characteristic habit is a considerable aid in identification. Some common examples are:

Acicular	Hematite, stibnite, jamesonite, rutile
Lath shaped	Ilmenite, hematite
Tabular	Covellite, molybdenite, graphite, hematite
Rhombic	Arsenopyrite, marcasite
Skeletal	Magnetite (due to fast crystallization), galena
Isometric forms	Cube—galena, pyrite Octahedron—chromite, spinel, pyrite, magnetite, galena Pentagonal dodecahedron—pyrite, bravoite

Crystal habit can also be used to advantage in the identification of commonly associated gangue minerals (e.g., octahedra or cubes of fluorite, rhombs of dolomite and siderite, quartz euhedra of characteristic trigonal morphology).

3.4.2 Cleavage and Parting

Although cleavage or parting is a mineral property often readily seen in hand specimen or in a thin section of a translucent mineral, it is not as commonly observed in a polished section. It may often be more readily seen at an early stage of section preparation (say after polishing with 6 μm diamond paste) than in the final specimen. Cleavage or parting is seen in polished section as one or more sets of parallel cracks, and if three or more cleavage directions are present, parallel rows of triangular pits may be observed. Such pits are particularly characteristic of galena (see Figures 3.5*a* and 7.19) and may also be observed in magnetite, pentlandite, gersdorffite, and other minerals. The development of such pits depends not only on the method of polishing but also on the orientation of the section surface relative to the cleavage directions. A prismatic cleavage gives diamond-shaped, triangular, or rectangular patterns; a pinacoidal cleavage gives a set of parallel cracks (see Figure 3.5*b*).

The cleavage of a mineral may not be evident at all in a carefully polished grain and, if the material is fine grained, it is unlikely to be evident at any stage of polishing. Cleavage is often more evident at the margins of grains and in certain cases may be beautifully brought out by weathering, some other form of alteration, or by etching. Fine examples of this include the alteration along cleavages of galena to cerussite, sphalerite to smithsonite, chalcopyrite to covellite (see Figure 7.10*b*).

3.4.3 Twinning

As further discussed in Section 7.6 three major types of twinning—growth, inversion, and deformation—can be observed in opaque minerals. Generally they

FIGURE 3.5 *(a)* Cleavage in galena shown by rows of triangular pits. (Width of field = 2000 μm.) *(b)* Cleavage in stibnite. (Width of field = 2000 μm.)

will not be seen in isotropic minerals unless the surface is etched (although they may often be evident in early stages of polishing) and are best observed in anisotropic minerals under crossed polars. Sometimes twinning may be evident from abrupt changes in the orientation of cleavages or lines of inclusions or in cubic phases that show internal reflections under crossed polars. The crystallographic planes involved in twinning are usually not determinable in polished section; nevertheless, the twin patterns in some minerals are characteristic and of considerable value in identification. Examples include the "arrowhead" twins (growth twins) in marcasite, the lamellar twins (deformation twins) seen in hematite and chalcopyrite, and the inversion twins seen in stannite and acanthite (see Figure 7.16).

3.5 OTHER AIDS TO IDENTIFICATION (PHASE EQUILIBRIA AND MINERAL ASSEMBLAGES)

The properties discussed in this chapter as the basis for identification by visual inspection under the ore microscope are all essentially the properties of individual grains. However, the samples normally studied in polished section are made up of assemblages of minerals and identification is greatly helped by considering the minerals as an assemblage rather than isolated individual phases. Ore mineral assemblages can be profitably considered in terms of (1) known phase equilibria and (2) characteristic ore types. For example, numerous studies of sulfide phase equilibria have demonstrated that most sulfide ores have reequilibrated on cooling and now occur in assemblages stable at less than 100–200°C. The refractory sulfides—pyrite, sphalerite, arsenopyrite—are notable exceptions and often retain compositional and textural features from higher formational temperatures. For most copper, lead, silver, and nickel ores, however, the low temperature phase diagrams provide a hint as to likely mineral associations. Thus using the copper–iron–sulfur system (Figure 8.14) as an example, if pyrite and chalcopyrite are identified in a polymineralic mass, one might expect to also find one of the adjacent phases—either pyrrhotite or bornite—but not both (as this would violate the Phase Rule—see Vaughan and Craig, 1978). Oxide minerals, though more refractory than sulfides, also generally occur in assemblages similar to those shown in phase diagrams. Thus one can reasonably expect to find hematite with magnetite or ilmenite, or both, but not with ulvöspinel (see Figure 9.12). The phase diagrams therefore provide a guide as to possible, but not necessary, mineral assemblages. They also provide an understanding of some textures (e.g., pentlandite exsolved from pyrrhotite, bornite exsolved from chalcopyrite) by revealing the extent and temperature dependence of mineral solid solutions.

The characteristic ore mineral associations and textural relationships discussed in Chapters 7–10 also provide valuable guides to mineral assemblages. For example, galena, sphalerite, and pyrite constitute a common assemblage in carbonate hosted deposits but tin minerals are virtually unknown in such ores. Pentlandite and chalcopyrite commonly occur with pyrrhotite, magnetite, and pyrite but sphalerite and galena are rarely present. The recognition of one unusual mineral (e.g., a telluride or selenide) should prompt the observer to be on the lookout for others.

The weathering of ores may result in equilibrium assemblages (e.g., hexagonal pyrrhotite altering to monoclinic pyrrhotite or bornite altering to covellite) or disequilibrium ores (e.g., pentlandite altering to violarite or chalcopyrite altering to covellite). In the former case, phase diagrams and characteristic assemblages are useful guides; in the latter case, equilibrium phase diagrams are less useful but the characteristic assemblages are still useful.

3.6 CONCLUDING STATEMENT

The qualitative methods of identification discussed in this chapter enable many minerals to be recognized without recourse to other methods, given a little experience on the part of the microscopist. It is important to emphasize that undue reliance should not be placed on any single property, or any single grain, but all the information should be considered in attempting to reach a decision. To confirm an identification or to choose between a number of remaining alternatives, the quantitative methods discussed in Chapters 4 and 5 may be used. Before turning to these methods, the origin of the optical effects seen using the reflected-light microscope will be considered.

BIBLIOGRAPHY

Cameron, E. N. (1961) *Ore Microscopy.* Wiley, New York.

Galopin, R. and Henry, N. F. M. (1972) *Microscopic Study of Opaque Minerals.* McCrone Research Association, London.

Ramdohr, P. (1976) *The Ore Minerals and their Intergrowths.* Pergamon, Oxford.

Uytenbogaardt, W. and Burke, E. A. J. (1971) *Tables for the Microscopic Identification of Ore Minerals.* Elsevier, Amsterdam.

Vaughan, D. J. and Craig, J. R. (1978) *Mineral Chemistry of Metal Sulfides.* Cambridge University Press, Cambridge, England.

4

Reflected Light Optics

4.1 INTRODUCTION

Light is a form of *electromagnetic radiation* that may be emitted by matter that is in a suitably "energized" (*excited*) state (e.g., the tungsten filament of a microscope lamp emits light when "energized" by the passage of an electric current). One of the very interesting consequences of the developments in physics in the early part of the twentieth century was the realization that light and other forms of electromagnetic radiation can be described both as waves and as a stream of particles (*photons*). These are not conflicting theories but complementary ways of describing light; in different circumstances either one may be the more appropriate. For most aspects of microscope optics, the "classical" approach of describing light as waves is more applicable. However, particularly (as outlined in Chapter 5) when the relationship between the reflecting process and the structure and composition of a solid is considered, it is useful to regard light as photons.

The electromagnetic radiation detected by the human eye is actually only a very small part of the complete *electromagnetic spectrum*, which can be regarded as a continuum from the very low energies (and long wavelengths) characteristic of radio waves to the very high energies (and short wavelengths) of gamma rays and cosmic rays. As shown in Figure 4.1, the more familiar regions of the infrared, visible light, ultraviolet and X-rays fall between these extremes of energy and wavelength. Points in the electromagnetic spectrum can be specified using a variety of energy or wavelength units. The most common energy unit employed by physicists is the electron volt* (eV), whereas wavelengths may be expressed in terms of *Angström units* (Å, where 1 Å $= 10^{-8}$ cm) or *nanometers* (nm, where 1 nm $= 10^{-7}$ cm $=$ 10 Å). Another unit employed in the literature of physics and chemistry is the wave number or reciprocal centimeter (cm^{-1}), which unlike the wavelength units varies linearly with energy. The nanometer is most commonly used in mineralogical literature and will be used in this book. In Figure 4.1, the relationships between these units are indicated for different

*One electron volt is the energy acquired by an electron accelerated through a potential difference of one volt.

regions of the spectrum. Shown in detail is the visible light region, which extends between approximately 390 and 770 nm; particular wavelength regions within this range are characterized by the eye as the different colors. Seven primary colors were recognized by Sir Isaac Newton (the "colors of the rainbow") and their wavelength limits are shown in Figure 4.1. *White light* from the sun or an artifical light source is comprised of contributions from all these wavelengths. Light of a very limited range of wavelengths, such as the characteristic yellow light from a sodium vapor lamp which consists chiefly of wavelengths 589.0 and 589.6 nm, is described as *monochromatic.*

In further developing the wave description of light, transmission can be considered to occur by a transverse wave motion in which the vibrations are perpendicular to the direction of travel of the energy. A wave resulting from this type of (simple harmonic) motion is shown in Figure 4.2*a* and has the shape of a sine curve. For such a wave,

$$c = \nu\lambda \tag{4.1}$$

where c = velocity of the wave, ν = frequency of the wave, λ = wavelength. As Figure 4.2*a* shows, the ray's vibration directions within any plane perpendicular to its path may be represented by a semicircle of radius equal to the vibration of the wave within that plane, except at points *a, b,* or *c* where the vibration is nil. In addition to this, a single wave of this type has a kind of spi-

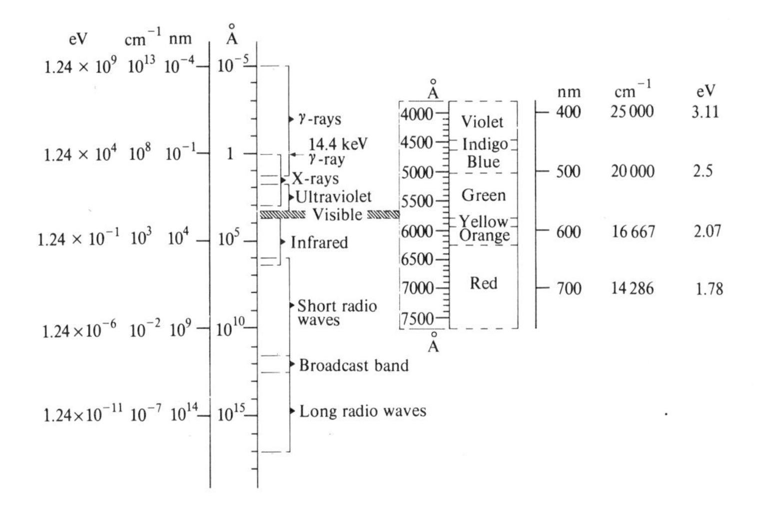

FIGURE 4.1 The electromagnetic spectrum between 10^{-5} and 10^{15} Å. Energies within this range are also shown as wavenumbers (cm^{-1}) and electron volts (eV). Energies of the visible light range are shown on an expanded scale. (Reproduced from *Mineral Chemistry of Metal Sulfides* by D. J. Vaughan and J. R. Craig, 1978, Cambridge University Press, with the publisher's permission).

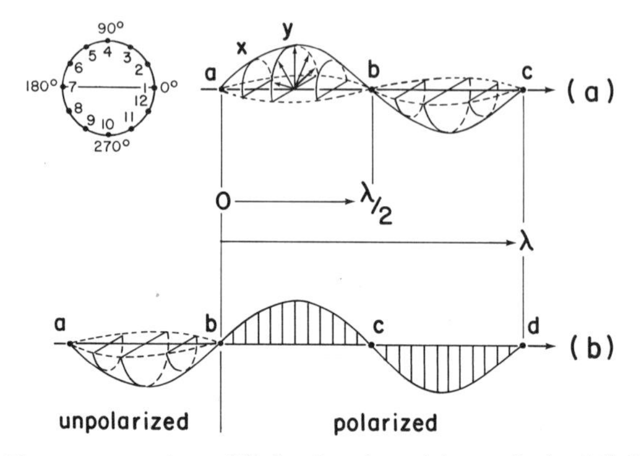

FIGURE 4.2 The wave motion of light showing: *(a)* unpolarized light with the wave motion represented along the propagation direction and normal to it (on the left); *(b)* unpolarized light in contrast to light that is linearly polarized.

ralling motion that can best be explained by describing the behavior of a single point on the wave which on progressing from $a \rightarrow b$ changes its direction of vibration with reference to the circular section normal to the line *a–b–c.* At point *a,* it can be considered as being at point 1 on circular section in Figure 4.2; by point *X* it will be at 2, point *Y* at 4 and so on. The position of this single point on the wave is termed its *phase.* Finally, the *amplitude* of the wave is the maximum vibrational displacement (i.e., the radius of the semicircular section at point *Y* in Figure 4.2).

It is possible to restrict the vibration of the light wave illustrated in Figure 4.2*a* to a single plane in which case the light is said to be *plane polarized,* although the more correct term *linearly polarized* is less confusing in discussing reflected-light optics and will be used in this text. The plane of vibration of this polarized light is that plane parallel to both the path of the ray and the vibration direction.

4.1.1 Interaction of Light with Transparent Media

The velocity of a light wave (c) or other electromagnetic wave is constant in a vacuum (299,793 km/sec) but changes if the wave passes into another transmitting medium, a change expressed by the concept of the *refractive index* of that medium:

$$n = \frac{c}{c_m} \tag{4.2}$$

where n = refractive index of the medium, c and c_{m} represent the velocity of light in a vacuum and in the medium. Since c_{m} is always less than c, n is always greater than 1.0, although for air $n = 1.0003$ (≈ 1). Since the refractive index is a ratio of two velocities, it is a dimensionless number.

Those materials through which monochromatic light travels at the same speed regardless of the direction of light vibration relative to the medium are optically *isotropic.* A vacuum, all gases, most liquids, glasses, and cubic (isometric) crystalline substances are isotropic; other materials (chiefly nonisometric crystals) are optically anisotropic and light rays may travel through them at different speeds depending on the direction of light vibration within them. The *optical indicatrix* shows how the refractive index of a transparent material varies with the vibration direction of the (monochromatic) light wave in the material. If an infinite number of vectors are imagined radiating in all directions from one point within the substance and each vector has a length proportional to the refractive index for light vibrating parallel to that vector direction, then the indicatrix is the surface connecting the tips of these vectors. For an isotropic substance, therefore, the indicatrix is a sphere of radius n (Figure 4.3*a*). However, it should be noted that the value of n does still vary as a function of wavelength of light. For anisotropic crystals, the refractive index varies with vibration direction in the crystal even for monochromatic light so that the indicatrix is not a sphere but an ellipsoid.

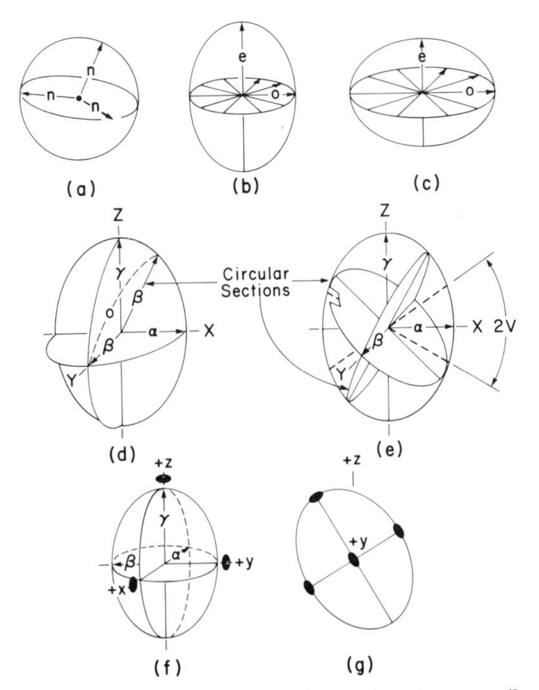

FIGURE 4.3 The optical indicatrix for: *(a)* an isotropic substance; *(b, c)* a uniaxial positive and negative crystal; *(d, e)* a biaxial crystal showing principal vibration axes and optic axes; *(f, g)* crystallographic orientation of the biaxial indicatrix in orthorhombic and monoclinic crystals.

The optical indicatrix for crystalline substances with hexagonal (used here as including the trigonal system) or tetragonal crystal symmetry is such that one cross section through the ellipsoid is circular and all other sections are elliptical. The indicatrix is termed *uniaxial* and the direction normal to this unique circular section is the *optic axis* and is parallel to the *c* axis of the crystal. The value of the refractive index in this circular section through the ellipsoid (termed the *ordinary* vibration direction and often symbolized ω or o) may be either the maximum or the minimum value of the refractive index so that the indicatrix may be either prolate or oblate as shown in Figures 4.3*b* and 4.3*c*. Conversely, the value of the refractive index of the crystal for the vibration along the direction of the optic axis (crystal *c* axis) and termed the *extraordinary* vibration direction (often symbolized ϵ or e) must also be either a maximum or minimum value. Materials in which the value (e-o) is positive (a prolate indicatrix) are termed *uniaxial positive*; those with (e-o) negative (an oblate indicatrix) are termed *uniaxial negative* in their optic signs.

The optical indicatrix for crystalline substances with orthorhombic, monoclinic, or triclinic crystal symmetry, although ellipsoidal (being geometrically termed the "general ellipsoid"), is such that there are two cross sections that are circular, all others being ellipsoidal. Again, the two unique directions normal to these sections are optic axis so that the indicatrix is termed *biaxial* (Figure 4.3*e*). The biaxial crystal has three principal refractive indices that are commonly symbolized α, β, γ with α and γ being the smallest and largest, respectively. The vector directions corresponding to the α and γ refractive indices are unique directions termed *principal vibration axes* and symbolized X and Z as shown in Figure 4.3*d*. These are always orthogonal axes of maximum and minimum length, respectively, and the principal vibration axis (Y), corresponding to the vector direction of the β refractive index, is also always normal to the X and Z axes. However, since the value of β is intermediate between α and γ, there must be an equivalent vector in the arc between the Z and X vibration axes as shown in Figure 4.3*d*. There must in fact be a complete series of vectors of length β between these two that give a circular section through the indicatrix. Examination of the figure will show that two such circular sections must exist intersecting along the Y axis as shown in Figure 4.3*e*; as in the uniaxial indicatrix, the normals to these circular sections define the optic axes (hence *biaxial*), which must always lie in the plane XZ (*optic axial plane*). The angle between these planes of circular section, and hence between the optic axes, varies depending on the relative values of α, β, and γ and by convention the acute angle between the optic axes is the *optic axial angle* (or $2V$). When Z (γ) is the acute bisectrix, the crystal is *biaxial positive* in optic sign and when X (α) is the acute bisectrix, it is *biaxial negative*. The special case when $2V = 90°$ is optically signless.*

Although in uniaxial crystals the optic axis always coincides with the crys-

*Some problems arising with regard to the determination of optic sign have been discussed by Galopin and Henry (1972, p. 277).

tallographic c-axis direction, in biaxial crystals the relationship between the orientation of the indicatrix and the crystallographic axes is not so straightforward. For orthorhombic crystals, the X, Y, and Z (principal vibration) axes of the indicatrix all coincide with the a, b, c crystallographic axes but although the relationship $a = X$, $b = Y$, $c = Z$ is possible (see Figure 4.3*f*) any one of the five other combinations ($a = Y$, $a = Z$, etc.) is also possible. In monoclinic crystals, a principle axis X, Y, or Z coincides with the single twofold axis normally selected as the b axis. The two other principal axes then lie in the plane perpendicular to this twofold axis (see Figure 4.3*g*). Finally, in the triclinic crystal, none of the X, Y, or Z axes coincides with a crystallographic axis unless by chance.

It is important to note that as the value of $2V$ approaches zero in a biaxial negative crystal it approaches a uniaxial negative indicatrix; in a similar way, a biaxial positive crystal approaches a uniaxial positive when $2V$ approaches zero. Also as e and o approach the same value, the uniaxial anisotropic crystal approaches the isotropic. For anisotropic materials, the difference between their two most divergent refractive indices ($e - o$ for uniaxial, $\gamma - \alpha$ for biaxial) at a particular wavelength is known as *birefringence.*

4.1.2 Interaction of Light with Opaque Media

The refractive index of an opaque substance is a complex number defined as follows:

$$N = n + ik \tag{4.3}$$

where N is the *complex refractive index,* n is the *refractive index* (or ratio of the velocities of light in the two adjoining media), k is the *absorption coefficient,* and i is the complex conjugate (Jenkins and White, 1976).

As with transparent crystalline solids, the interaction of light with opaque crystalline materials depends on the physical state of the material. Under the reflected light microscope, a flat polished surface is subjected to light at normal incidence, a certain percentage of which is reflected directly back as the *specular* component. Such specular reflectivity is normally expressed as a percentage of the incident light and relates to the optical constants, n and k, at normal incidence through the Fresnel equation:

$$R = \frac{(n - N)^2 + k^2}{(n + N)^2 + k^2} \tag{4.4}$$

where n = refractive index of the substance
N = refractive index of the medium (commonly, air or immersion oil)
k = absorption coefficient of the substance
R = reflectance (when $R = 1$ corresponds to 100% reflectance)

When the medium is air, for which $N \sim 1$, the Fresnel equation becomes

$$R = \frac{(n - 1)^2 + k^2}{(n + 1)^2 + k^2} \tag{4.5}$$

Equation 4.5 makes it clear that when the medium has a refractive index above 1 (e.g., water, $N = 1.33$; index oil, $N = 1.515$) the reflectance is reduced from the value in air.

If an opaque material is crushed to a fine powder, the nature of its reflecting properties changes because some light can now pass through particles before being reflected back. Such *diffuse reflectance* is analogous to the *streak* of a mineral and is the color exhibited by the mineral when light is transmitted through it (e.g., in thin or polished thin sections of certain minerals).

The physical nature of the reflecting process and the significance of variations in n and k in substances of different composition will be further discussed in Chapter 5. In the remainder of this chapter, the classical treatment of optics is applied to understanding the behavior of flat polished surfaces of opaque materials when examined under the ore microscope.

4.2 REFLECTION OF LINEARLY (OR "PLANE") POLARIZED LIGHT

4.2.1 Monochromatic Linearly Polarized Light Reflected from an Isotropic Surface

An incident beam of linearly polarized light is unchanged in its polarization when reflected from a perfectly flat isotropic surface (whether of a transparent or an opaque mineral). The value of the percentage reflectance ($R\%$) is given in terms of the optical constants by the Fresnel equation (see equation 4.4). It has been shown for isotropic transparent crystals, that the uniform value of refractive index (n) for any orientation of the vibration direction in the crystal is represented by the spherical optical indicatrix (Figure 4.3*a*). However, the refractive index of an absorbing substance (N) contains a number couple (equation 4.3) so that this "complex indicatrix" cannot be represented by a three-dimensional surface. It is possible to represent the variation of the component refractive index (n) and absorption coefficient (k) terms as a function of vibration direction in the crystal, and each of these *indicating surfaces** is a sphere as shown in Figure 4.4*a*. The reflectance is also a real number and can be represented by a single surface that for isotropic materials will, of course, be a sphere (Figure 4.4*a*). Consequently, isotropic opaque minerals have only one reflectance value (R) whatever the orientation. As in transparent materials all cubic (isometric) crystals, as well as basal sections of hexagonal (including trigonal)

*The term indicatrix should be restricted to the application explained in Section 4.1.1.

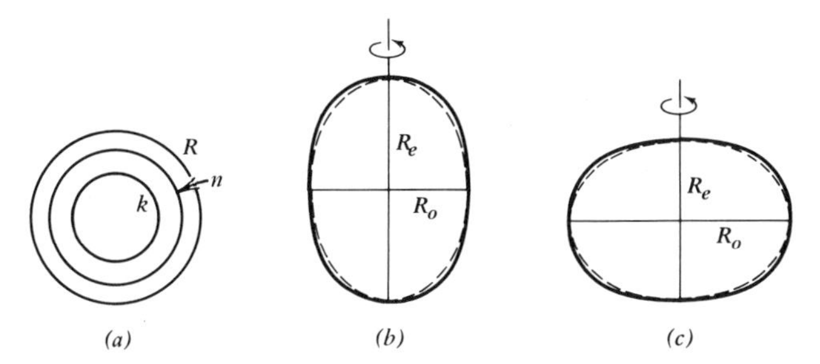

FIGURE 4.4 Indicating surfaces for opaque substances: *(a)* R, n, and k for an isometric substance; *(b)* the R-surface only for a uniaxial positive crystal; *(c)* the R-surface only for a uniaxial negative crystal. Note that the dashed line in *(b)* and *(c)* shows the outline of a regular ellipsoid drawn on the same axis and that these figures represent sections through three-dimensional spheres or ellipsoids.

and tetragonal crystals, are isotropic. Another useful way of expressing these relationships is in terms of *optical symmetry planes.* Every plane through an isometric crystal (or straight line on a polished flat surface of that crystal) is a plane (or line) of optical symmetry. The same is true of basal sections of hexagonal and tetragonal crystals so that linearly polarized light incident on such surfaces is still linearly polarized on reflection.

4.2.2 Monochromatic Linearly Polarized Light Reflected from an Anisotropic Surface

As has been discussed for the simpler case of transparent materials, crystalline opaque substances with hexagonal or tetragonal symmetry are optically *uniaxial.* However, for absorbing uniaxial crystals, the indicating surfaces for n, k, and R depart in shape from the regular ellipsoid familiar in transparent materials and are mathematically more complex. This is illustrated in Figures 4.4*b*, and 4.4*c* and further discussed in the following section. Nevertheless, as the indicating surfaces for R illustrated in Figures 4.4*b* and 4.4*c* show, the figure outlined has a single circular section with a unique direction normal to it (i.e., the optic axis). As in transparent materials, the vibration direction parallel to this axis is the (principal) *extraordinary* vibration direction, whereas that in the circular section is the *ordinary* vibration direction. Reflectance values measured for light vibrating parallel to these directions are consequently labeled R_e and R_o and each one is generally either the maximum or minimum value. However, as in transparent crystals, either $R_e > R_o$, resulting in the prolate figure of *uniaxial positive* crystals, or $R_o > R_e$, resulting in the oblate figure characteristic of *uniaxial negative* crystals* (Figures 4.4*b* and 4.4*c*). As in the true uniaxial indica-

*These are the *bireflectance signs* of the crystals.

trix, the optic axis is parallel to the crystallographic *c*-axis so the ordinary vibration and the circular ("isotropic") section lies in the basal plane. The values of both R_o and R_e can only be obtained from a prismatic section (parallel to *c*), and the difference between these values ($R_e - R_o$) is a measure of the *bireflectance* of the mineral. As noted in Chapter 3, bireflectance is a clearly observable qualitative property of many opaque minerals and one that can be quantified (see Chapter 5).

When a beam of linearly polarized light is reflected at normal incidence from a flat polished surface of an opaque uniaxial mineral (but *not* the basal section), the light beam can be considered as split up into two mutally perpendicular linearly polarized beams. These correspond to the two lines of optical symmetry present in any section of a uniaxial crystal (and analogous to the extinction directions observed in a transparent mineral in thin section). For the two positions of the polished surface corresponding to alignment of the polarization direction of the incident beam with these vibration directions, the light is reflected with its polarization state unchanged. However, for orientations intermediate between these (i.e., when the microscope stage is rotated away from either of these unique directions), components of the two vibrations will combine and where there is a phase difference between them the resulting beam will be *elliptically polarized.*

Further discussion of reflected light optics requires development of the ideas of elliptical polarization. In fact, there are three principal types of polarized light: the familiar *linearly* (or *plane*) polarized, *elliptically* polarized, and *circularly* polarized. Linearly polarized light has already been described (Section 4.1 and Figure 4.2). In order to understand the other two types of polarization it is necessary to consider two linearly polarized waves with vibration planes at right angles to one another traveling along the same ray path. The wave motion which results will be a single wave produced by combination of the two waves, but the nature of this resultant wave will depend on the relative phase and amplitude of the two original waves. The simplest case is illustrated in Figure 4.5*a* where the waves are of the same amplitude and are "in phase" (i.e., the nodes of zero vibrational amplitude, like *a, b, c* of Figure 4.2, perfectly coincide). Here, the resultant wave produced by interference between the two waves is still linearly polarized and its direction of polarization is at 45° to the two "parent" waves. The relationship between parent waves and the resultant can also be illustrated by the cross-sectional view normal to the direction of wave propagation, as shown in Figure 4.5*d*. If the two parent waves differ only in amplitude, the resultant wave is still linearly polarized but is not at 45° to the parent waves. Resolution of the component vectors shows it will be closer to the vibration direction of greater amplitude (Figure 4.5*e*).

Complications occur when the two linearly polarized waves differ in phase. If this is a phase difference of 90° or, to put it another way, if the nodes of zero vibration are displayed by $1/4\ \lambda$ [or $3/4\ \lambda$, $5/4\ \lambda \ldots (2n + 1)/4\lambda$ so that the point of zero vibration of one parent wave coincides with the point of maximum vibration (amplitude) of the other parent wave, the resultant wave will be circularly polarized. This is illustrated in Figure 4.5*b*, from which it can be seen that the resultant vibration vectors are of constant length but variable azimuth so that they progress in a spiral like the thread of a screw. Viewed as the cross section normal to the direction of propagation, the vectors outline a circle (Figure 4.5*f*)—hence, circular polarization. Clearly, like the combination of two linearly polarized waves that are "in phase," a phase difference of 90° resulting in circular polarization is a special case. When the phase difference is not 90° [i.e., when the nodal points of the two waves are displaced by differences other than $(n + 1)/4\lambda$, the vibration vectors progress in a spiral but are not of constant length (Figure 4.5*c*)]. Consequently, the cross section shown in Figure 4.5*g* is an ellipse rather than a circle—hence, elliptical polarization.

The kind of phase difference noted in the preceding can result from the interaction of linearly polarized light with a material such as a uniaxial crystal, the optical constants (n, k and hence N) of which vary with direction. Hence elliptical polarization results for orientations other than those corresponding to vibration along directions of optical (as well as crystallographic) symmetry. In the isotropic crystal, of course, every direction is a direction of optical symmetry and there is no variation in the values of the optical constants with orientation. Consequently, the incident linearly polarized light beam is unchanged in its polarization after reflection.

Opaque crystalline materials with orthorhombic, monoclinic, or triclinic symmetry can also, like the transparent substances, be termed *biaxial.* However, the differences between the indicatrix of transparent materials and the indicating surfaces of the opaques is much greater than between uniaxial transparent and opaque crystals. The biaxial indicatrix derives its name from the fact that for two directions (the optic axes) normal to circular sections of the indicatrix, linearly polarized light is transmitted without any change in polarization. The same is not true of biaxial opaque crystals because here the "axes" for the indicating surface for n do not coincide with the surface for k. There are singular directions but they do not coincide for the three surfaces involved (n,k,R); hence there is no section analogous to the circular section in transparent biaxial minerals. Furthermore, in the opaque crystals, only certain planar sections are geometrically representable as indicating surfaces. In the orthorhombic system, it is the sections in the *xy, yz,* and *xz* crystallographic planes as shown in Figure 4.6*a*. The surfaces all have the symmetry of the crystallographic point group *mmm* as illustrated in Figure 4.6*b* and they are the optical symmetry planes. In the monoclinic system, only the section perpendicular to the diad axis is representable (Figure 4.6*c*) and the surface, which is a crystallographic and optical symmetry plane has the symmetry of the point group $2/m$ (Figure 4.6*d*). In the triclinic system, *no* surface is geometrically representable as the indicating surface for n, k, and R (and there are *no* optical symmetry planes).

For the absorbing biaxial crystals, light normally remains linearly polarized only for vibration directions parallel to a plane of optical symmetry (or cut nor-

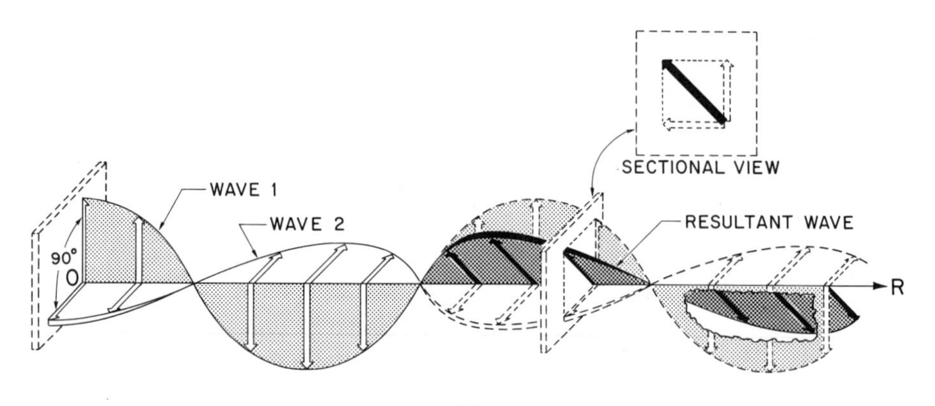
SECTIONAL VIEW
WAVE 1
WAVE 2
RESULTANT WAVE
90°
O
R

(a)

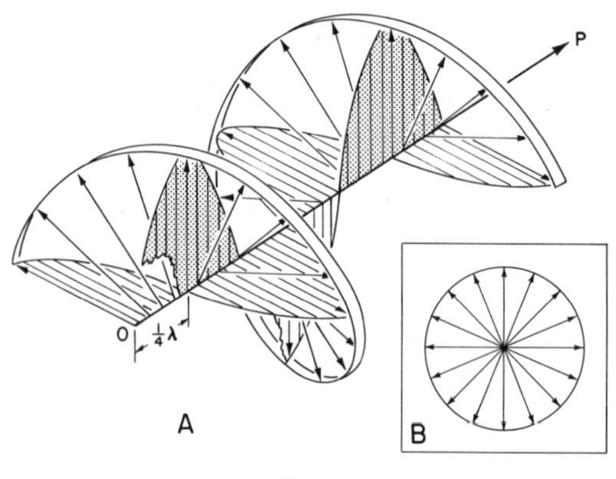
P
O
$\frac{1}{4}\lambda$
A
B

(b)

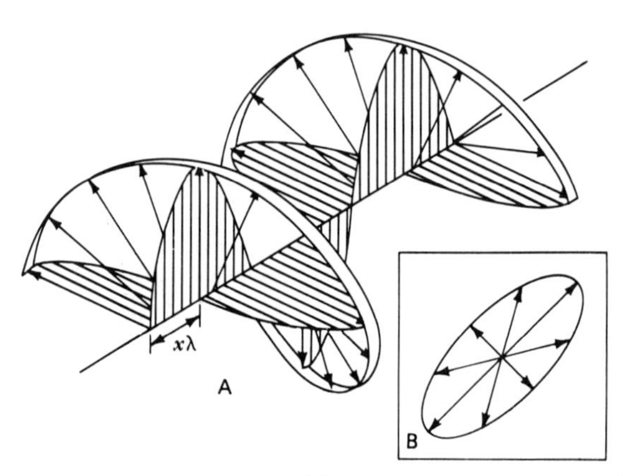
$x\lambda$
A
B

(c)

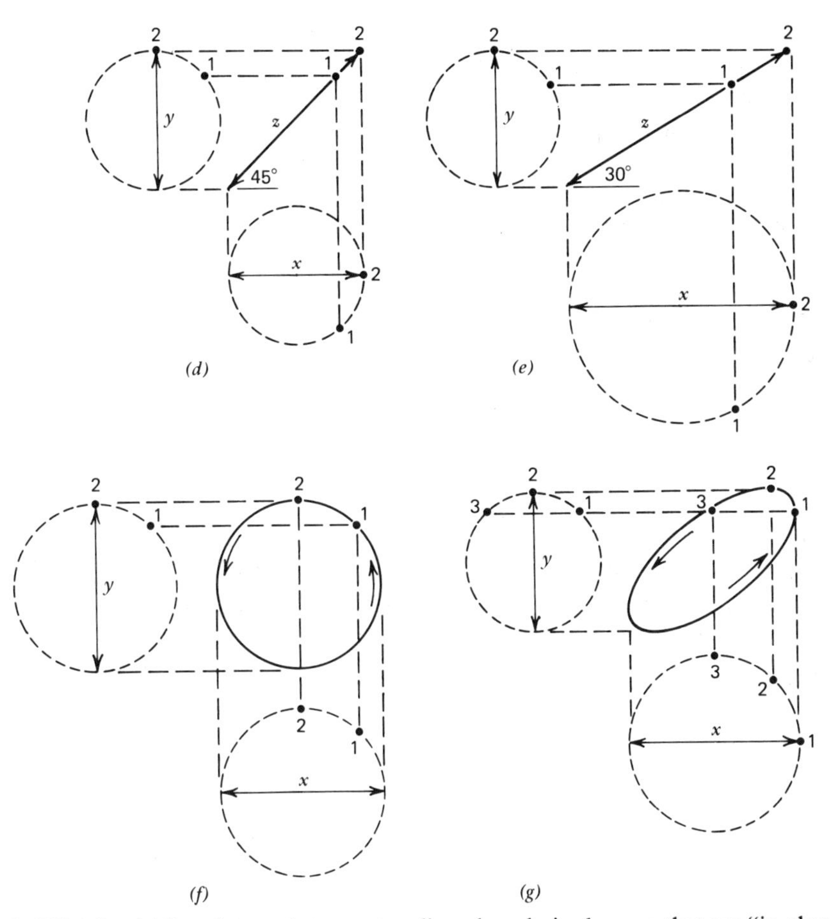

FIGURE 4.5 *(a)* Interference between two linearly polarized waves that are "in phase" and oriented at right angles to each other. The resultant wave motion is still linearly polarized and produced by adding the vibration vectors of the two parent waves as illustrated in the sectional view. (Reproduced from *An Introduction to the Methods of Optical Crystallography* by F. D. Bloss, Holt, Reinhart and Winston, with the publisher's permission.) *(b)* Interference between two mutually perpendicular linearly polarized waves that are out of phase by $\frac{1}{4}\lambda$. The resultant wave motion is the spiralling vibration shown in *(A)*, which actually has a circular cross section *(B)* so that the wave is circularly polarized. (Reproduced, as above, from Bloss (1961), with permission of the publisher). *(c)* Interference between two mutually perpendicular linearly polarized waves that are out of phase (by $x\lambda$). The resultant wave motion is again a spiralling vibration shown in *(A)* but this has an elliptical cross section *(B)* so the wave is elliptically polarized. *(d)* Interference of two linear vibrations *(x,y)* normal to each other, equal in phase and amplitude. Resultant is the linear ("plane polarized") vibration *z*. *(e)* Interference of two linear vibrations *(x,y)* normal to each other, equal in phase but not in amplitude. Resultant is linear (plane polarized) vibration *z*. *(f)* Interference of two linear vibrations *(x,y)*, normal to each other, equal in amplitude but differing in phase by 90°. Resultant is a circular vibration. *(g)* Interference of two linear vibrations *(x,y)* normal to each other, differing in amplitude and in phase by 45°. Resultant is an elliptical vibration.

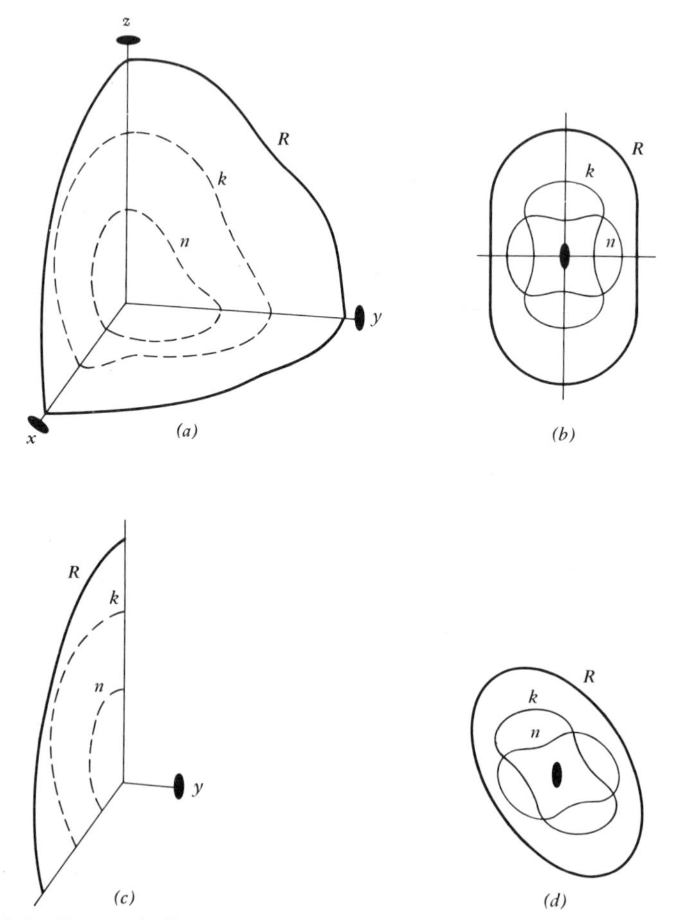

FIGURE 4.6 Geometrically representable sections of indicating surfaces in the biaxial systems: *(a, b)* orthorhombic in which the three pinacoidal sections are representable; *(c, d)* monoclinic in which only the section perpendicular to the diad axis is representable.

mal to an optic axis). For an orthorhombic crystal, there are three values of the optical constants (n_α, n_β, n_γ; k_α, k_β, k_γ) and of the reflectance (R_α, R_β, R_γ) corresponding to light vibrating along the X, Y, and Z directions of optical symmetry, which coincide with one or the other of the a, b, c crystallographic axes. As in the biaxial indicatrix, the R_β value that corresponds to the Y axis is the intermediate value, so that the bireflectance is given by the difference between R_α and R_γ. In practice, however, since specially oriented sections are required to make such measurements on orthorhombic crystals (when values may be reported as R_a, R_b, R_c corresponding to a, b, c crystal axes), reflectance values are normally reported as minimum and maximum values and symbolized R_1 and R_2, since although one may start with $R_2 > R_1$, this relationship may change with the wavelength of the light. In this case, of course, the bireflectance is the difference between R_1 and R_2. In the monoclinic system, as shown in Figures 4.6*c* and 4.6*d* although two of the values R_α, R_β, and R_γ are contained in the symmetry plane normal to the diad (*b* crystallographic) axis, the only practical measurements are again of maximum and minimum values (R_2 and R_1) and the same is true of triclinic crystals that show no relationship between crystal axes and directions of optic axes.

4.2.3 "White" Linearly Polarized Light Reflected from an Isotropic Surface

As already outlined, "white" light consists of contributions from light of wavelengths (or energies) throughout the whole of the visible region of the electromagnetic spectrum. The fact that many opaque materials do not uniformly reflect back all the component wavelengths of incident "white" light is what produces the phenomenon of color as detected by the human eye or some other system capable of recording reflected intensities throughout the visible region. Such measuring systems will be discussed much more fully in Chapter 5; it is sufficient for the present to note that the measurement of the reflectance of a variety of isotropic homogeneous crystalline materials generally yields curves such as those illustrated in Figure 4.7. The reflectance of a perfectly white material will be independent of the wavelength of the incident light; blue materials will show greater values of $R\%$ towards the 400 nm (or blue) region of the spectrum and the red materials greater $R\%$ values toward the 700 nm (or red) region. The curves shown in Figure 4.7 are termed *spectral dispersion* curves and are both a quantitative representation of the color of the opaque material and

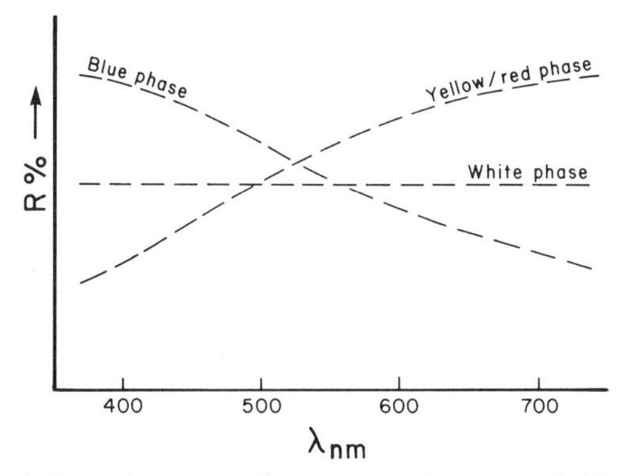

FIGURE 4.7 Spectral dispersion curves for opaque substances of different color.

an aid to identification (see Section 5). Many such curves have been measured and show that for most materials, the optical constants (n and k) vary as a function of the wavelength of light.

4.2.4 "White" Linearly Polarized Light Reflected from an Anisotropic Surface

As already discussed, in anisotropic substances n, k, and R vary as a function of orientation; thus for such materials in "white" light, the combined effects of orientation and wavelength must be considered. For uniaxial opaque minerals, spectral dispersion curves can be plotted for R_o and R_e as illustrated in Figure 5.5 for the mineral covellite (curves for other orientations commonly, but not always, lie between these). These curves, in which R_o and R_e are plotted against wavelength, show the bireflectance of the mineral, which is the separation between R_o and R_e at a particular wavelength; the differing shapes of the curves as a function of wavelength (dispersion) illustrate the property of *reflection pleochroism.* For biaxial minerals, the same data can be presented through plots of R_1 and R_2.

The variation of n, k, and R as a function of wavelength will naturally result in changes in their indicating surfaces as a function of wavelength. For isotropic materials, this only means a variation in the sizes of the spheres (Figure 4.4*a*) but for anisotropic crystals the changes will be much more complex. This can be illustrated by considering the changes in the reflectance (R) surface for covellite (CuS) for a series of wavelengths. The data in Figure 4.8 show that from 656 to 678 nm, R_o grows relative to R_e and at 700 nm, $R_o = R_e$ (von Gehlen and Piller, 1964; Galopin and Henry, 1972).

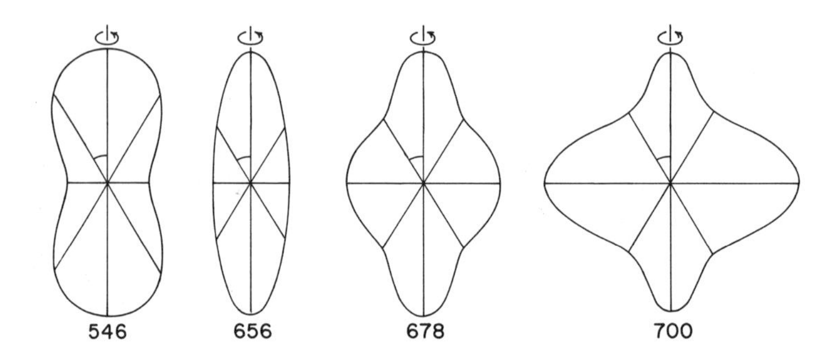

FIGURE 4.8 Sections through the reflectivity surface for covellite (CuS) showing its variation as a function of wavelength (in nm) from 546 to 700 nm. Note that the shape of the surface shows that R_o or R_e is not always a maximum or minimum value at a particular wavelength. The optic axis is vertical.

4.3 REFLECTION BETWEEN CROSSED POLARS

4.3.1 Monochromatic Linearly Polarized Light

All sections of isotropic substances as well as basal sections of uniaxial crystals are such that every vibration direction of a beam of linearly polarized light at normal incidence coincides with a plane of optical symmetry. The light is therefore reflected as linearly polarized light with the direction of polarization unchanged. If the analyzer is inserted in the cross (90°) position relative to the polarizer, reflected light is completely blocked whatever the position of the section on the stage (i.e., the section will be *optically isotropic*). Light may still reach the observer from internal reflections, surface scratches, or imperfections, but none is reflected by the flat polished surface. Also, faint illumination (particularly at high magnification) may be observed from slight ellipticity produced when the incident beam is not perfectly normal to the surface; this does not alter in intensity on rotating the microscope stage.

Apart from the basal section discussed previously, all other sections of uniaxial crystals are perpendicular to two optical symmetry planes. This is also true of sections of the type (Ok1), (hO1) and (hkO) in the orthorhombic system, and the (hO1) sections of the monoclinic system. In such *symmetric* sections, the two vibrations along these directions are linearly polarized. When the crossed polars are aligned with these directions, which will occur every 90° of rotation of the stage, the section will extinguish.* In other positions, the resultant wave of these two vibrations (which may differ in amplitude, in phase, or both) is not parallel to the vibration direction of the analyzer and the section will not be completely dark (i.e., it will exhibit *optical anisotropy*). The case where the two vibrations differ only in amplitude (Figure 4.5*e*) produces a resultant that is still linearly polarized but not at 45° to the parent waves. In Figure 4.9, the type of anisotropy that results from observing this effect under crossed polars is illustrated. When the stage is moved from the extinction position (e.g., when the section is in the 45° position halfway between two extinction positions), the resultant vibration is rotated and the reflected beam is rotated through the angle ω, so that a component (shown by OT in the figure) can be transmitted by the analyzer to the observer. This rotation angle can be measured by rotating the analyzer ($OP \rightarrow OR$) to restore extinction and recording the angle.† Clearly, the larger this rotation, the greater the component of light transmitted by the analyzer and the greater the anisotropy observed under the microscope. This is the case when the two vibrations differ only in amplitude; if they differ only in phase (Figure 4.5*g*), the resultant light wave is elliptically polarized although it is not rotated (Figure 4.9*b*). Although the section will again extinguish every 90° of rotation of the

*This is analogous to the "straight extinction" observed in certain minerals in transmitted light.

†This is the anisotropic rotation (symbolized ω or sometimes A_r).

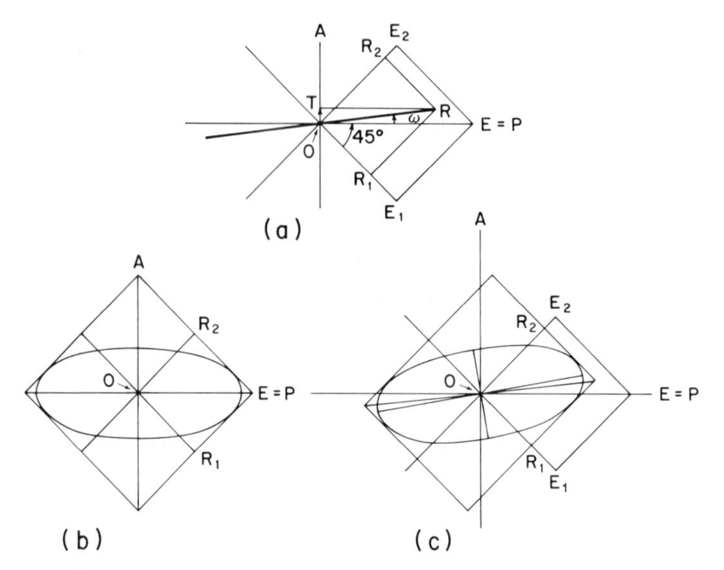

FIGURE 4.9 The origins of anisotropy when a linearly polarized beam parallel to polarizer *(P)* is reflected with the analyzer *(A)* inserted. *(a)* When the two component vibrations (R_1,R_2) differ only in amplitude, the resultant *(R)* linearly polarized beam is rotated through the angle ω when the section is, as here, in the 45° position. The amplitude transmitted by the analyzer *(A)* is given by the length *OT*. *(b)* When the two component vibrations (R_1,R_2) differ only in phase, the resultant wave is elliptically polarized although with the major axis of the ellipse along the vibration direction of the incident beam (i.e., the polarizer, *P*). *(c)* When the two component vibrations (R_1,R_2) differ in both phase and amplitude, the reflected beam is elliptically polarized and the major axis of the ellipse is rotated. (*OE* = incident vibration direction; OE_1, OE_2 = component incident vibration directions.)

stage under crossed polars (because the light is still linearly polarized in these orientations parallel to optical symmetry planes), some light is transmitted in the intermediate positions where the light is elliptically polarized and this will be a maximum in a certain position (although not necessarily the 45° position, see Galopin and Henry, 1972, p. 282). Furthermore, in this position, the section cannot be extinguished by rotation of the analyzer since the elliptically polarized light will still have a component that can be transmitted by the analyzer. When the two vibrations reflected from a symmetric section differ in both amplitude and phase, then the reflected beam is elliptically polarized and the major axis of the ellipse is rotated (towards the vibration of greater amplitude). This combination of the first two phenomena described is illustrated in Figure 4.9*c*. Again, of course, the section will extinguish every 90° of rotation of the stage under exactly crossed polars. However, in the 45° position, although the analyzer can be rotated to coincide with the major axis of the ellipse, since the light is still elliptically polarized, the section cannot be extinguished. It is important to remember here that we have been discussing how anisotropy is produced when the stage is rotated to the 45° position and that the same phenomena occur between the 45° positions and the extinction (90°) positions. Since the rotation angle or ellipticity of the vibration is gradually decreasing as the section is rotated towards extinction, the intensity of the anisotropy also decreases gradually.

Sections of orthorhombic and monoclinic crystals other than those discussed already and all sections of triclinic crystals (which have no planes of optical symmetry) are termed *asymmetric sections.* Here, both vibrations that result from reflection of a beam of plane-polarized light are elliptically polarized with the major axes of the ellipses normal to each other. Rotation of the microscope stage under crossed polars does not produce complete extinction at *any* position but there are four minima of illumination at 90° intervals. These correspond to light being dominantly contributed by only one of the two elliptical vibrations. In the 45° positions (which approximately correspond to illumination maxima), the resultant vibration is also elliptically polarized and extinction cannot be achieved by rotation of the analyzer.

4.3.2 "White" Linearly Polarized Light

It has been explained already that for many opaque minerals the optical constants (n, k) and hence reflectance (R) vary as a function of wavelength throughout the visible region (i.e., they exhibit spectral *dispersion* and are therefore colored when observed in "white" light). Under crossed polars with white linearly polarized light illumination, isotropic minerals or sections remain in extinction but other sections do not, and the light transmitted through the analyzer may show color as well as intensity variations. These "anisotropic-rotation colors" can be of diagnostic use in certain minerals, so it is useful to examine their origin.

In the case of symmetric sections discussed in Section 4.3.1 and illustrated in Figure 4.9*a*, we saw that if the two vibrations produced on reflecting a wave of linearly polarized light differ in amplitude, the resultant wave is rotated and the extent of this rotation is a function of the difference in amplitudes (and in turn in the difference in reflectances). Since the reflectances in such sections can vary as a function of both orientation and wavelength, the angle of rotation can vary as a function of wavelength resulting in "dispersion of the rotation" (and hence of the anisotropy). This is illustrated in Figure 4.10*a* in which the R_2 reflectance value for both blue and red light is the same (nondispersed), whereas the R_1 reflectance value is very much greater for red light (strong dispersion). This results in greater rotation of the resultant reflected vibration for blue than red light so that more blue light is transmitted through the analyzer (so a blue anisotropic tint would be observed under the microscope). If the analyzer in this case is rotated anticlockwise, so as to partially uncross it, more red light is transmitted relative to blue light so that the anisotropic tint systematically changes (Figure 4.9*b*). In symmetric sections it is also true that the tint observed for a given angle

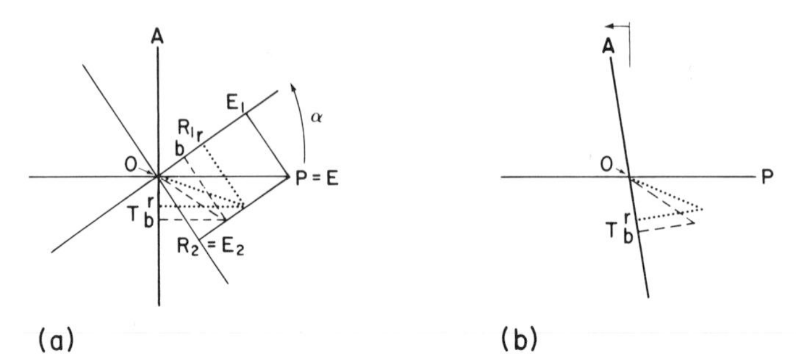

FIGURE 4.10 Dispersion of the rotation in symmetric sections. *(a)* In the position with the section rotated $\alpha°$ from extinction, the blue light *(b)* is more rotated than the red *(r)* light so that the amplitude transmitted *(OT)* through the analyzer *(A)* is greater for blue light in this section. *(b)* On uncrossing the analyzer, the amount of red light transmitted by the analyzer increases relative to the amount of blue light. (A = analyzer; P = polarizer; OE = incident vibration direction; OE_1, OE_2 = component incident vibrations directions; R_1, R_2 = reflectances in these directions.)

of clockwise turn of the stage is the same as that observed for the same angle of turn anticlockwise. In asymmetric sections, all vibrations are elliptically polarized (see Section 4.3.1) but again, variation in reflectance and hence amplitude with orientation of the section and with wavelength means that the rotation of the major axis of the ellipse can be wavelength dependent. The intensity of light transmitted by the analyzer is again different for different wavelengths, producing the anisotropic tints that again will show variation as the stage is rotated or as the polars are uncrossed. However, for the asymmetric section, the tint observed for a given angle of clockwise turn of the stage is not the same as that observed for the same angle of turn anticlockwise.

4.3.3 Convergent Light

Employing the usual viewing conditions under the reflected light microscope, the light is reflected back from the surface of the section at normal (or near normal) incidence. However, if an objective of large numerical aperture is employed (see Section 1.2), *convergent* light enters the microscope system as shown in Figure 4.11*a*. Through insertion of the Bertrand lens (or removal of the objective if no Bertrand lens is fitted on the microscope), the image in the back focal plane of the objective can be brought into focus for the observer. In this mode of observation, the light seen in the field of view is not reflected at normal incidence except near the center of the field. If linearly polarized light is reflected in this way from an isotropic surface (i.e., an isometric crystal or basal section of a uniaxial crystal), then apart from the center of the field and the N–S, E–W cross-wire directions, the oblique angle of incidence itself causes the linearly polarized beam to be rotated. This *reflection rotation* increases away from the N–S, E–W crosswires (see Figure 4.11*b*). If the polars are then crossed, a black cross (see Figure 4.11*c* 1) is observed along the crosswires where the reflected light is extinguished by the analyzer, but elsewhere the field is illuminated. The black cross, the arms of which are correctly termed *isogyres*, remains stationary when the stage is rotated (cf. the uniaxial interference figure in transmitted light microscopy although the figure here is observed on reflection from an *isotropic* material).

Reflection of linearly polarized light even at normal incidence from an anisotropic surface results in effects of rotation of the incident vibration direction (anisotropic rotation) and in elliptical polarization when the section is not in the extinction position. In examining anisotropic sections in convergent light, the reflection rotation described above is superimposed on these effects. Thus when the anisotropic section is examined under crossed polars in the extinction position, a black cross is observed as for the isotropic sections (see Figure 4.11*c* 1), but on rotating the stage the isogyres break up and move outwards to form a pair of isogyres in opposite quadrants (cf. the acute bisectrix figure in transmitted light) (see Figure 4.11*c* 3). The separation of the isogyres with the stage at the 45° position is a general indication of the amount of anisotropy of the section. This can be explained if we consider the 45° position illustrated in Figure 4.11*c* 3. If the vibration direction of the larger reflectance value (R_2) is in the northeast quadrant as shown, then the anisotropic rotation will be towards R_2 (anticlockwise), whereas the reflection rotation (Figure 4.11*b*) will be clockwise. At a certain point these two effects will cancel each other and this is the position of the isogyre. Similarly, in the southwest quadrant, anticlockwise anisotropic rotation and clockwise reflection rotation will cancel, producing the isogyre. However, in the two remaining quadrants, both rotation effects are anticlockwise; thus no cancellation (and no isogyre) occurs.

Although the observations described above resemble the study of interference figures in transmitted light microscopy, with these surface effects there is no path difference in the substance and, therefore, no interference effects. They are correctly termed *convergent-light figures*. The amount of information that is obtainable from these figures is also rather limited. It is not possible to distinguish between an isometric crystal or a basal section of a uniaxial crystal, or between a nonbasal uniaxial section and the general section of a biaxial crystal. Convergent-light figures can be used for setting a section to extinction or determining the vibration direction of greater reflectance (R_2 in Figure 4.11*c* 3) in an anisotropic section although both can be obtained without resorting to such techniques.

Convergent-light figures are also useful in the study of dispersion effects. Isotropic sections in convergent "white" light can show dispersion of the reflection rotation (i.e., the degree of rotation can vary with the wavelength). The result will be color in the quadrants near the edge of the field; red patches indicate rotation is greater for red than blue light and vice versa for blue patches. This is only observed for strong dispersion but weaker dispersion can be studied by

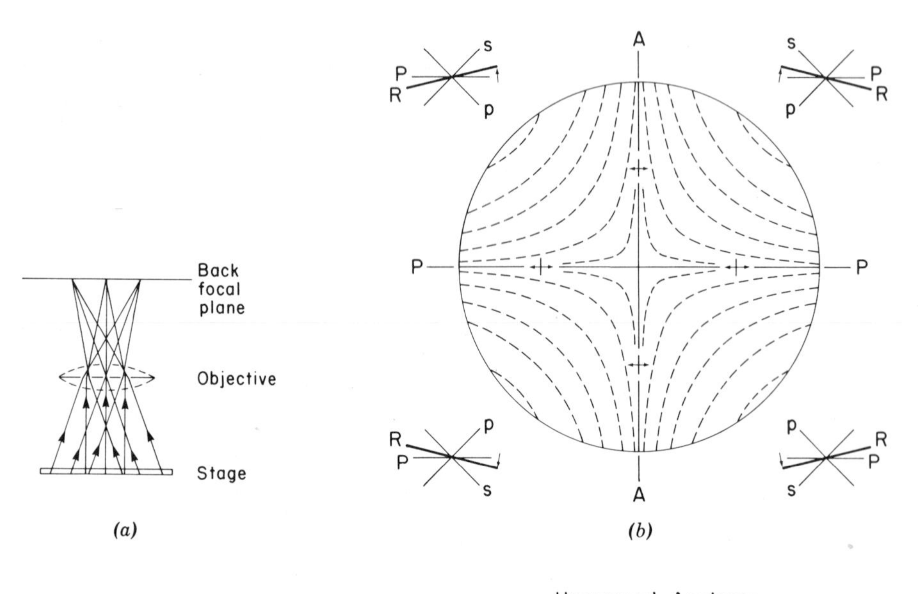

(a) *(b)*

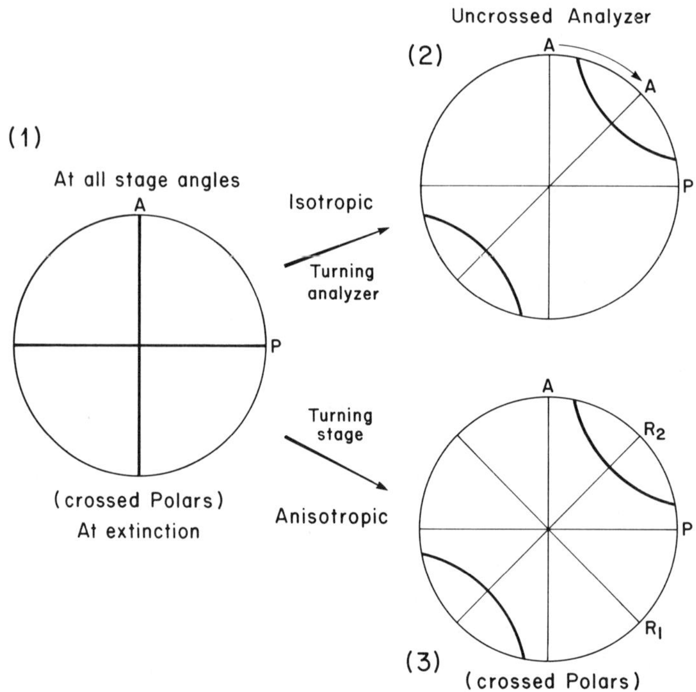

(c)

FIGURE 4.11

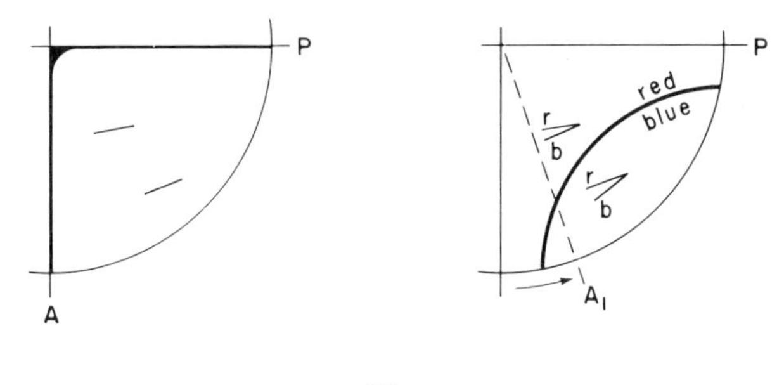

(d)

FIGURE 4.11 *(a)* Formation of a convergent-light figure. *(b)* Convergent-light figure for an isotropic surface showing vibration directions for points on the figure and (around the figure) the rotation of the resultant (*R* of "*p*" and "*s*") relative to the polarizer *(P)*. (*A* = analyzer.) *(c)* Convergent-light figures: (1) isogyres seen at all stage positions for isotropic sections and at extinction for anisotropic sections; (2) effect of uncrossing analyzer by 45° (section still in 90° position) in isotropic sections; (3) effect of rotating stage to 45° position for anisotropic section under crossed polars. *(d)* One quadrant of an isotropic section in convergent "white" light showing the effect of uncrossing the analyzer (*A* to position A_1) forming two isogyres with colored fringes. Lines labelled *r* and *b* show rotation of the red and blue light. (*P* = polarizer direction.)

partly uncrossing the analyzer, in which case the isogyres will part and colored fringes will be observed on either side of the isogyres (Figures 4.11*c* 2 and 4.11*d*). For the case shown in Figure 4.11*d*, the isogyres are colored red on the convex and blue on the concave side and the reflection rotation is dispersed with rotation greater for blue light. This means that a spectral dispersion profile of this phase would show $R\%$ greater at the red end. A phase with a spectral dispersion curve showing $R\%$ greater at the blue end would have red fringes on the concave side of the isogyre.

Dispersion of the reflection rotation in anisotropic sections can be studied in exactly the same way when the section is in the extinction position (isogyres crossed) and the analyzer is uncrossed. When the anisotropic section is examined in the 45° position (two isogyres under crossed polars), the effects observed are due to a combination of dispersion of reflection rotation *and* of anisotropic rotation. In certain cases, it is possible to make some deductions about the contribution of anisotropic rotation as well as some general observations about the overall dispersion (i.e., whether weak or strong).

4.4 CONCLUDING REMARKS

In this chapter, the nature of light and its reflection from a flat polished surface of an opaque substance have been considered. In particular, the reflection of linearly polarized light, its direct examination, and its examination under crossed polars have been discussed, as have the effects associated with convergent light. In this discussion, "classical" optical theory has been used to explain the phenomena qualitatively observed under the ore microscope and employed in identification. It will be appreciated from this discussion that, whereas for microscopic observations in transmitted light a whole range of diagnostic optical properties are readily determined (relative value of n, nature of indicatrix, optic sign, $2V$, birefringence, etc.), such data cannot be readily (if at all) determined for opaque phases. Attempts have been made to use measurements of anisotropic rotation properties as diagnostic parameters (see Cameron, 1961, for further discussion) but these methods have proved to be of limited usefulness and not been widely used. The two properties that have been adopted as quantitative parameters for identification are reflectance and hardness. The next two chapters are devoted to discussing these methods.

BIBLIOGRAPHY

Bloss, F. D. (1961) *An Introduction to the Methods of Optical Crystallography.* Holt, Rinehart & Winston, New York.

Cameron, E. N. (1961). *Ore Microscopy.* Wiley, New York.

Galopin, R. and Henry, N. F. M. (1972). *Microscopic Study of Opaque Minerals.* McCrone Research Associates, London.

Jenkins, F. and White, H. E. (1976). *Fundamentals of Optics,* 4th ed. McGraw-Hill, New York.

von Gehlen, K. and Piller, H. (1964) Zur Optik von Covellin. *Beit. Mineral. Petrogr.* **10**, 94.

5

Quantitative Methods—Reflectance Measurement

5.1 INTRODUCTION

The percentage of light at *normal* incidence reflected back to an observer (or instrumental observation system) from a flat polished surface of a particular ore mineral is the *reflectance** (R or $R\%$) of that mineral. It has already been explained that this parameter is directly related to the optical constants, n and k, through the Fresnel equation (see Section 4.1.2), which is restated here because of its importance:

$$R = \frac{(n - N)^2 + k^2}{(n + N)^2 + k^2} \tag{5.1}$$

where n = refractive index of the mineral
N = refractive index of the medium into which reflection takes place (when this is air, $N = 1$)
k = absorption coefficient of the mineral
R = reflectance ($R = 1$ when $R\% = 100\%$)

From the discussions in Chapter 4, it is clear that for many minerals or other solid materials, reflectance varies as a function of the wave-length of the incident light. Hence although reflectances have been determined in white light in the past (before the development of more sensitive photosensors), a reflectance value for a substance should be given at a specified wavelength to be meaningful. Furthermore, for all crystalline substances that are not isotropic (i.e., all noncubic minerals), reflectance will commonly vary as a function of crystallographic orientation of the polished surface relative to the vibration direction of the linearly polarized incident light. Thus although cubic minerals have a single value of re-

*The term *reflectivity* is used in some articles.

flectance ($R\%$) at a specified wavelength of light, Section 4.2.2 shows that other minerals show maximum and minimum values of reflectance with all possible intermediate values. The origin and significance of the terminology has already been explained but it is worth recalling. Uniaxial (hexagonal and tetragonal) minerals have two reflectances; R_o, R_e. Biaxial (orthorhombic, monoclinic, and triclinic) minerals have, in theory, three reflectances; sometimes symbolized R_p, R_m, R_g (where $R_p < R_m < R_g$) but since only two reflectances are readily measured and one or the other need not be a maximum throughout the visible region, the symbols R_1 and R_2 are preferred. This chapter is concerned with the quantitative measurement of these reflectance values, their physical significance, and their applications in mineral identification.

The "direct" measurement of reflectance requires relatively large specimens and is the kind of method employed in the calibration of standards used in other methods. The intensity of the stabilized light source beam is measured by photometer; then the intensity of this beam is measured when reflected from a relatively large polished surface of the material at angles close to 90°. The plot of R against angle permits such measurements to be extrapolated to obtain the value at 90°. It is not a technique readily modified for use with the microscope, so that reflectance measurements in ore microscopy have centered on comparison with a standard of known reflectance (known from measurement by the direct method).

The methods first developed and marketed by microscope manufacturers for quantitative reflectance measurement under the ore microscope relied on visual comparison of unknown and standard. In 1937, Berek developed a *slit microphotometer* employing a field of view divided into two, with the mineral grain on one side and the capability of varying the intensity of illumination on the other. The variable illumination was achieved by an analyzer that could be rotated to progressively cross or uncross polarizer and analyzer; the angle of rotation of the analyzer when the observer had matched intensity with the sample was then read off. Settings were calibrated against standards of known reflectance. The microphotometer developed in 1953 by Hallimond also relied on visual matching by the observer, this time matching light intensity reflected from a small mirrored surface located in the center of the field of view. The intensity of illumination of this area could be varied until it matched the unknown phase surrounding it in the field of view. Again, the instrument was calibrated using standards of known reflectance. However, as early as 1927, Orcel was experimenting with the use of a photoelectric device to measure light reflected from a polished surface under the microscope. This method, in which a photometer reading for a beam of light reflected by the unknown is compared with a reading made under the same conditions for a standard, has proved the most suitable for use with the ore microscope. The early measurements were unreliable, however, due to the primitive electronic systems and problems of specimen preparation and standardization. Subsequent developments, particularly the use of the selenium barrier-layer photoelectric cell by Bowie and Taylor (1957) in a series of systematic measurements in white light, established that reliable reflectance measurements can be made in this way. Most modern microphotometers are a development of this work started by Orcel but employ photomultipliers for light measurement, devices a million times more sensitive than the selenium cell. The discussion of reflectance measurement in the rest of this chapter will, therefore, center on these modern instruments.

The selenium photocell is still used in certain routine work and operates using a photovoltaic effect in which the current generated is directly proportional to the light intensity. The cell is simply arranged in circuit with a galvanometer. The photomultiplier operates from a photoemissive effect such that when incident photons fall on a cathode, electrons are ejected from it and are attracted to the first of a series of dynodes, positively charged relative to the cathode, at which many more electrons are ejected. This process continues at the other dynodes in the series. The resulting current amplification is again recorded by the deflection of a galvanometer; both analogue and digital readout systems are available. The size of specimen area illuminated and the size of the field sampled by the photomultiplier are limited by a series of stops or diaphragms.

Before considering the techniques of modern microscope photometry, a number of other general observations should be made, particularly concerning the relationship between the illumination system, the photometer, and the human eye. The phenomenon of spectral dispersion has already been explained (Section 4.2.3.) and illustrated in Figure 4.7. Such spectral dispersion curves assume that the intensity of illumination is uniform throughout the visible region and that the response of the eye or artificial monitoring system is independent of the wavelength. Neither of these is actually true. The intensity of a microscope lamp, or even of light from the sky, is not uniform across the visible region, as illustrated in Figure 5.1. Light from the sky is appreciably more intense towards the blue end of the spectrum. Microscope lamps differ, of course, depending on their construction and operating conditions (factors normally defined by speci-

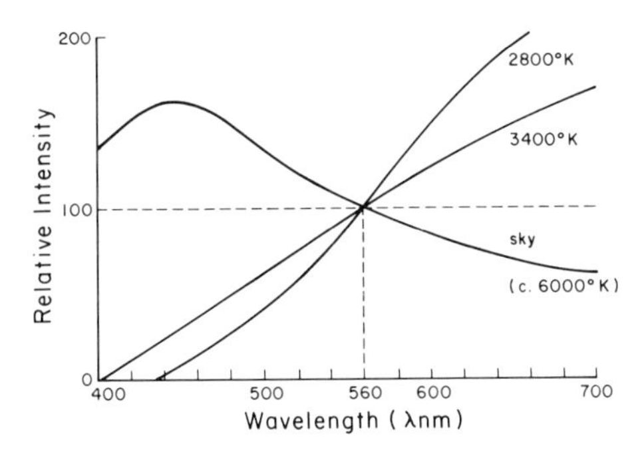

FIGURE 5.1 Spectral curves of the relative intensity (normalized for 560 nm) of the clear sky and for microscope lamps with color temperatures of 2800 and 3400°K. (After R. Galopin and N. F. M. Henry, 1972.)

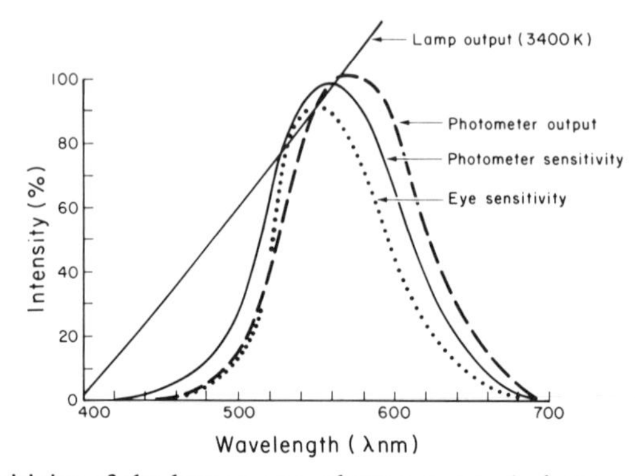

FIGURE 5.2 Sensitivity of the human eye, photometer and photometer measuring system as a function of wavelength. (Modified after R. Galopin and N. F. M. Henry, 1972.)

fying their *color temperature*, see Section 1.2.4. Two examples are given in Figure 5.1. from which it can be seen that microscope lamps show a greater intensity towards the red/yellow end of the visible spectrum. In practice, this means that to a single observer, ore mineral colors will appear slightly different when observed using different microscopes with different illumination systems. The human eye is also far from being uniformly sensitive as a detector throughout the visible region. It is extremely insensitive at the blue and red ends of the spectrum, with a sensitivity maximum near the center of the visible (~550 nm) as shown in Figure 5.2. When the eye is replaced by a photoelectric device, this too varies in the sensitivity of its response, commonly with a curve similar to that of the eye (Figure 5.2.). However, this device in turn is coupled to a galvanometer, which can introduce a further displacement in the zone of maximum sensitivity. In Figure 5.2 are shown the relationships between the absolute intensity of light from a microscope lamp (color temperature 3400°) reflected from a perfectly "white" surface and the sensitivity of the eye and photometer. Fortunately, of course, in measuring reflectances against standards for which absolute values are known by direct measurement, nonlinearity in source and sensor equally affect sample and standard and therefore cancel. They do, however, affect the sensitivity and accuracy of measurements, which are most reliable in the central region of the visible spectrum.

5.2 MEASUREMENT TECHNIQUES

As already outlined, reflectance measurements are now performed using photoelectric devices which, for research work, consist chiefly of a photomultiplier tube mounted on the ore microscope. For teaching and certain routine work, the selenium photocell is still employed in simple systems although robust and inexpensive photomultipliers are now available and give much better results. Whatever photoelectric device is employed, the principal components and their arrangement will be as illustrated in Figure 5.3.

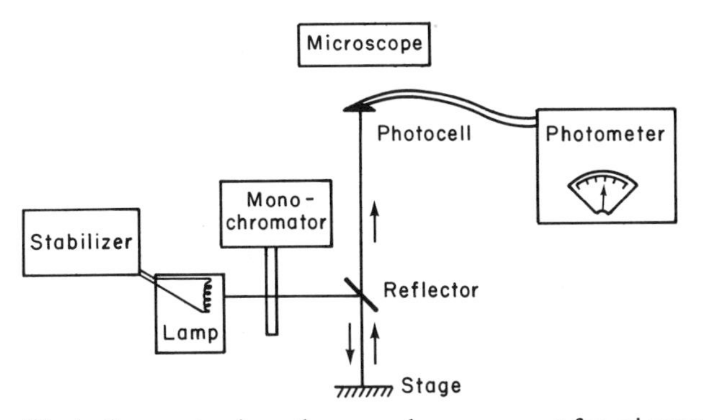

FIGURE 5.3 Block diagram to show the general arrangement for microscope photometry. (After R. Galopin and N. F. M. Henry, 1972.)

The lamp, which is built into most modern microscope systems, must be run at a high filament temperature (~3500°K) and should be stabilized for accurate work. In some systems, the *monochromator* is inserted between the light source and specimen, but in others it is placed immediately in front of the photocell or photomultiplier. The most convenient and commonly employed monochromator is the interference filter, of which there are two kinds. The "band-type" allows to pass through a range of wavelengths with half-height bandwidth of ~20 nm whereas for the "line-type" this is only ~10 nm. Both types are available as single "plate filters" for specified wavelengths or as "running filters" which cover the whole of the visible spectrum. The latter (already calibrated by the manufacturer) is most commonly used on commercial instruments. Even for research work, no greater accuracy is achieved by using a monochromator with a half-height bandwidth less than ~10 nm.

The microscope used in the reflectance measuring system, commonly a modified modern teaching or research ore microscope, must be fitted with stops or diaphragms to limit the beam (which must be at normal incidence) and the size of area illuminated on the polished section. Most systems are also arranged so that the light from the specimen can pass into the photoelectric device or be deflected by operating a simple lever and pass through an ocular, enabling the operator to view the specimen.

5.2.1 Measurement Procedure

Obviously, precise measurement procedures will vary between different systems, but a typical routine measurement might involve the following sequence for a reflectance measurement at a specific wavelength in air:

1 Standard and specimen are both carefully cleaned and leveled.

2 An objective is selected (for coarse-grained material a magnification of 8× to 16× is most suitable) and the specimen placed on the stage and sharply focused.

3 A wavelength for measurement is selected (commonly 546 or 589 nm, see Section 5.2.2) and the monochromator adjusted accordingly.

4 The photometer field stop and illuminator field stop are adjusted so that the former is about half the diameter of the latter, which in turn covers a homogeneous area of the specimen.

5 The photometer is adjusted so that readings for specimen and standard are both on scale by inserting each in turn and passing the beam through to the photomultiplier photocell.

6 The reading is taken for the specimen by placing it on the stage and passing light to the photomultiplier and taking the galvanometer reading (G'_{sp}).

7 The reading is taken for the standard, which is also carefully focused with conditions being maintained exactly the same as for the specimen (G'_{st}).

8 Again with the same conditions maintained, a reading is taken with a black box held over the front of the objective. This is a reading of *primary glare*, which is light reflected from the back surface of the objective before reaching the specimen (C) (see Figure 5.7).

9 The reflectance is simply calculated after the value for the glare measurement has been subtracted:

$$G_{sp} = (G'_{sp} - C) \text{ and } G_{st} = (G'_{st} - C)$$

then:

$$R\%_{specimen} = \frac{G_{sp}}{G_{st}} \times R\%_{standard}$$

The above procedure produces one reflectance value for a specified wavelength which is all that is required for a cubic (or isotropic noncrystalline) material. If a material is known to be uniaxial, it is possible to prepare oriented polished sections of single crystals cut perpendicular and parallel to the *c* crystallographic axis so as to measure R_o and R_e and determine the bireflectance and reflectance sign. Much more commonly, however, the measurements are being made on a randomly oriented aggregate, in which case it is necessary to search for suitable basal and prismatic sections for measurement. Examination under crossed polars to look for a section as near isotropic as possible, and another showing maximum anisotropy, aids the selection. In the latter case, the two extinction positions indicate the two vibration directions; the reflectance is measured for both, one being R_o and the other R'_e (not R_e since we do not know if it is the extreme value). If it is uncertain whether the material is uniaxial, one constant reflectance value in all reliable measurements (i.e., the R_o value present in all general sections) will confirm a uniaxial phase. For a biaxial material of relatively high symmetry (orthorhombic and possibly monoclinic), it is possible to prepare oriented polished sections of coarse single crystal material. In this way, it is possible to determine R_p, R_m, and R_g (or R_a, R_b, R_c) for an orthorhombic crystal. This is specialized work and more commonly the problem is one of determining reflectances of random grains in a polished section. Any such grain of a material that is at least moderately bireflecting will show two reflectance values that will lie between certain extremes. Unlike the uniaxial material, neither value will be constant from one grain to the next, and it is necessary to search the polished section in order to find values for both R_p and R_g (or R_1 and R_2). In any of these determinations, it is important to make an appreciable number of repeat measurements to ensure consistency of results.

Since most ore minerals exhibit at least some spectral dispersion (Section 4.2.3, Figure 4.7), reflectance measurement is much more valuable in identification and characterization if readings are taken at several wavelengths through the visible region or if a whole series of measurements are made at regular (say 20 nm) intervals between 400 and 700 nm so that a spectral reflectance curve can be constructed. In Figure 5.4 the spectral reflectance curve for pyrite is shown. This is an isotropic phase with a much greater reflectance at the yellow/red end of the visible region, as would be expected from its yellow color. For a uniaxial mineral, it is possible to plot a whole family of curves between the extreme values of R_o and R_e, but in practice only R_o and R_e values are recorded. The example of covellite is shown in Figure 5.5 where spectral curves measured both in air and using an oil immersion objective are shown. In air, the color of the *e*-vibration is pale bluish white, pale because the rise at the blue end occurs where the sensitivity of the eye is falling rapidly (see Figure 5.2). The *o*-vibration is deep blue because the proportion of blue light reflected is large compared with yellow light and the steep rise at the end comes where the sensitivity of the eye is declining rapidly. When the covellite specimen is immersed in oil, the shift in R_o to shorter wavelengths is sufficient to introduce an observable red component and the mineral appears a bluish purple.* Covellite provides an excellent example of the value of spectral reflectance measurements in understanding the colors of ore minerals in air and in oil immersion. The third example of a spectral reflectance curve shown is that of arsenopyrite, one of the few biaxial minerals for which single crystal oriented data are available (Figure 5.6). The three vibration directions are specified by crystallographic axes since inspection of the curves will show that assignments of R_g, R_m, and R_p over the whole visible range would be meaningless.

The procedure for making a single measurement that was outlined at the start of this section applies also to the measurement of spectral curves. The modern photomultiplier system is well suited to making measurements throughout the

*This phenomenon does not occur in modifications of CuS that differ in details of their crystal structures. These phases were first observed because of their distinctive behavior under oil immersion and initially were called "blue remaining" (or bleiblaubender) covellite. Two minerals have now been characterized by careful work and named spionkopite ($Cu_{1.4}S$) and yarrowite ($Cu_{1.12}S$).

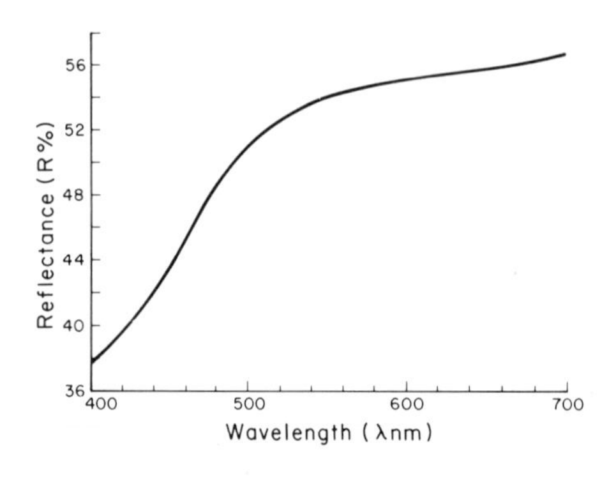

FIGURE 5.4 Spectral reflectance curve for pyrite (FeS_2).

visible spectrum even on very small grains. In making such measurements, it is possible to measure the standard at all wavelengths initially and then measure the specimen or to measure each consecutively at each wavelength. The first procedure makes considerable demands on the stability of the apparatus and requires carefully matched setting of the monochromator at each wavelength. The second procedure is potentially subject to focusing and leveling errors in addition to problems of returning to the same areas on specimen and standard. However, it is possible to use a specially designed mechanical specimen changer stage, which can be mounted on the microscope. An example of such a device is the *Lanham Specimen-changer Stage* on which specimen and standard can be

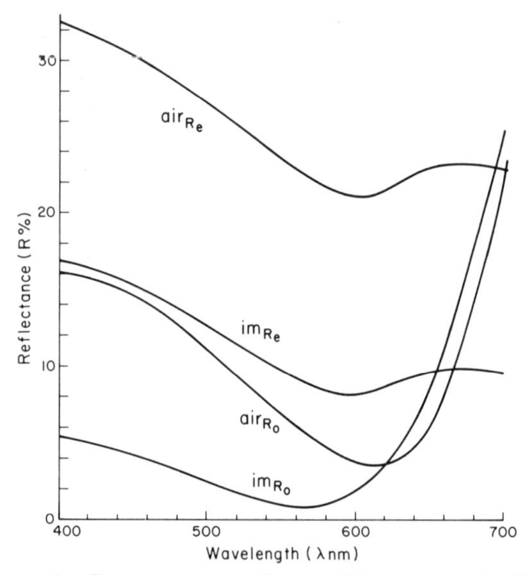

FIGURE 5.5 Spectral reflectance curves for covellite in air and oil immersion (im). (Data from IMA/COM Quantitative Data File, 1977.)

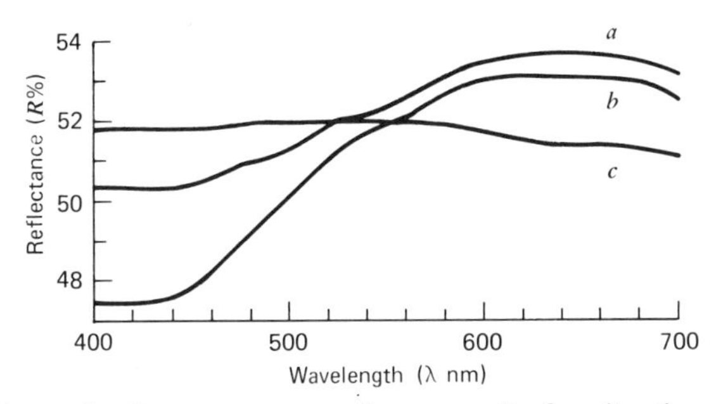

FIGURE 5.6 Spectral reflectance curves of arsenopyrite for vibrations parallel to the main crystallographic directions. (Data from IMA/COM Quantitative Data File, 1977.)

mounted, leveled, and focused, after which switching between the two is a trivial mechanical operation and both always return to the same area at the same focusing position.

5.2.2 Standards

Since the widely used methods of reflectance measurement depend on direct comparison of the unknown with a standard of known reflectance, accurate and reliable standards are of the utmost importance. Such matters are the concern of the Commission on Ore Microscopy (hereafter abbreviated COM) of the International Mineralogical Association, which selected the following materials as standards for reflectance measurement:

Black glass	$R\%$ (air) ~4.5% at 546 nm
Silicon carbide (SiC)	$R\%$ (air) ~20% at 546 nm
Tungsten titanium carbide (WTiC)	$R\%$ (air) ~50% at 546 nm

These standards, which have been chosen because they can take and maintain a good polish and exhibit little spectral dispersion, are obtainable from manufacturers of equipment for microscope photometry, each standard being individually calibrated in air and immersion oil (Cargille D/A oil has been accepted by the COM for measurements made in oil). It is also desirable to employ a standard close in reflectance to the unknown being measured. The standards are relatively expensive, and for routine work it is quite adequate to use a secondary standard, itself calibrated against one of these. The COM has also considered the question of a small number of standard wavelengths for reporting of discrete reflectance values. Such measurements must be made at 470, 546, 589, and 650 nm. If only a single measurement is made this should be at 546 nm. (This wavelength has been chosen because reflectance values at 546 nm are remarkably close to the luminance value, $Y\%$, for most ore minerals; see Section 5.5).

5.2.3 Errors in Measurement and Their Correction

In making reflectance measurements it is important to be aware of the errors that can arise and how they can be avoided. These aspects are discussed in much greater detail by Galopin and Henry (1972). Clearly the light source and photometer must be stable and the photometer reading linear before any measurements can be attempted. The specimen should be as well polished as possible and both specimen and standard must be clean. The reliable calibration of the reflectance standards and of the monochromator are beyond the control of the operator but the following errors are the responsibility of the user of the instrument:

1 Leveling errors. It is very important that both specimen and standard are normal to the axis of the microscope and for detailed work, a special specimen stage with leveling screws (e.g., the Lanham stage) can be employed. It is possible to check whether the specimen is level by using the *conoscopic leveling test* in which conoscopic (or convergent) illumination (preferably using the Bertrand lens, see Section 4.3.3.) is employed. When the illuminator aperture diaphragm is closed down, an image of a spot of light will form in the back focal plane of the objective. The objective used should be of low magnification ($<5\times$) to obtain a low numerical aperture. When the stage is turned, this image will remain stationary only when the specimen surface is perfectly level.

2 Focusing errors. Accurate focusing is also extremely important for good results and the position of focus must be the same for specimen and standard. The change of height of the reflecting surface over which no loss of image sharpness occurs is called the depth of focus and within this range, photometer readings do not change. Outside this, a small change in the specimen position produces a large change in the reading. The problem is greatest, therefore, with an objective of large numerical aperture that will have a small depth of focus.

3 Errors due to setting of the microscope. The various settings of stops and diaphragms with different objectives can be critical, particularly regarding the problems of *glare* (see Figure 5.7). In the measurement procedure (Section 5.2.1), it was shown how a black box measurement can be used to cor-

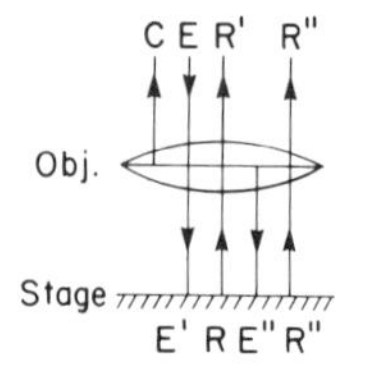

FIGURE 5.7 Primary and secondary glare. Primary glare *(C)* causes loss of intensity of the incident beam *(E)* becoming E', which is reflected as R. Secondary glare (E'') produces additional reflection (R''). (After R. Galopin and N. F. M. Henry, 1972.)

rect for the effect of primary glare. *Secondary glare* arises from light on its way from the specimen being reflected partly back down by the objective. This light forms a secondary component but one that varies according to the reflectance of the specimen and hence can differ between specimen and standard. This effect can be reduced by using good quality objectives, by keeping the illuminator field stop fairly small and the photometer field stop only about half this size, and by using objectives of low numerical aperture wherever possible.

5.2.4 Reflectance Measurements in Oil Immersion and the Determination of the Optical Constants, *n* and *k*

As already indicated in Sections 5.2.1 and 5.2.2, reflectance measurements can be made in oil. This gives another series of values that may be an aid to identification but commonly such measurements are undertaken to enable the optical constants, n and k, to be derived by simultaneous substitution in the Fresnel equation (equation 5.1). Detailed procedures for reflectance measurement in oil and for solution of the equations to derive optical constants are given in Galopin and Henry (1972), although this method for the determination of n and k suffers from serious error problems (Embrey and Criddle, 1978).

5.3 APPLICATIONS TO MINERAL IDENTIFICATION

The first objective of reflectance measurement is the identification of an opaque mineral. In this regard a reflectance measurement is the single most useful quantitative optical parameter and is rapidly obtainable using equipment that is easily operated and relatively inexpensive to set up. Standards for reflectance determinations are now readily available, so the only other requirements are comprehensive data for ore mineral reflectances and search procedures for utilizing these data. Compilations of reflectance data, often combined with data on Vickers Hardness numbers (see Chapter 6), have been published by Bowie and Taylor (1958) Gray and Millman (1962), McLeod and Chamberlain (1968), Galopin and Henry (1972), and Uytenbogaart and Burke (1971). However, as compilations of reflectance data, all of these sources have now been superceded by the data cards published by the COM. An example of one of the COM Data Cards is reproduced in Figure 5.8 from which it is seen that as well as the basic information on mineral name, formula, and crystal symmetry, the reflectance ranges in air and oil are given at the four standard wavelengths and detailed spectral curve readings given at 20 nm intervals from 400 to 700 nm. When available, data are also given on Vickers hardness number and quantitative color measurements (see Section 5.4). Ancillary information on the standard used, the polishing method employed, and chemical composition of the material may be provided as well as a reference to the available X-ray data. Reflectance data at

1.0520.1	Name Arsenopyrite		Chemical formula FeAsS	Symmetry Monoclinic
λnm	470	546	589	650
R% air	48.7-51.85	51.85-52.2	51.7-53.2	51.3-53.6
R% oil	33.9-36.95	37.2-37.55	37.05-38.5	37.0-39.1

VHN: 1081 sf on (001) section — Load gf 100

λnm	Air			Cargille oil D/A 58.884		
	R_a	R_b	R_c	R_a	R_b	R_c
400	50.3	47.45	51.8	33.2	31.95	36.55
420	50.3	47.45	51.8	33.42	32.15	36.7
440	50.3	47.55	51.8	33.9	32.55	36.8
460	50.6	48.15	51.8	34.55	33.4	36.95
480	51.0	49.1	51.9	35.3	34.4	37.0
500	51.4	50.1	51.9	36.0	35.4	37.05
520	51.85	51.0	51.9	36.75	36.3	37.15
540	52.15	51.65	51.9	37.4	37.1	37.2
560	52.5	52.2	51.9	37.9	37.7	37.15
580	53.05	52.65	51.8	38.3	38.15	37.05
600	53.45	52.95	51.65	38.65	38.5	37.05
620	53.6	53.0	51.5	38.95	38.8	37.05
640	53.6	53.0	51.3	39.05	38.85	37.05
660	53.6	52.95	51.3	39.1	38.9	37.0
680	53.45	52.75	51.15	39.1	38.8	37.0
700	53.15	52.4	51.0	38.9	38.6	37.0

Standard: SiC, NPL No. 3AR68

Filter or other monochromator: Schott interference line filter

Polishing method: Grinding: alumina on glass; Finishing: alumina on lead, microcloth

Chemical composition: Quantitative XRF and optical emission spectrographic analysis

Element	
As	43.1
Fe	34.3
S	20.5
Ba	0.1
Si	0.05
Al	0.02
Mg	0.0015
Ca	0.001
	98.1

X-ray data: Diagram corresponds to PDF No. 14-218

a = 9.55Å
b = 5.69
c = 6.43
$\beta \approx 90°$

Provenance: Unknown, M. Sc. P. R. S. 4

Author(s): P. R. Simpson; VHN M. J. Cope; U. K. L3(b)

Additional information: Colour values (C)

a	.315	.320	52.5	air
	.320	.327	37.6	oil
b	.318	.325	51.8	air
	.322	.330	37.4	oil
c	.310	.316	51.8	air
	.310	.317	37.1	oil

FIGURE 5.8 Example of a card from the IMA/COM Quantitative Data File (First Issue, 1977) showing complete reflectance, Vickers microhardness, and quantitative color data for arsenopyrite. Additional information on X-ray data and chemical composition are provided. (Reproduced with permission)

546 and 589 nm (air) taken from the COM tables are provided in Appendix 1 for all the common ore minerals. A list of these minerals arranged in order of increasing reflectance is also provided in Appendix 2.

The first systems designed to utilize reflectance data in a scheme for ore mineral identification were plots of reflectance against Vickers Hardness number. Bowie and Taylor (1958) plotted average reflectance in *white* light against average microindentation hardness so that each mineral was represented by a point on the chart. Gray and Millman (1962) plotted reflectance in white light against hardness, representing reflectance variations due to bireflectance or compositional variation so that each mineral appeared as a line on the chart. McLeod and Chamberlain (1968) produced a chart with all available published data and showing both reflectance and hardness variations by intersecting horizontal and vertical lines. However, the original literature has to be consulted to determine whether the data are for white or monochromatic light reflectances. In the book by Galopin and Henry (1972), a series of charts for different major ore mineral groups show hardness plotted against reflectance, but again in white light. The most recent determinative chart of this type is that of Tarkian (1974) in which reflectance values at 589 nm (air) are plotted against Vickers Hardness number, with ranges in both incorporated so that the minerals occupy a "box" on the chart. A chart of this type is provided in Appendix 4 but based on the COM Data File.

A number of other determinative schemes have recently been introduced including various semiautomated and computerized search procedures. By far the most significant advance for the student, however, has been the introduction of the *Bowie-Simpson System* which is available as a students issue (Bowie and Simpson, 1978) containing 33 common ore minerals. Four charts that show the minerals ordered as to increasing reflectance at 546 nm, and then in the same order for the other standard COM wavelengths of 470, 589, and 650 nm, form the nucleus of the system. The style of presentation is illustrated in Figure 5.9 which shows that the reflectance range for each mineral is represented by a horizontal line. Ticks and markings on this line distinguish isotropic, uniaxial, and

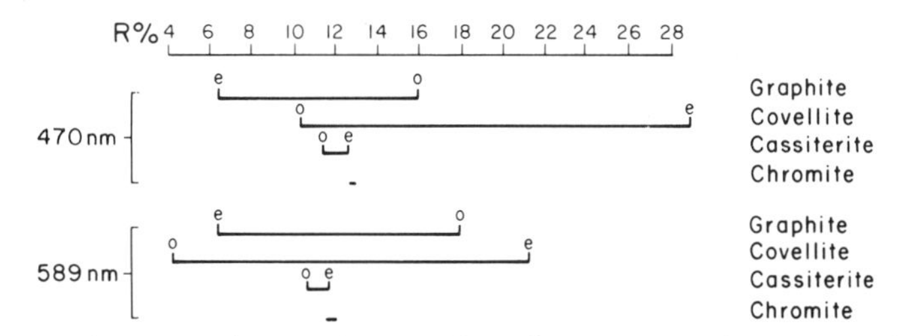

FIGURE 5.9 The Bowie-Simpson System. The reflectance range is represented by a horizontal line, unmarked for an isotropic phase, with ticks (*o* and *e*) for a uniaxial phase, ticks (and no letters) for a biaxial phase. (After S. H. U. Bowie and P. R. Simpson, 1978.)

biaxial species and whether the range shown is due to bireflectance or compositional variation between individual measured grains (see Figure 5.9). Accompanying tables contain data on Vickers hardness and qualitative properties observable under the microscope. Unless the mineral has very distinctive qualitative properties enabling immediate identification, the procedure involves taking a reflectance measurement at 546 nm that may (in combination with the qualitative properties) conclude the identification or reduce the possibilities to a few minerals. In the latter case, measurements at one or more of the other standard wavelengths should permit identification. Only in some cases is it necessary to use the second quantitative technique of Vickers hardness measurement.

Two other recent determinative schemes that should be mentioned are the Delft scheme, which utilizes punched "property cards" to systematically limit the possible mineral species (Kühnel, Prins, and Roorda, 1978), and the Nottingham Interactive System for Opaque Mineral Identification (NISOMI), which is a series of computer programs that undertake search routines based on input of reflectance and microhardness data (Atkin and Harvey, 1978).

5.4 APPLICATIONS TO THE COMPOSITIONAL CHARACTERIZATION OF MINERALS

Reflectance variations and variations in color (which may be treated quantitatively as described in Section 5.5) may be sensitive to variations in mineral composition as well as structural differences. An important and not yet fully explored field is the correlation of reflectance variations with compositional variations in minerals. A number of contrasting examples of this application of reflectance measurements will be considered. The first is the use of a reflecto-

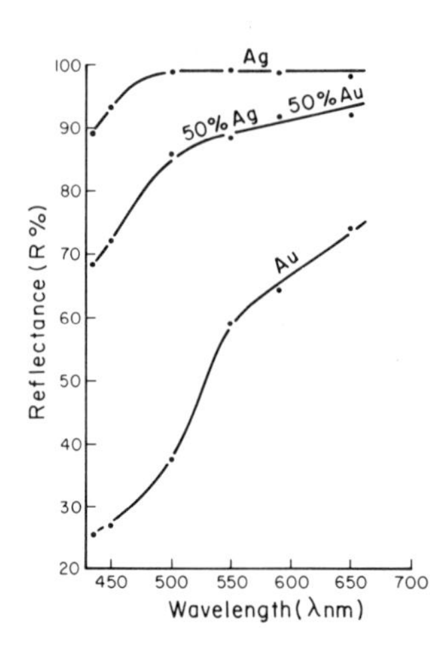

FIGURE 5.10 Spectral reflectance curves of pure gold, silver, and an artificial alloy of 50% gold and silver. (After H. Squair, 1965.)

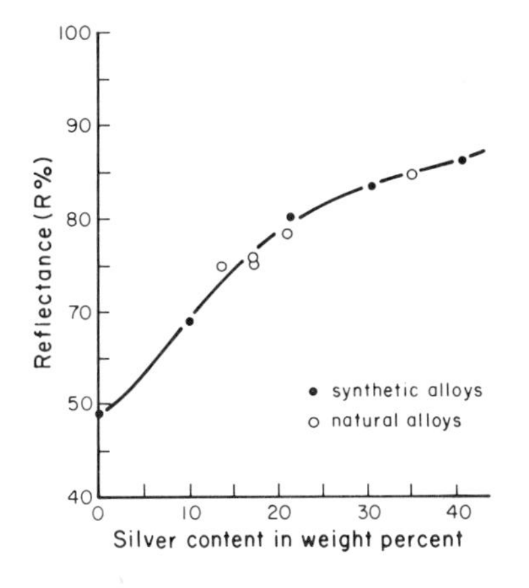

FIGURE 5.11 Plot of reflectance (at $\lambda = 550$ nm) against silver content for synthetic gold-silver alloys. (After H. Squair, 1965.)

metric method for determining the silver content of natural gold-silver alloys described by Squair (1965) and by Eales (1967). As can be seen from the spectral curves in Figure 5.10, reflectances for pure gold and pure silver show a good separation throughout the visible range with an alloy containing 50% silver having an intermediate reflectance and spectral curve. On the basis of the separation in reflectances of the endmembers and the range of maximum sensitivity of the microphotometer used, Squair (1965) chose a wavelength of 550 nm at which to measure a series of synthetic alloys and produce a determinative curve such as that shown in Figure 5.11. Here, the synthetic alloys have been used to plot a curve for the mineralogically important range from 0–40% Ag. The reflectances of natural alloys of known chemical composition are also plotted on the figure and give an indication of the accuracy of this method. An important factor in the application of this technique, however, is that the gold should contain no other metal than silver in significant quantities. In particular copper is known to occur in fairly large amounts in some natural samples and would invalidate the determinative curve. Eales (1967) has also pointed out that gold reflectances are very sensitive to polishing technique.

A number of opaque oxide mineral systems exhibit solid solution behavior, which causes systematic variations in reflectance. An important example is the substitution of magnesium in ilmenite ($FeTiO_3$) which forms a complete solid solution to geikielite ($MgTiO_3$). Cervelle, Lévy, and Caye (1971) have studied the effect of magnesium substitution on reflectance and developed a rapid method for determination of magnesium content of ilmenite. A series of spectral curves for ilmenites with increasing MgO content are illustrated in Figure 5.12, which shows not only the pronounced effect of this substitution on the reflectances, but also that these materials exhibit very little dispersion. This means that a rapid method for magnesium determination can function with a simple white light source. The determinative curve that relates ilmenite reflectance (actually the R_o value present in all sections) to MgO content is shown in Figure 5.13. This rapid

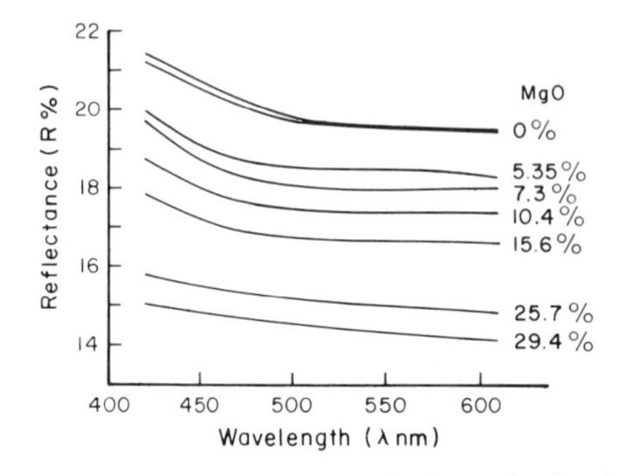

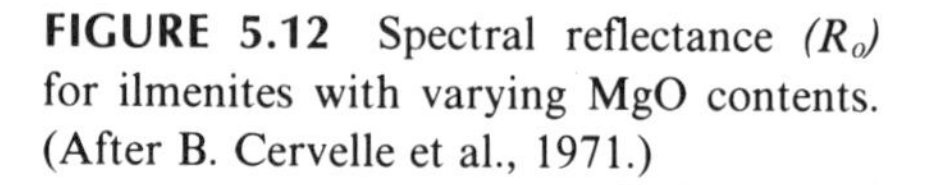
FIGURE 5.12 Spectral reflectance *(R_o)* for ilmenites with varying MgO contents. (After B. Cervelle et al., 1971.)

method for the determination of magnesium in ilmenites has practical applications in diamond prospecting, since ilmenites from kimberlites characteristically carry ~10% MgO.

In the group of iron sulfide minerals generally termed the "pyrrhotites" and including the stoichiometric endmember troilite (FeS) and the metal-deficient "hexagonal" ($\sim Fe_9S_{10}$) and monoclinic ($\sim Fe_7S_8$) pyrrhotites, systematic reflectance variations have been observed as a function of composition. Vaughan (1973) observed a general increase in average reflectance with increasing metal deficiency in a series of synthetic samples. Carpenter and Bailey (1973) demonstrated that measurements of R_o (at 546 nm) could be used to distinguish troilite from "hexagonal" or intermediate pyrrhotite and either of these from monoclinic pyrrhotite.

In the mineralogically very complex series of compositions related to the tennantite-tetrahedrite group, Charlat and Lévy (1976) have examined the reflectivities as a function of complex chemical substitutions in compositions of the type $(Cu,Ag)_{10}(Cu,Fe,Zn,Hg)_2(As,Sb)_4S_{13}$. Although optical properties alone are insufficient to determine chemical composition, they can be used to predict possibilities, (e.g., whether the material contains any silver).

Other examples of this systematic approach to the relationship between reflec-

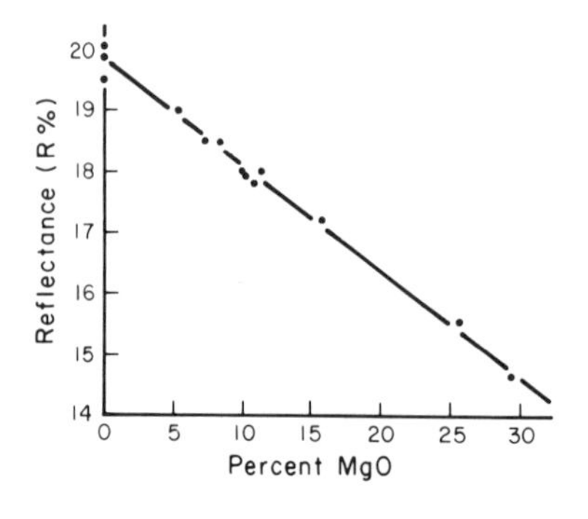

FIGURE 5.13 Plot of reflectance *(R%)*, at about 460 nm, against percent MgO content for ilmenites ($FeTiO_3$–$MgTiO_3$ series). (After B. Cervelle et al., 1971.)

tance and composition include studies of the sulfides and sulfosalts of copper (Lévy, 1967), of the platinum minerals (Stumpfl and Tarkian, 1973) and of the silver sulfosalts (Pinet et al., 1978).

5.5 QUANTITATIVE COLOR

The colors of ore minerals or other opaque materials observed in reflected linearly polarized light are clearly amongst their major diagnostic features (see Chapter 3). Qualitative descriptions of color are inevitably imprecise and somewhat subjective, so that a system for quantitatively specifying color has obvious advantages. The first publication to deal with color measurements specifically in ore microscopy was that of Piller (1966), although it is only much more recently that the potential of such measurements has been widely appreciated by ore microscopists. In an introductory text, it is not appropriate to go into the details of quantitative color determinations, except to say that the parameters can be easily derived from spectral reflectance data by straightforward calculations. In this section, the theory of quantitative color specification is outlined to give a clearer impression of this property and to explain the terminology. Quantitative color data are given, when available, on the COM data cards and in Appendix 1 of this book.

The sensation of color, in terms of its subjective perception by the eye, can be described by three attributes—*hue, saturation,* and *brightness.* The spectral colors of the visible region (Figure 4.1) are given names associated with particular wavelength ranges (630–780 nm, red; 450–490 nm, blue, etc.). Each member of this continuous series of colors is a hue; the pure color has maximum (100%) saturation for that hue, whereas white light is regarded in this system of color quantification as having zero saturation. Mixing increasing proportions of white light with a particular hue results in colors of decreasing saturation. The *brightness,* or light intensity of the color, also affects color perception.

It is possible to match any color, as perceived by the eye, by mixing only three spectral colors (e.g., red, green and blue) in appropriate proportions. The amounts of each needed to match a particular color are called the *tristimulus* values for that color. If these values are expressed as fractions summing to unity, they are called *chromaticity* coordinates. The science of color measurement is, of course, well established in other fields, and in 1931, the Commission Internationale de l'Eclairage developed the CIE system based on hypothetical primary colors (X, Y, Z) obtained mathematically from experimental data. The chromaticity coordinates (x, y, z) for any color are positive fractions that sum to unity so that only two need be specified. Commonly, x and y are plotted on orthogonal axes (see Figure 5.14) when the pure spectrum colors fall on an inverted U-shaped curve (the *spectrum locus*), the ends of which are joined by a straight line (the *purple line*). This line is a locus of colors not found in the spectrum but the area enclosed by the purple line and spectrum locus encloses chromaticity coordinates for *all* possible colors. The boundary of this area represents the pure colors (100% saturation), whereas a central point within the area represents white (0% saturation). The color sensation produced by an object depends, of course, on the light source used to illuminate it, so the point representing 0% saturation varies with the light source. "Ideal" white light would show the same energy at every wavelength and would plot at coordinates $x = 0.3, y = 0.3$ on the chromaticity diagram (Figure 5.14). However, real light sources differ appreciably from this ideal, so that the C.I.E. system specifies several standard white light sources. Source *A* (coordinates $x = 0.4476$, $y = 0.4075$) corresponds to a tungsten filament microscope lamp and source *C* with average daylight ($x = 0.3101, y = 0.3163$) or the microscope lamp modified by a conversion filter. The color temperatures of these sources are 2854 and 6770° K respectively, and both are plotted on Figure 5.14.

In determining the chromaticity coordinates for an opaque mineral, it is only necessary to measure the reflectance at a series of wavelengths in the visible range and subject these values to certain mathematical manipulations. However, the chromaticity coordinates are calculated relative to one of the standard illuminants (*A, C*) as a reference achromatic point, and spectral energy distribution values for this source (available from the literature) are needed for the calculation. Although chromaticity coordinates can be calculated for a standard illuminant even when measurements are not made using one of the standard illuminants, conversion of data from one reference illuminant to another requires complete recalculation.

The chromaticity coordinates are a precise means of specifying a color but do not immediately convey any visual impression of the color concerned. From this point of view, two parameters that can be used to specify exactly a point on the chromaticity diagram are more useful. The parameters—*dominant wavelength* (λ_d) and *excitation purity* (P_e)—comprise part of the *monochromatic* or *Helmholtz* system of color specification. These can be illustrated with reference to two examples, *R* and *S*, plotted in terms of their chromaticity coordinates on Figure 5.14. Using the standard *C* illuminant, a line drawn from the point *C* representing this illuminant and passing through the sample point, will intersect the spectrum locus at the dominant wavelength ($\lambda_d = 540$ nm for sample *R*). This is the spectrum color which will match the specimen color when mixed with the "white" of the standard *C* illuminant, and therefore conveys an immediate impression of the color concerned. For an example like sample *S*, the corresponding parameter is the *complementary wavelength* (λ_c) and is given by extending the line from *S* to *C* to intersect the spectrum locus (i.e., $\lambda_c = 500$ nm). In this system, the excitation purity is a measure of saturation. It is the distance of the sample point from the achromatic (in this case *C* illuminant) point expressed as a percentage of the distance to the point on the spectrum locus representing the dominant wavelength (or distance from the achromatic point to the purple line). P_e% values for samples *R* and *S* are shown as contours on Figure 5.14 and are

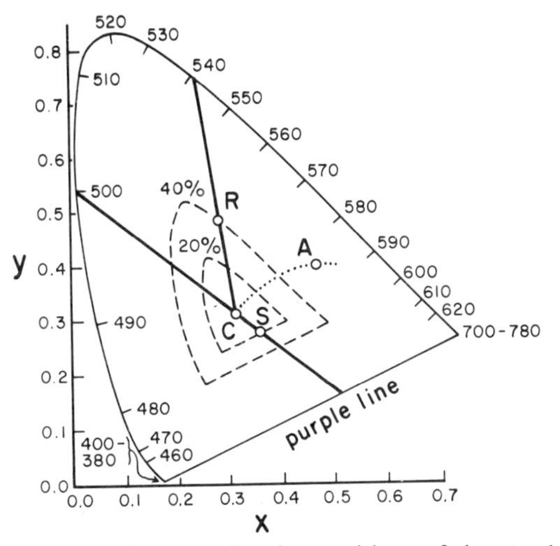

FIGURE 5.14 Chromaticity diagram showing positions of the standard sources *(A,C)* and the $P_e\%$ and λ_D values for two phases *(R,S)*. (After W. Htein and R. Phillips, 1973.)

40 and 20%, respectively. The one attribute of color not specified by giving chromaticity coordinates or the dominant wavelength and excitation purity is the brightness. A parameter to define this can be incorporated by considering the chromaticity diagram of Figure 5.14 as the base of a three-dimensional figure in which the height above the base (i.e., normal to the x-y plane) represents increasing brightness on an arbitrary percentage scale. An equivalent parameter is termed the luminance ($Y\%$) in the Helmholtz system where $Y = 100\%$ corresponds to a white light reflectance measurement made with a standard C illuminant.

Two other questions that are of particular interest in ore microscopy are: (1) What range of values do the common ore minerals show on a chromaticity diagram? (2) What is the minimum difference in color coordinates that can be detected by the average observer? Both questions can be answered by reference to Figure 5.15, which shows that nearly all of the ore minerals plot in the central portion of the chromaticity diagram. This is a vivid illustration of why many beginning students of ore microscopy see most minerals as grey or white. However, under ideal conditions, the experienced observer can discriminate down to the level shown by the *discrimination ellipse*, an example of which is also plotted in Figure 5.15. Points plotting inside such an ellipse are indistinguishable from the central color and from each other even under ideal viewing conditions.

This discussion of quantitative color has necessarily been brief, but enough to illustrate the importance of the topic in ore microscopy. Further information is available in articles by Piller (1966), Atkin and Harvey (1979a,b) and a book by Henry and Phillips (1981). Clearly the description of the mineral pyrite as λ_d

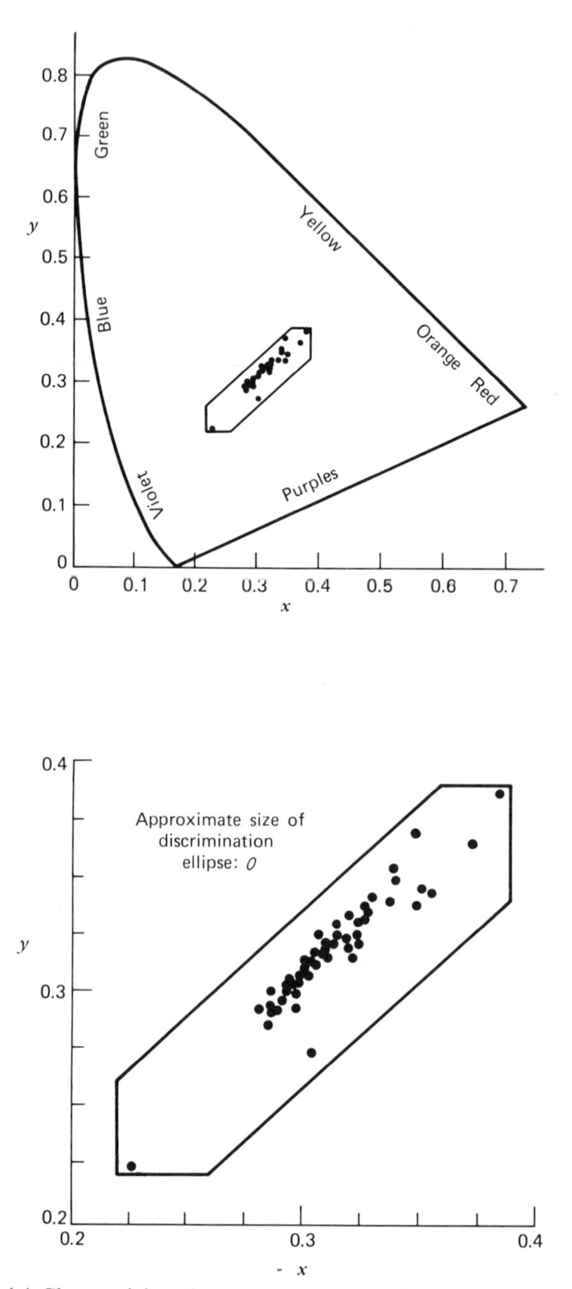

FIGURE 5.15 *(a)* Chromaticity diagram showing the distribution of the chromaticity of all opaque minerals represented in the first issue of the COM Quantitative Data File for which appropriate data are availabile. *(b)* Enlargement of the hexagonal area in Figure *3a* that encloses all chromaticity points of opaque minerals. (After B. P. Atkin and P. K. Harvey, *Canad. Mineral.* **17**, 639, 1979; used with permission.)

= 573, P_e = 13%, $Y\%$ = 52.9 rather than as color light yellow is an important quantitative advance that will be further stimulated by the availability of reliable quantitative color data in the COM tables.

5.6 THE CORRELATION OF ELECTRONIC STRUCTURE WITH REFLECTANCE VARIATION

Discussions of the theory of light reflection from polished surfaces generally utilize a classical approach based on the Maxwell equations for electromagnetic waves. This approach, essential to the understanding of refractive index and polarization phenomena, is outlined by Galopin and Henry (1972). However, the classical approach could not explain why, for example, there is a systematic decrease in reflectance (at 496 nm) in the pyrite structure series $FeS_2 \rightarrow CoS_2 \rightarrow NiS_2 \rightarrow CuS_2$. Variations of this type are a result of changes in the *electronic structures* of minerals and materials and require an explanation related to the description of light as photons rather than waves (Section 4.1). A simplified approach of this type based on work by Burns and Vaughan (1970) and Vaughan (1979) can provide a physical picture of the reflecting process as well as making predicitions of reflectance variation with composition possible.

The values of n, k, and hence R (see equation 5.1) for a solid depend on the interaction of light photons with electrons of the atoms in the solid, and therefore on the distribution of electrons in the solid (or its electronic structure). Electrons surrounding an atomic nucleus occur in *orbitals* that are sometimes represented visually as spheres, dumbbell-shaped regions, and so on, centered on the nucleus and within which the probability of the electron occurring is much greater. In simple terms, an electron in an orbital also has a clearly defined energy and successive electrons added to the system will be at higher energies occupying higher energy orbitals. It is also possible for an electron to be promoted from its normal *(ground state)* orbital energy level to a higher energy empty orbital in a process of *excitation.* This excitation requires energy, which may be a beam of light. When atoms come together to form compounds or minerals, the inner orbitals remain essentially unchanged, but the outermost orbitals overlap to form more complex *molecular orbitals.* Also, the extensive overlap between orbitals in solids causes a broadening of the formerly discrete, clearly defined, energy levels into *bands* of closely spaced energy levels. The optical and other electronic properties of materials depend on the nature of the highest energy orbitals or energy bands that contain electrons and the lowest energy empty orbitals. Three important cases can be recognized and are illustrated in the energy level diagrams of Figure 5.16—insulators, semiconductors, and metals.

In an insulator such as pure quartz, the highest orbitals containing electrons (or *valence band* of Figure 5.16) are completely filled with electrons and the only empty orbitals (*conduction band*) into which these electrons could be excited are at much higher energy. It requires a lot more energy than that provided by a beam of visible light to cause such an excitation, and light passes through without being absorbed. In this case, k is zero but $n > 1$ because the light frequency is affected by interaction with the bound (core) electrons. In a metal, the valence band and conduction band can be envisaged as overlapping (Figure 5.16). Electrons in the highest energy filled orbitals can readily move to and from the unfilled orbitals given a small amount of energy (such as visible light energy). A light wave incident on a metal surface may therefore be appreciably absorbed ($k > 0$) as well as slowed down ($n > 1$). In this case, the reflectance is high because light is reemitted when the excited electrons return to the ground state. This is the case for many metals throughout the visible light range and for many opaque minerals. The semiconductor (Figure 5.16) in its simplest form can be considered an intermediate case between the metal and insulator. The energy required to excite electrons into the conduction band is greater than in a metal, but much less than in an insulator. Frequently, metal sulfides and oxides are semiconductors that require energies of the order of visible light to produce such excitation.

Electrons excited into the conduction band in a metal or semiconductor are delocalized and not located on a single atom; that is, they are effectively free and are responsible for the conduction of electricity. The effective number of free electrons (n_{eff}) can be determined and Burns and Vaughan (1970) plotted values of n_{eff} for the pyrite type FeS_2, CoS_2, NiS_2 and CuS_2 against $R\%$ (at 496 nm) and found a linear relationship. Data available for $CuSe_2$, $CuTe_2$, and Ag also conform with this correlation, which shows that reflectance increases with the effective number of free electrons. In the pyrite-type compounds (see Figure 5.17) the highest energy levels that contain electrons (the 3d orbitals of the metals) are split into a group of lower energy and a group of higher energy orbitals, which are labeled t_{2g} and e_g orbitals, respectively. In FeS_2, the t_{2g} orbitals are filled with electrons but the e_g orbitals are empty. The successive addition of electrons across the transition series to CuS_2 results in increased occupancy of the e_g orbitals (see Figure 5.17). These e_g orbitals form a band by overlap through the crystal (not shown in Figure 5.17), so that electrons excited into the e_g band are delocalized (i.e., become effectively free). As Figure 5.17 shows, both values of n_{eff} and $R\%$ systematically decrease in the series $FeS_2 > CoS_2 > NiS_2 > CuS_2$, and the values of n_{eff} are roughly proportional to the number of empty e_g levels into which t_{2g} electrons may be excited. The reduction in reflectance in the

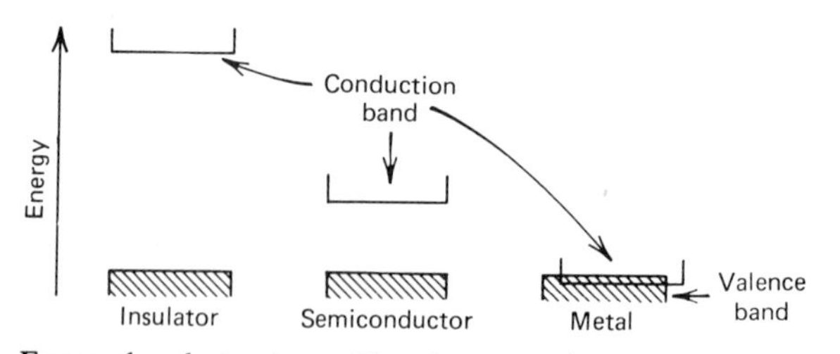

FIGURE 5.16 Energy band structure of insulator, semiconductor, and metal.

	FeS_2	CoS_2	NiS_2	CuS_2
R% (496 nm)	52	34	27	17
n	2.5	2.5	1.7	2.0
k	3.0	1.8	1.4	0.8
n_{eff}*	2.97	2.08	1.11	0.62

(* for a volume of ~100 Å³)

FIGURE 5.17 Electron occupancy of t_{2g} and e_g energy levels in pyrite-type compounds and the value of reflectance (R%), the optical constants (n and k) and the effective number of free electrons (n_{eff}). (After R. G. Burns and D. J. Vaughan, 1973.)

series $FeS_2 > CoS_2 > NiS_2 > CuS_2$ is due to filling of the e_g orbital levels making fewer of them available for excited electrons that reemit light energy on returning to the ground state.

The reflectances of these ore minerals are therefore interpreted in terms of their electronic structures and the transitions of electrons, which can occur between ground and excited states resulting in the absorption and reemission of energy. Certain transitions of this type may or may not occur depending on the relative orientation of a beam of linearly polarized light and the reflecting crystal. In this way, an interpretation can be offered for the property of bireflectance in substances like graphite and molybdenite. The interpretation of reflectance data for isotropic and anisotropic phases is further discussed by Vaughan (1979).

5.7 CONCLUDING REMARKS

Reflectance is the single most important quantitative parameter that can be used to identify or characterize an opaque mineral. The availability of efficient and relatively inexpensive instrumentation, good standards, and reliable data for most ore minerals makes reflectance measurement a powerful technique for the ore microscopist. Further development of correlations between compositional variation and reflectance variation should increase the value of the technique, as should the more widespread use of quantitative color determinations. Interpretation of ore mineral reflectances also offers a challenge to those concerned with understanding the fundamental structures of minerals and a source of further data.

BIBLIOGRAPHY

Atkin, B. P. and Harvey, P. K. (1979a) Nottingham Interactive System for Opaque Mineral Identification: NISOMI. *Trans. Inst. Min. Metal.* **88**, 1324–1327.

Atkin, B. P. and Harvey, P. K. (1979b) The use of quantitative color values for opaque-mineral identification. *Can. Mineral.* **17**, 639–647.

Bowie, S. H. U. and Simpson, P. R. (1978) *The Bowie-Simpson System for the Microscopic Determination of Ore Minerals: First Students Issue.* Applied Mineralogy Group, Mineralogical Society, London.

Bowie, S. H. U. and Taylor, K. (1958) A system of ore mineral identification. *Mining Mag. (Lond.)* **99**, 265.

Burns, R. G. and Vaughan, D. J. (1970) Interpretation of the reflectivity behavior of ore minerals. *Am. Mineral.* **55**, 1576–1586.

Carpenter, R. H. and Bailey, H. C. (1973) Application of R_O and A_r measurements to the study of pyrrhotite and troilite. *Am. Mineral.* **58**, 440–443.

Cervelle, B., Lévy, C., and Caye, R., (1971) Dosage rapide du magnésium dans les ilménites. *Mineral. Deposita* **6**, 34–40.

Charlat, M. and Lévy, C. (1976) Influence des principales substitutions sur les propriétés optiques dans la série tennantite tétraédrite. *Bull. Soc. Franc. Mineral. Crist.* **99**, 29–37.

Commission on Ore Microscopy (1977). *IMA/COM Quantitative Data File.* Edited by N. F. M. Henry. Applied Mineralogy Group, Mineralogical Society, London.

Eales, H. V. (1967) Reflectivity of gold and gold-silver alloys. *Econ. Geol.* **62**, 412–420.

Embrey, P. G. and Criddle, A. J. (1978) Error problems in the two media method of deriving the optical constants *n* and *k* from measured reflectances. *Am. Mineral.* **63**, 853–862.

Galopin, R. and Henry, N. F. M. (1972) *Microscopic Study of Opaque Minerals.* McCrone Research Associates, London.

Gray, I. M. and Millman, A. P. (1962) Reflection characteristics of ore minerals. *Econ. Geol.* **57**, 325–349.

Henry, N. F. M. (1979) *Reflected-Light Microscopy in the Study of Opaque Minerals: A First Course.* Applied Mineralogy Group, Mineralogical Society, London.

Henry, N. F. M. and Phillips, R. (1981) *Quantitative Color* (in press).

Htein, W. and Phillips, R. (1973) Quantitative specification of the colours of opaque minerals. *Mineral. Mater. News Bull. Quant. Microsc. Methods* No. 1, pp. 2–3, No. 2, pp. 5–8.

Kuhnel, R. A., Prins. J. J. and Roorda, H. J. (1976) The Delft System for Mineral Identification. I. Ore Minerals. *Mineral. Mater. News Bull. Microsc. Methods* No. 1, pp. 2–3.

Lévy, C. (1967) Contribution à la minéralogie des sulfures de cuivre du type Cu_3. *Mem. Bur. Rech. Géol. Min.* No. 54.

McLeod, C. R. and Chamberlain, J. A. (1968) Reflectivity and Vickers microhardness. Paper 68-64, *Geol. Surv. Can.*

Piller, H. (1966) Colour measurements in ore microscopy. *Mineral. Deposita* **1**, 175–192.

Pinet, M., Cervelle, B. and Desnoyers, C. (1978) Reflectance, indice de réfraction et expression quantitative de la couleur de proustites et pyrargyrites naturelles et artificielles: interpretation génétique. *Bull. Mineral.* **101**, 43–53.

Singh, D. S. (1965) Measurement of spectral reflectivity with the Reichert Microphotometer. *Trans. Inst. Min. Metall.* **74**, 901–916.

Squair, H. (1965) A reflectometric method for determining the silver content of natural gold alloys. *Trans. Inst. Min. Metall.* **74**, 917–931.

Stumpfl. E. F. and Tarkian, M. (1973) Natural osmium-iridium alloys and iron-bearing platinum: new electron probe and optical data. *Neues Jb. Miner. Monat.*, 313–322.

Tarkian, M. (1974) A key diagram for the optical determination of common ore minerals. *Mineral. Sci. Eng.* **6**, 101.

Vaughan, D. J. (1978) The interpretation and prediction of the properties of opaque minerals from crystal chemical models. *Bull. Mineral.* **101**, 484–497.

6

Quantitative Methods—Microindentation Hardness

6.1 INTRODUCTION

The hardness of an ore mineral has been estimated or measured by mineralogists in one of three ways: through examination of polishing hardness, scratch hardness, or microindentation hardness. Polishing hardness, discussed in detail in Chapter 3, only enables the hardness of a phase to be estimated relative to other phases in a polished section. Scratch hardness may also be qualitatively estimated by visual examination of the relative intensity of surface scratches on a polished section (see Chapter 3). The *Mohs Scale* of scratch hardness, universally employed in the study of minerals in hand specimen, is a simple quantification of this property. Early attempts at measuring mineral hardness under the microscope involved drawing a scribe across the surface while applying a known load as in the work of Talmage (1925). Such methods have been superseded by the more accurate techniques of microindentation hardness measurement in which a static indenter is lowered onto the mineral surface under a known load and the size of the resulting impression is determined.

The measurement of hardness on the microscopic scale has involved a variety of instruments and types of indenter, the most common indenters being the Vickers (a square-based pyramid) and the Knoop (an elongated pyramid). Most systematic studies of ore minerals (Bowie and Taylor, 1958; Young and Millman, 1964) have employed Vickers microhardness determination and this technique has been widely adopted in ore microscopy. It is important to note that although the various types of hardness (polishing, scratch, indentation) will produce very similar results for a series of minerals, differences in relative hardness may be observed on detailed examination since the methods (and nature of "deformation" involved) are not equivalent.

62. VICKERS HARDNESS MEASUREMENT

6.2.1 Theory

The measurement of Vickers hardness provides a Vickers Hardness Number (VHN) for a material. The hardness number is defined as the ratio of the load applied to the indenter (gram or kilogram force) divided by the contact area of the impression (square millimeters). The Vickers indenter is a square-based diamond pyramid with a 130° included angle between opposite faces, so that a perfect indentation is seen as a square with equal diagonals (although it is actually a pyramidal hole of maximum depth one seventh the diagonal length).

The area of the Vickers indentation can be expressed in terms of the length of the diagonal d (in μm) as:

$$\tfrac{1}{2} d^2 \operatorname{cosec} 68° \qquad (6.1)$$

The VHN, being the ratio of load L (in gram force, gf) to area of indentation, is given by

$$\text{VHN} = \frac{2 \sin 68° \times L}{d^2} = \frac{1.8544 \times L}{d^2} \quad \text{g}/\mu^2$$

or

$$\frac{1854.4 \times L}{d^2} \quad \text{g/mm}^2 \qquad \text{(the normal units employed)} \qquad (6.2)$$

Microindentation hardness testers normally employ loads of 100–200 gf, which result in indentations of diagonal lengths approximately 5–100 μm, commonly a few tens of microns in diameter. The load employed in a VHN determination must be stated since, as discussed later, values obtained are not independent of load.

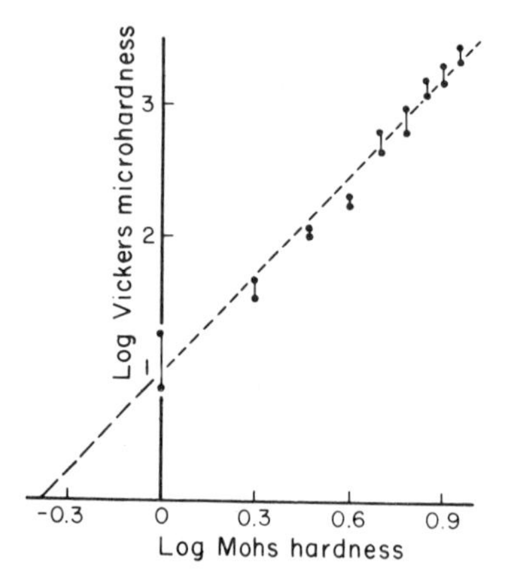

FIGURE 6.1 Correlation of Vickers microhardness with Mohs Scale of hardness showing the virtually linear relationship between log Mohs and log VHN. (After B. B. Young and A. P. Millman, 1964.)

A number of workers have compared the scale of VHN values with the more familiar Mohs Scale. As shown in Figure 6.1, Young and Millman (1964) suggest an approximately linear relationship when log Mohs is plotted against log VHN, a relationship expressed by the equation:

$$\log \text{VHN} = 2.5 \log \text{Mohs} + 1.00 \tag{6.3}$$

A log/linear relationship was determined by Bowie and Simpson (1977):

$$\log \text{VHN} = \text{Mohs} \log 1.2 + C \tag{6.4}$$

6.2.2 Instrumental Techniques

A variety of instruments for measuring Vickers hardness are commercially available; some are marketed as complete instruments and others designed as attachments to a standard ore microscope. In all instances, a reflected-light microscope forms the basis of the instrument, with a permanent or detachable indenter head that can be rotated into position in place of a normal objective; the second major addition is a micrometer eyepiece to enable the indentation diagonals to be accurately measured. The load may be applied directly when a plunger is depressed to allow the indenter to descend onto the sample surface, or it may be applied pneumatically via an air line from depressing a lever on an impeller drum.

An example of the typical sequence of operations in a single hardness determination with the Vickers Company instrument would be:

1 Selection of a suitable grain or area for measurement using a low power objective (on the surface of a *leveled* polished section).

2 Selection of a high power (×40) objective, focusing, and checking that the area is free from flaws.

3 Rotation of indenter nosepiece into position (and in this instance focusing through this objective-mounted indenter nosepiece).

4 Selection of a suitable load (commonly 100 gf, although a range of loads from 10 to 200 gf is available for softer or harder materials).

5 Depression of lever to initiate indenting process, after a set time (say 15 sec) the lever is raised to withdraw indenter.

6 Examination of indentation under high power objective.

7 Measurement of the length of each diagonal using the micrometer eyepiece, when the indentation has been suitably positioned.

The data on diagonal lengths (averaged for each indentation) can be used in equation 6.2 to calculate the VHN although tables are normally provided by manufacturers for direct conversion of diagonal measurements to VHN values. An experienced operator can obtain satisfactory results from a couple of indentations on each of three or four grains except when a wide range of hardness is exhibited by the material.

VHN values have been obtained for most ore minerals and the accepted standard values have been published by the Commission on Ore Microscopy (COM) in the IMA/COM Data File (see Section 5.3, Figure 5.8). In Appendix 1 of this book, VHN values are provided for the more common ore minerals, and in Appendix 2 these minerals are listed in order of increasing VHN.

6.2.3 Sources of Error and Accuracy and Precision of Measurement

Assuming that the instrument is working correctly and has been properly calibrated (a standard block of known VHN is normally provided to check this), errors may still arise from a number of sources including vibration effects, inconsistent indentation times, and misuse of the micrometer eyepiece. Vibration may be a serious source of error and the system is best mounted on a bench designed to minimize this effect. Indentation time is important because a certain amount of creep occurs during indentation and there may be some recovery after indentation. A loading time of 15 sec has been approved by the Commission on Ore Microscopy. Errors arising in the actual measurement of the indentation depend partly on factors outside the control of the operator such as the resolving power of the objective and quality of the indentation (which may always be poor in certain materials). However, accurate focusing and careful location of the ends of the diagonals are important for acceptable results. Precise measurement can also be improved by using monochromatic illumination (Bowie and Simpson, 1977 suggest using light of $\lambda = 500$ nm and a dry objective of N.A. $= 0.85$ for relatively rapid and precise results).

Producing an average hardness number for a particular mineral is easier than establishing a true range of hardness, and reproducing the same average values (to within 5%) on the same instrument is fairly simple. Bowie and Simpson (1977) have noted that if $H_1, H_2, H_3 \ldots H_n$ are the values obtained in n tests on the same specimen and the respective differences from the mean ($\bar{H}$) are $(H_1 - \bar{H}), (H_2 - \bar{H}) \ldots (H_n - \bar{H})$ then the standard error of any individual observation is

$$\frac{\Sigma (H - \bar{H})^2}{n}$$

and the standard error of the mean is:

$$\frac{\Sigma (H - \bar{H})^2}{n^2}$$

Relatively large variations may result using different types of indenter and as a result of applying different loads as discussed below. Generally, a load of 100 gf should be employed unless the specimen is too soft, too brittle, or too small for this to be acceptable.

6.3 SHAPES OF HARDNESS MICROINDENTATIONS

A perfectly square, clear indentation very rarely results from hardness testing of minerals; the shape of the impression and any fracturing characteristics that may result from indentation can provide useful additional information regarding the identity (and sometimes orientation) of minerals. Young and Millman (1964) have examined the shape and fracture characteristics of indentations in a large number of ore minerals. These authors have noted the shape characteristics as consisting of combinations of four distinct indentation types: (1) straight edge, (2) concave edge, (3) convex edge, and (4) sigmoidal edge (see Figure 6.2). The extent to which curvature is developed in those indentations without straight edges can also be classified as weak or strong. Fracturing or deformation may also occur during indentation, particularly if the mineral has a distinct cleavage or fracture. Again, Young and Millman (1964) have classified the types observed: (1) star radial fractures, (2) side radial fractures, (3) cleavage fractures, (4) parting fractures, (5) simple shell fractures, (6) cleavage shell fractures, and (7) concentric shell fractures. The various shape, deformation and fracture characteristics of indentations are illustrated in Figure 6.2 and their development in the major ore mineral groups is shown in Table 6.1. Some data on fracture characteristics are also provided in the tables of Appendix I.

The shapes and fracture characteristics of minerals may vary with orientation

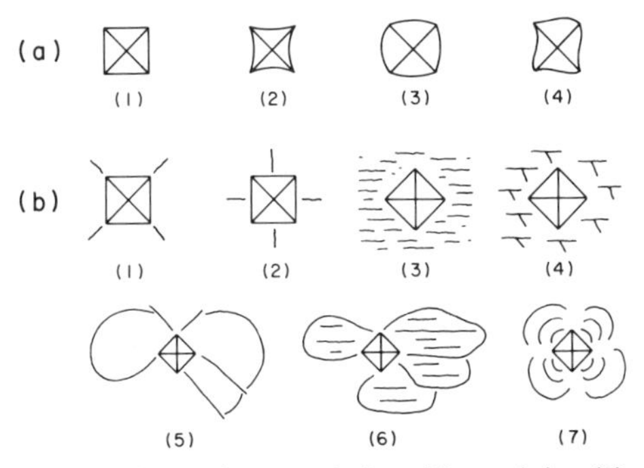

FIGURE 6.2 *(a)* Indention shape characteristics: (1) straight; (2) concave; (3) convex; (4) sigmoidal. *(b)* Indentation fracture characteristics: (1) star radial; (2) side radial; (3) cleavage; (4) parting; (5) simple shell; (6) cleavage shell; (7) concentric shell. (Based on Young and Millman, 1964.)

TABLE 6.1 Categories of Indentation Shape, Fracture Characteristics, and Anisotropy Associated With Mineral Structural Groups*

Mineral Structural Group	Indentation Anisotropy	Characteristics of Indentation				Characteristics of Fractures
		Straight	Concave	Convex	Sigmoidal	
Metals and semimetals	Mod. to st.	—	Wk. to st.	Wk. to st.	Wk. to st.	Wk. GPT
Simple sulfides	Wk. to st.	—	Wk. to st.	Wk. to st.	Wk. to st.	Wk. rad. and GPT
Co-Ni-Fe sulfides						
Pyrite type	Iso.	Com.	Wk.	—	—	St. rad. shells
Spinel type	Iso.	Com.	Wk.	—	—	Wk. rad. shells
Marcasite type	Mod. to st.	Com.	Wk.	Wk.	—	St. rad. shells
Complex sulfides						
Sphalerite type	Wk. or iso.	—	Wk. to st.	—	—	Com. rad. and shells
Wurtzite type	Mod. to st.	Com.	St.	Wk.	—	St. rad. shells
Niccolite type	Mod. to st.	—	St.	Wk.	Wk.	Com. rad. and GPT
Sulfosalts						
Chain type	Mod. to st.	Com.	Wk. to st.	Wk. to st.	—	Com. rad. and shells
Sphalerite type	Iso.	Com.	Wk. to st.	Wk. to st.	—	Com. rad. and shells
Spinels, isometric oxides, rutile-type oxides, metamict oxides	Wk. or iso.	Com.	Wk. to st.	—	—	Wk. rad.
Hydrated oxides, hex. hematite and zincite type oxides	Wk. to st.	—	Wk. to st.	—	—	Occ. rad. shells
Silicates, sulfates, phosphates, carbonates	Wk. or iso.	—	Wk.	—	—	St. rad. shells

**Abbreviations:* wk., weak; mod., moderate; st., strong; iso., isotropic; com., common; rad., radial fractures; shells, shell fractures; occ., occasional; GPT, glide plane traces.

of the grain, particularly in materials such as covellite or molybdenite, which have highly anisotropic structures.

6.4 FACTORS AFFECTING MICROINDENTATION HARDNESS VALUES OF MINERALS

6.4.1 Variation with Load

The actual VHN value determined for a mineral is not independent of the load used in measurement; generally an increase in hardness is shown with decrease in the applied load. For this reason, measurements are often made at a standard load of 100 gf, although Young and Millman (1964) have pointed out the advantages of using a series of standard loads (15, 25, 50, and 100 gf) depending on the hardness of the mineral under examination. These authors undertook a systematic study of hardness variation with load in a variety of minerals and found, for example, that a mean percentage increase in VHN in going from 100 gf load to 15 gf is ~24% for a mineral in the 600–1200 kg/mm^2 hardness range and ~12% for a mineral in the 60–120 kg/mm^2 hardness range. They also examined the effects of both load and orientation on the VHN values of galena as shown in Figure 6.3 and discussed in the following.

One source of this load dependence of hardness is the deformation of the surface layer of a mineral during polishing. The hard layer (~10–20 μm thick) produced by some polishing methods may become important when small loads are used but should have less influence at loads of 100 gf as shown by tests using different polishing techniques (Bowie and Taylor, 1958). The load dependence may also be related to the actual mechanism of deformation during indentation or may even originate from instrumental effects (Bowie and Simpson, 1977).

6.4.2 Variation with Mineral Texture

Ideally, relatively large and well crystallized grains are required for hardness determination. When masses of microcrystalline or cryptocrystalline material are measured the hardness may be markedly lower (e.g., microcrystalline hematite and goethite give VHN values ~70% of values for coarse crystals). On single grains, excessive fracturing normally occurs and values are unreliable if the indentation diagonal exceeds one-third the grain size. In practice, the minimum grain size is ~100 μm although use of smaller loads may make it possible to obtain results on smaller grains.

6.4.3 Variation with Mineral Orientation

Most minerals show some degree of hardness anisotropy and this effect may be considerable in fibrous, layered, or prismatic phases [e.g., molybdenite VHN

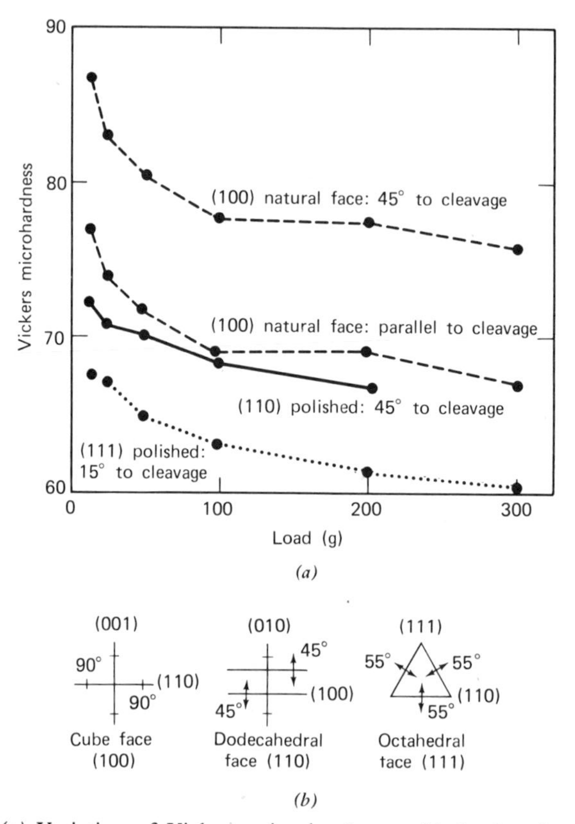

FIGURE 6.3 *(a)* Variation of Vickers microhardness with load and orientation in a crystal of galena. *(b)* Attitudes of (001) cleavage glide planes in galena in sections parallel to cube, dodecahedral and octahedral planes. (From *Mineral Chemistry of Metal Sulfides* by D. J. Vaughan and J. R. Craig, 1978, Cambridge University Press, after B. B. Young and A. P. Millman, 1964; used with permission.)

(0001) 33–74, VHN (1010) 4–10 at 100 gf load]. The measurements of Young and Millman (1964) involved determinations on oriented sections of 50 minerals and demonstrated that most minerals gave different hardness values on different crystal faces and different indenter orientations on the same face. Indentation shape and fracture characteristics often also exhibit variations with crystal orientation.

An interesting aspect of hardness anisotropy is that it is not limited to non-isometric minerals; cubic minerals such as galena, sphalerite, and native copper show considerable hardness anisotropy. In the case of galena, a detailed interpretation of the variation of hardness with orientation (illustrated in Figure 6.3) has been presented based on the attitudes of various glide planes in the structure. The relationship between hardness values for various faces was determined to be VHN (001) > > VHN (110) > VHN (111) (see Figure 6.3) and plastic deformation in galena is known to take place by gliding along (100) and (111) planes, the (100) planes being dominant. Hence on indenting the cube face (001, 100, 010) there is little tendency for movement along (100) planes which are perpendicular or parallel to the surface. On indenting the (110) face, deformation occurs readily by movement along two sets of (100) glide planes oriented at 45° to the indentation direction. On the (111) face, three (100) planes make angles of 35° with the indentation direction and are all capable of easy glide translation, resulting in the lowest hardnss of all. The relationships of glide planes to crystal faces are also shown in Figure 6.3.

6.4.4 Variation with Mineral Composition

The variation of hardness with mineral composition has been examined in a number of mineral series and solid solutions. In some cases hardness variation can be directly linked to variations in structure and bonding in the series. For example, in the series galena (PbS)-clausthalite (PbSe)-altaite (PbTe), which all have the halite-type structure, decreasing VHN (at 100 gf) follows roughly the increasing unit cell parameters (PbS, VHN 57–86, a_o = 5.94Å; PbSe VHN 46–72, a_o = 6.45Å; PbTe VHN 34–38, a_o = 6.45Å). In the series of isostructural disulfides, hauerite (MnS_2)-pyrite (FeS_2)-cattierite (CoS_2)-vaesite (NiS_2) hardness variations can be correlated not only with unit cell parameter but in turn with the electron occupancy of certain orbitals associated with the metals (see Vaughan and Craig, 1978). These variations in the electronic structures of the disulfides also explain differences in metal-sulfur bond strengths which, like the hardness values, decrease in the sequence $FeS_2 > CoS_2 > NiS_2$. An intimate relationship clearly exists between hardness, bond strength and the nature of bonding in minerals but its precise formulation is complex.

Detailed studies of hardness variation have been undertaken on a number of solid solution series. As summarized by Vaughan and Craig (1978) several authors have studied the variation of the hardness of sphalerite (Zn,Fe)S with iron content. As illustrated in Figure 6.4*a*, all these studies show a sharp increase in hardness on substitution of small amounts of iron (< 2 wt. %) and most studies show a subsequent decrease with further iron substitution. The cell parameters of sphalerites show a linear increase with iron substitution (Figure 6.4*a*) and possibly the initial hardness increase is due to some iron atoms filling defect sites before the expansion in the unit cell becomes dominant and hardness decreases. Young and Millman (1964) also report the complex hardness variations observed with compositional variation in the hübnerite-wolframite-ferberite $(Fe,Mn)WO_4$ series. Their measurements, illustrated in Figure 6.4*b*, were performed on oriented crystals.

Clearly the variation in hardness with composition in solid solution series is complex and could never be used to give more than a very crude estimate of

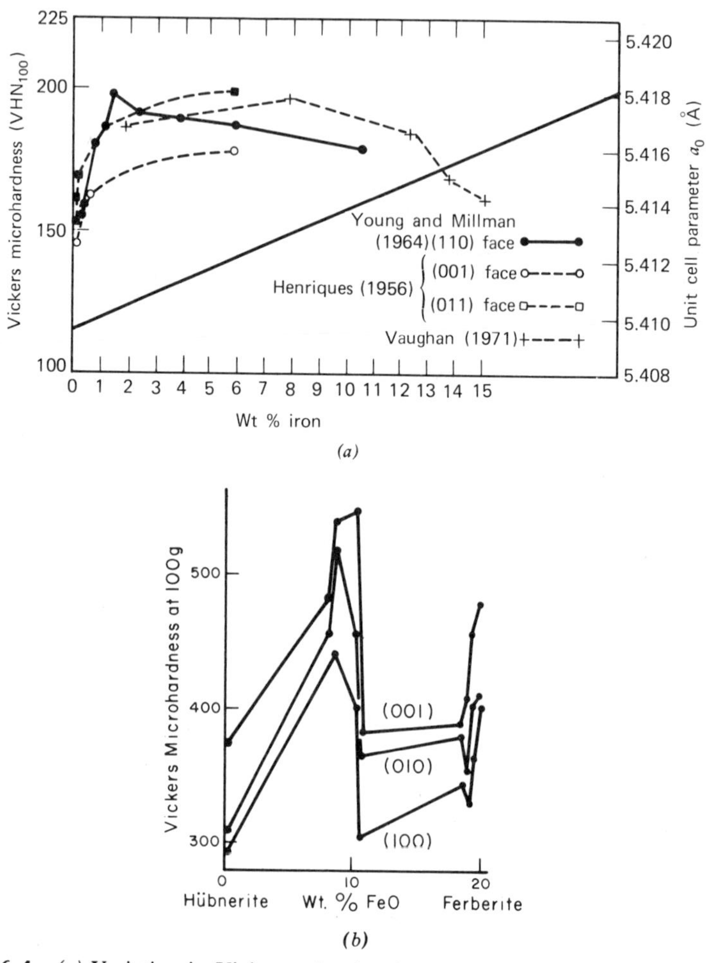

FIGURE 6.4 *(a)* Variation in Vickers microhardness (at 100 gf load) and cell size (Å) with iron content of sphalerite. (After D. J. Vaughan and J. R. Craig, 1978) *(b)* Variation of Vickers microhardness with iron content in the wolframite series (in various orientations). (From *Mineral Chemistry of Metal Sulfides* by D. J. Vaughan and J. R. Craig, 1978, Cambridge University Press, after B. B. Young and A. P. Millman, 1964; used with permission.)

composition. What may be of greater value is the information on crystal chemical variations implied by such studies.

6.4.5 Variation with Mechanical and Thermal History

The presence of structural imperfections in crystalline materials has a very marked influence on hardness properties. Defects, particularly the linear regions of mismatching of the crystal lattice known as *dislocations* are introduced by mechanical deformation (a process known as *work hardening*). This means that the cutting, grinding, and polishing of specimen preparation can increase hardness values by 5–30% compared to the values for untreated crystal or cleavage faces of some materials. Since most laboratories use both comparable and consistent methods of specimen preparation, the consequences generally do not seriously impair the use of the technique in mineral identification.

The formation of dislocations, the processes of work hardening, and the effects of both the conditions of initial crystallization and any subsequent heat treatment on defect formation have been extensively studied by metallurgists. In studying natural materials that have undergone mechanical deformation (and sometimes heat treatment) during tectonism and metamorphism, Stanton and Willey (1970, 1971) have demonstrated another important field of application of hardness determinations of ore minerals. For example, galena deformed by tectonic movement has commonly undergone a process of natural work hardening. This hardening may be eliminated by heat treatment and the softening results from two separate processes: recovery and recrystallization. Although recrystallization initially results in softening, complete recrystallization may ultimately lead to substantial hardening as Stanton and Willey (1971) have shown in their studies of the recrystallization of naturally deformed sphalerite and galena. In Figure 6.5, the softening-hardening behavior that accompanies the recovery-recrystallization of naturally deformed galena from Broken Hill (New South Wales, Australia) is illustrated. The curves show the distinct separation

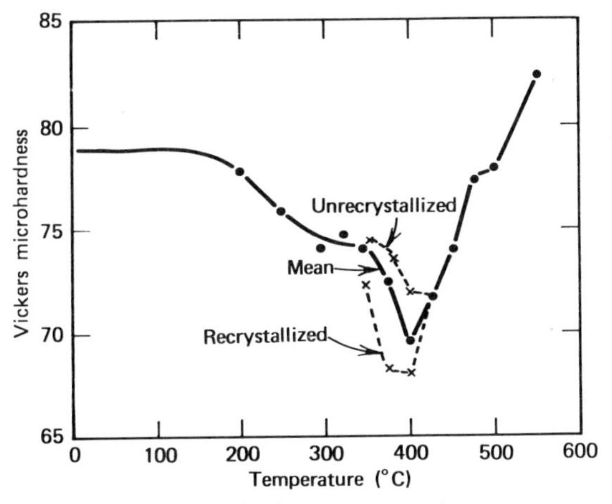

FIGURE 6.5 Softening-hardening behavior accompanying recovery-recrystallization of naturally deformed galena from Broken Hill, Australia. Specimens were heated for 1 day and the plotted values are an average of 70 indentations. (From *Mineral Chemistry of Metal Sulfides* by D. J. Vaughan and J. R. Craig, 1978, Cambridge University Press, after Stanton and Willey, 1971; used with permission.)

in hardness between naturally unrecrystallized and recrystallized material once experimental recrystallization has commenced, and the greater hardness of the completely recrystallized material. The careful heating and hardness testing of sulfides that have undergone natural deformation or annealing may give valuable information on their postdepositional histories. Such information as the upper temperature limits reached since deformation, whether recovery has occurred, whether a polycrystalline aggregate is a result of primary deposition or recrystallization, and to what extent this recrystallization has proceeded is obtainable from such studies.

In related studies, Kelly and Clark (1975) and Roscoe (1975) have examined the nature of deformation in chalcopyrite; the former authors have also summarized data for pyrrhotite, galena, and sphalerite. These workers, although not actually measuring hardness, have carefully examined the strengths of minerals as functions of temperature and confining pressure (see Figure 10.25).

6.5 CONCLUDING REMARKS

The measurement of Vickers hardness provides an established and tested method for quantitative determination of the hardness of minerals in polished section. This has been widely used as an aid to mineral identification (see Section 5.3), although in many laboratories hardness determination is now used only where optical methods fail to provide conclusive identification.

The hardness of a mineral is a complex property of both theoretical and practical interest. Detailed studies of hardness variation with orientation, compositional variation, and thermal and mechanical history of minerals should be a valuable field of future study with wider geological application than has been exploited so far.

BIBLIOGRAPHY

Bowie, S. H. U. and Simpson, P. R. (1977) Microscopy: reflected light. In J. Zussman, ed., *Physical Methods in Determinative Mineralogy,* (2nd ed. Academic, London, pp. 109–165).

Bowie, S. H. U. and Taylor K. (1958) A system of ore mineral identification *Min. Mag. (Lond.)* **99**, 237.

Kelly, W. C. and Clark, B. R. (1975) Sulfide deformation studies: III Experimental deformation of chalcopyrite to 2,000 bars and 500°C *Econ. Geol.* **70,** 431–453.

Roscoe, W. E. (1975) Experimental deformation of natural chalcopyrite at temperatures up to 300°C over the strain rate range 10^{-2} to 10^{-6} sec^{-1} *Econ. Geol.* **70**, 454–472.

Stanton, R. L. and Willey, H. G. (1970) Natural work hardening in galena and its experimental reduction. *Econ. Geol.* **65**, 182–94.

———(1971) Recrystallization softening and hardening in sphalerite and galena. *Econ. Geol.* **66**, 1232–8.

Talmage, S. B. (1925) Quantitative standards for hardness of the ore minerals. *Econ. Geol.* **20**, 535–53.

Vaughan, D. J. and Craig, J. R. (1978) *Mineral Chemistry of Metal Sulfides,* Cambridge University Press, Cambridge, England.

Young, B. B. and Millman, A. P. (1964) Microhardness and deformation characteristics of ore minerals. *Trans. Inst. Min. Metall.* **73**, 437–66.

7 Ore Mineral Textures

7.1 INTRODUCTION

Ore microscopy involves not only the identification of individual mineral grains but also the interpretation of ore mineral textures, that is, the relationships between grains. These textures may provide evidence of the nature of such processes as initial ore deposition, postdepositional reequilibration or metamorphism, deformation, annealing, and meteoric weathering. The recognition and interpretation of textures is thus an important step in understanding the origin and postdepositional history of an ore. The extent to which the ore minerals retain the compositions and textures formed during initial crystallization varies widely. Figure 7.1 illustrates this variability in terms of equilibration rates and shows that oxides, disulfides, arsenides, and sphalerite are the most refractory of ore minerals. These minerals are more likely to preserve evidence of their original conditions of formation than are minerals such as the pyrrhotites or Cu-Fe sulfides. Argentite, sulfosalts, and native metals are among the most readily reequilibrated ore minerals and thus are the least likely to reflect initial formation conditions. The textures observed in many polymetallic ores reflect the stages, sometimes numerous, in their development and postdepositional history. Textural information is also important in the milling and beneficiation of ores, an aspect discussed in Chapter 11. In this text, it is impossible to discuss completely the vast variety of textures observed in ores, but a number of common examples will be described and used to illustrate the principles involved in textural interpretation. Many additional textures found in specific ore associations are illustrated in Chapters 9 and 10. It is worth noting that many textures are still inadequately understood and that experienced ore microscopists still disagree as to their precise origin.

7.2 PRIMARY TEXTURES OF ORE MINERALS FORMED FROM MELTS

The growth of ore minerals in silicate melts generally results in the development of euhedral to subhedral crystals because there is little obstruction to the growth of faces. Thus primary chromite, magnetite, ilmenite and platinum minerals, phases that are refractory enough to retain original textures, often occur as well-developed equant euhedra interspersed in the plagioclase, olivine, and pyroxene of the host rock (Figure 9.1). Unobstructed growth, especially in rapidly cooled basalts, sometimes results in the formation of skeletal crystals (Figure 7.2*a*) that may be wholly or partially contained within subsequently crystallizing silicates. Poikilitic development of silicates in oxides or oxides in silicates is not uncommon. In oxide-rich layers, the simultaneous crystallization of mutually interfering grains results in subhedral crystals with widely variable interfacial angles. In contrast, the interfacial angles at triple-grain junctions of monomineralic masses that have been annealed during slow cooling or during metamorphism approach 120° (see Section 7.7 for further discussion).

Iron (plus nickel, copper)-sulfur (-oxygen) melts, from which iron-nickel-copper ores form (see Section 9.3) generally crystallize later than the enclosing silicates. The magnetite often present in these ores crystallizes while the iron sulfides are wholly or partly molten and thus tends to be euhedral or skeletal, whereas the much less refractory sulfides (mostly pyrrhotite) exhibit textures of

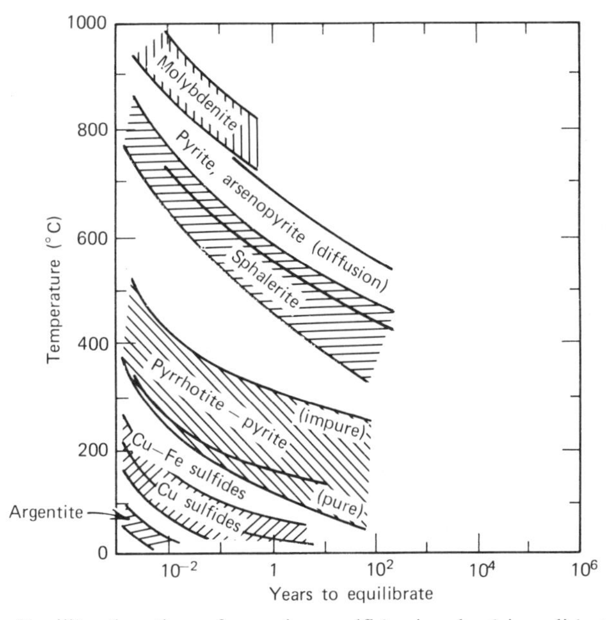

FIGURE 7.1 Equilibration times for various sulfides involved in solid-state reactions. The field widths represent differing rates in different reactions as well as changes in rates due to compositional differences in a phase and a great deal of experimental uncertainty. (From H. L. Barnes, *Geochemistry of Hydrothermal Ore Deposits*, 2nd ed., pg. 287, Wiley-Interscience, New York, 1979, p. 287; used with permission.)

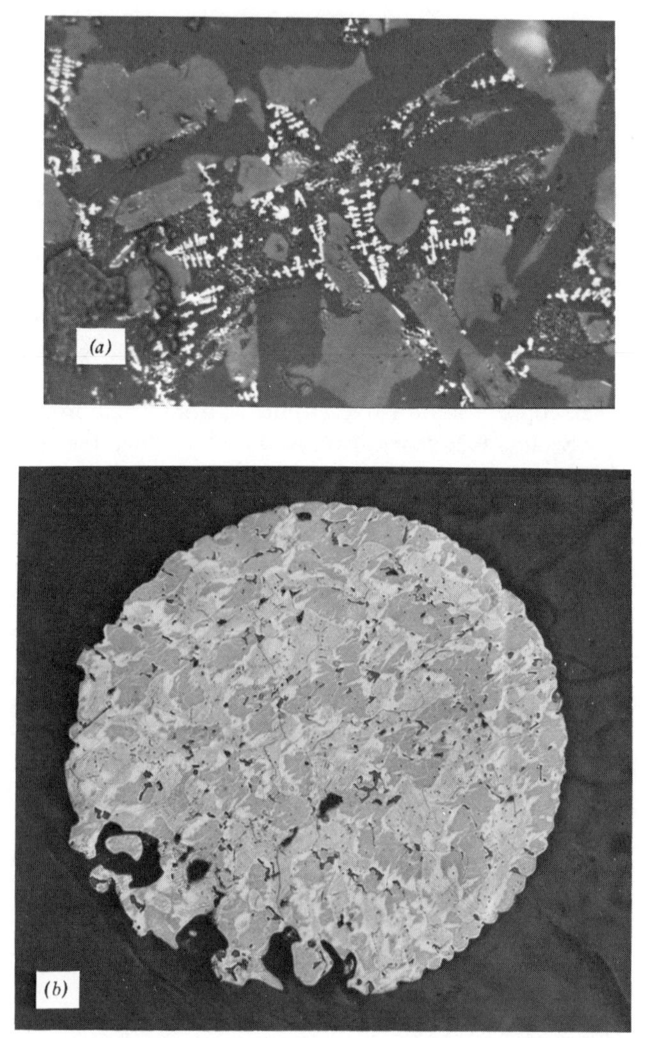

FIGURE 7.2 *(a)* Skeletal crystals of ilmenite in basalt from Hawaii. (Width of field = 150 μm.) *(b)* Sulfide droplet in Mid-Atlantic Ridge basalt, composed of $(Fe, Ni)_{1-x}S$ monosulfide solid solution (medium gray) and $(Cu,Fe)S_{2-x}$ intermediate solid solution (light gray) with rims and flames of pentlandite (bright). (Width of field = 250 μm.) (Reproduced from G. K. Czamanske and J. G. Moore, *Geol. Soc. Amer. Bull.* **88**, 591, 1977, with permission.)

cooling and annealing. Primary iron-sulfur (-oxygen) melts can also result in the formation of small (< 100 μm) round droplets trapped in rapidly cooled basalts and basaltic glasses (see Figure 7.2*b*).

7.3 PRIMARY TEXTURES OF OPEN-SPACE DEPOSITION

The growth of ore minerals in vugs and open veins is commonly characterized by the formation of crystals with well-developed faces, of crystals which exhibit growth-zoning (Figure 7.3), and of colloform or zoned monomineralic bands (Figures 7.4*a*–7.4*c*). All these features result from unobstructed growth of minerals into fluid-filled voids, the banding being the result of a change in the physico-chemical environment of mineralization with time. Deposition from hydrothermal solutions in open fissures can result in *comb structures* and in *symmetrically* and *rhythmically* crustified veins as shown in Figures 7.5*a* and 7.5*b*. Movement along such a vein may cause brecciation resulting in *breccia* ore (Figure 7.6). All of these textures may be developed on a scale ranging from macroscopic to microscopic. Open-space filling of this type is exemplified by the Cu-Pb-Zn(-Ag) vein deposits (discussed in Section 9.6) composed of pyrite, sphalerite, galena, chalcopyrite and silver-bearing sulfosalts, and by some Pb-Zn ores in carbonates (discussed in Section 10.6). Pyrite in these ores generally

FIGURE 7.3 Growth-zoning in a single crystal of sphalerite (from Creede, Colorado) viewed in transmitted light through a doubly polished thin section. (Width of field = 2 cm.)

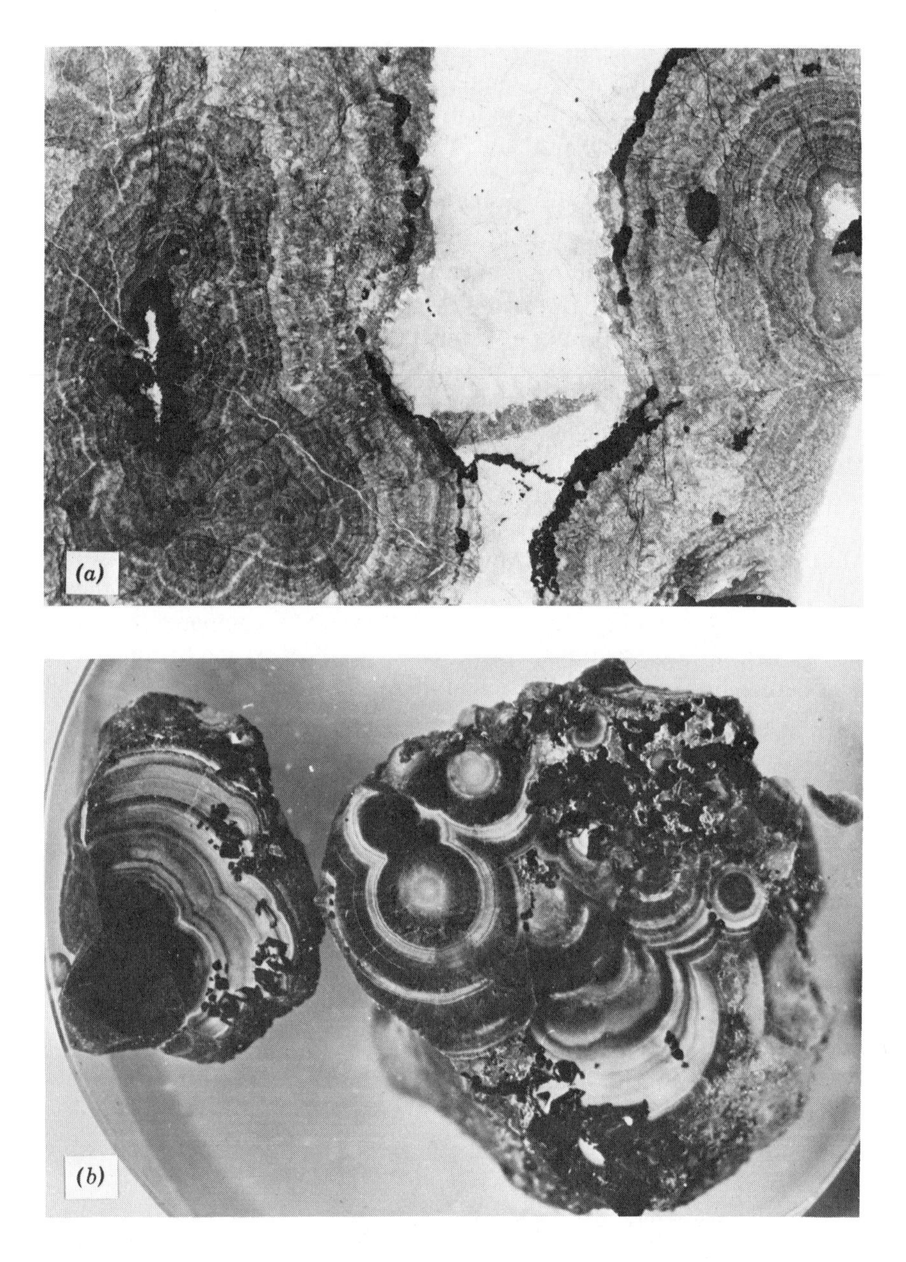

FIGURE 7.4 *(a)* Concentric growth-banding in sphalerite (from Austinville, Virginia) viewed in transmitted light through a doubly polished thin section. Black areas are pyrite, white areas are dolomite. (Width of field = 4 cm.) *(b)* Colloform growth-banding in sphalerite (from Pine Point, North West Territories, Canada) seen in reflected light. (Width of field = 2 cm.) *(c)* Growth-zoning in pyrite showing radial and concentric development (from Ainai Mine, Japan) seen in reflected light. (Width of field = 2000 μm.)

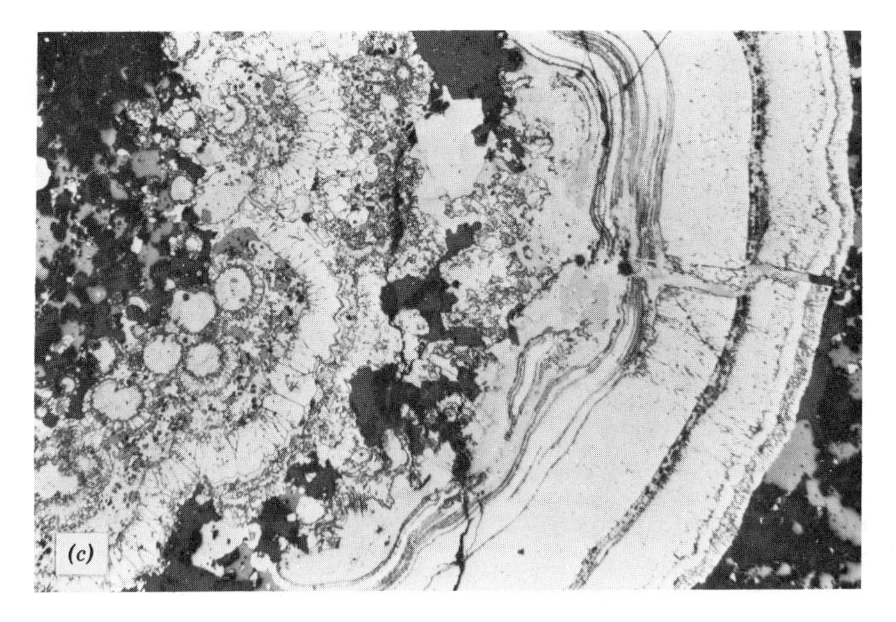

FIGURE 7.4 *(Continued)*

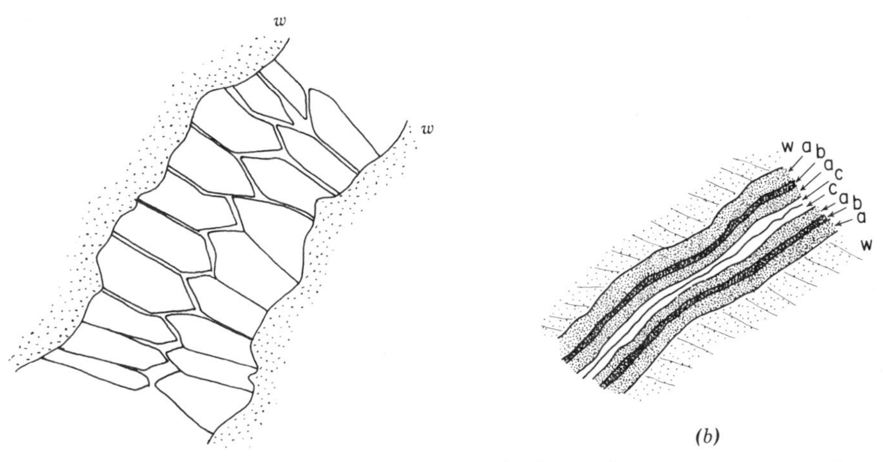

FIGURE 7.5 *(a)* Comb structure showing growth of crystals outward from open fracture walls. *(b)* Symmetrically crustified vein showing successive deposition of minerals inwards from open fracture walls. This vein is also rhythmically crustified in showing the depositional sequences *a-b-a-c* (*w* = wallrock). Scale variable from millimeter to several meters across vein.

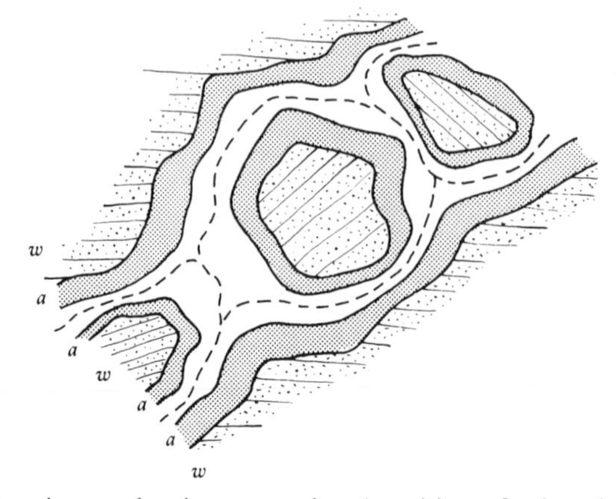

FIGURE 7.6 Breccia ore showing successive deposition of minerals on breccia fragments and other wallrocks. Scale variable from wallrock fragments of millimeter to several meters across breccia fragments.

forms as isolated cubes, more rarely as octahedra or pyritohedra, or as aggregates of interfering crystals along the walls of fractures. Sphalerite may occur as honey-yellow to black crystals or radiating colloform aggregates, which often contain a well-developed growth banding or zoning (Figure 7.3); this structure is clearly visible in doubly polished thin sections but is difficult to see in polished sections (see Section 2.5 and Figure 2.7). Darker sphalerite colors generally in-

FIGURE 7.7 Compositionally zoned crystals of bravoite (Fe,Ni,Co)S_2 from Maubach, Germany. (Width of field = 60 μm.)

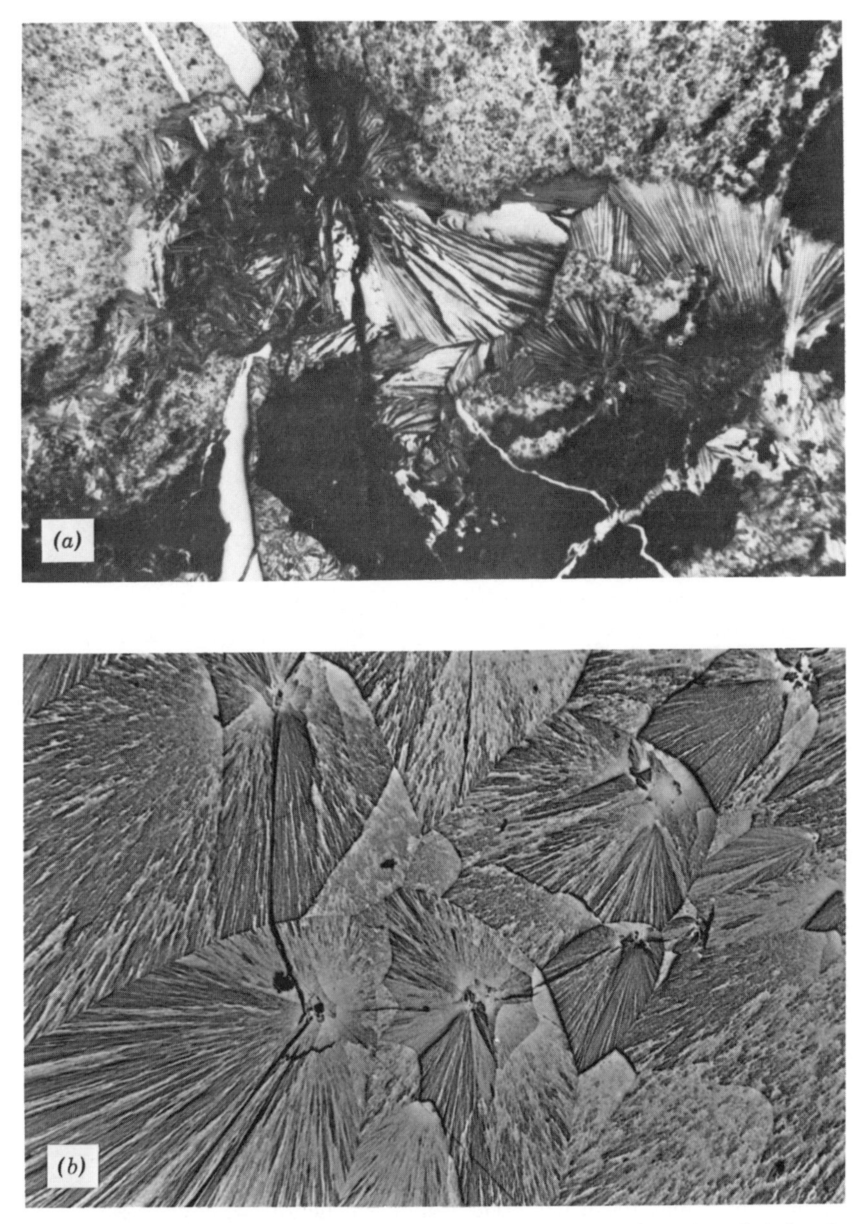

FIGURE 7.8 *(a)* Radiating fibrous crystals of the manganese oxide mineral chalcophanite infilling an open fracture, Red Brush Mine, Virginia. (Width of field = 2000 μm.) *(b)* Radiating clusters of fibrous goethite crystals forming a botryoidal aggregate in a weathering zone, Giles Co., Virginia. (Width of field = 2000 μm.)

dicate higher iron contents but this correlation is by no means consistent and is especially unreliable if iron contents are below 5%. Galena, commonly observed as anhedral intergranular aggregates in many types of ores, often occurs as well-formed cubes and less commonly as octahedral or skeletal crystals in open voids. Episodic precipitation, sometimes with intervening periods of leaching, often leaves hopperlike crystals of galena. The unimpeded growth of chalcopyrite, tetrahedrite, and the "ruby-silvers" (polybasite-pearcite) commonly results in the formation of euhedral crystals.

Sequential deposition from cobalt- and nickel-bearing solutions may result in the development of concentrically growth-zoned pyrite-bravoite crystals (Figure 7.7), which often reveal changing crystal morphology (cube, octahedron, pyritohedron) during growth. A similar process of sequential deposition from metal- and sulfur-bearing fluids circulating through the intergranular pore spaces in sediments may leave concentric sulfide coatings on the sediment grains.

Iron and manganese oxides and hydroxides often form in open fractures as a result of meteoric water circulation. These minerals (e.g., goethite, lepidocrocite, pyrolusite, cryptomelane) may form concentric overgrowths inward from vein walls or complex masses of fibrous (brushlike) crystals (Figures 7.8*a* and 7.8*b*) radiating from multiple growth-sites along an open fracture.

"Colloform" textures (see Figure 7.4*b*) have often been cited as evidence for initial formation by colloidal deposition; however, Roedder (1968) has shown that many colloform sphalerites in Pb-Zn ores (see Section 8.2.3) grew as tiny fibrous crystals projecting into a supersaturated ore fluid.

7.4 SECONDARY TEXTURES RESULTING FROM REPLACEMENT (INCLUDING WEATHERING)

Replacement of one ore mineral by another or by a mineral formed during weathering is common in many types of ores. However, a major problem in textural interpretation is the recognition of replacement when no vestige of the replaced phase remains. Probably the most easily recognized replacement textures are those in which organic materials, such as wood fragments (Figures 7.9*a* and 10.6) or fossil shells, have been pseudomorphed by metal sulfides (commonly pyrite, marcasite, chalcocite) or oxides (commonly hematite, goethite, "limonite," uranium minerals). Pyrite cubes and marcasite laths that have been replaced by iron oxides during weathering (Figure 7.9*b*) are also readily identified.

Replacement may result from one or more of the following processes: (1) dissolution and subsequent reprecipitation, (2) oxidation, and (3) solid state diffusion. The resulting boundary between the replaced and the replacing mineral is commonly either sharp and irregular (a *careous* = corroded texture) or *diffuse.*

Edwards (1947), Bastin (1950), and Ramdohr (1969) have described a wide variety of replacement geometries—rim, zonal, frontal, and so on—but they all appear to represent variations of the same process. Replacement textures depend chiefly for their development on three features of the phase being replaced.

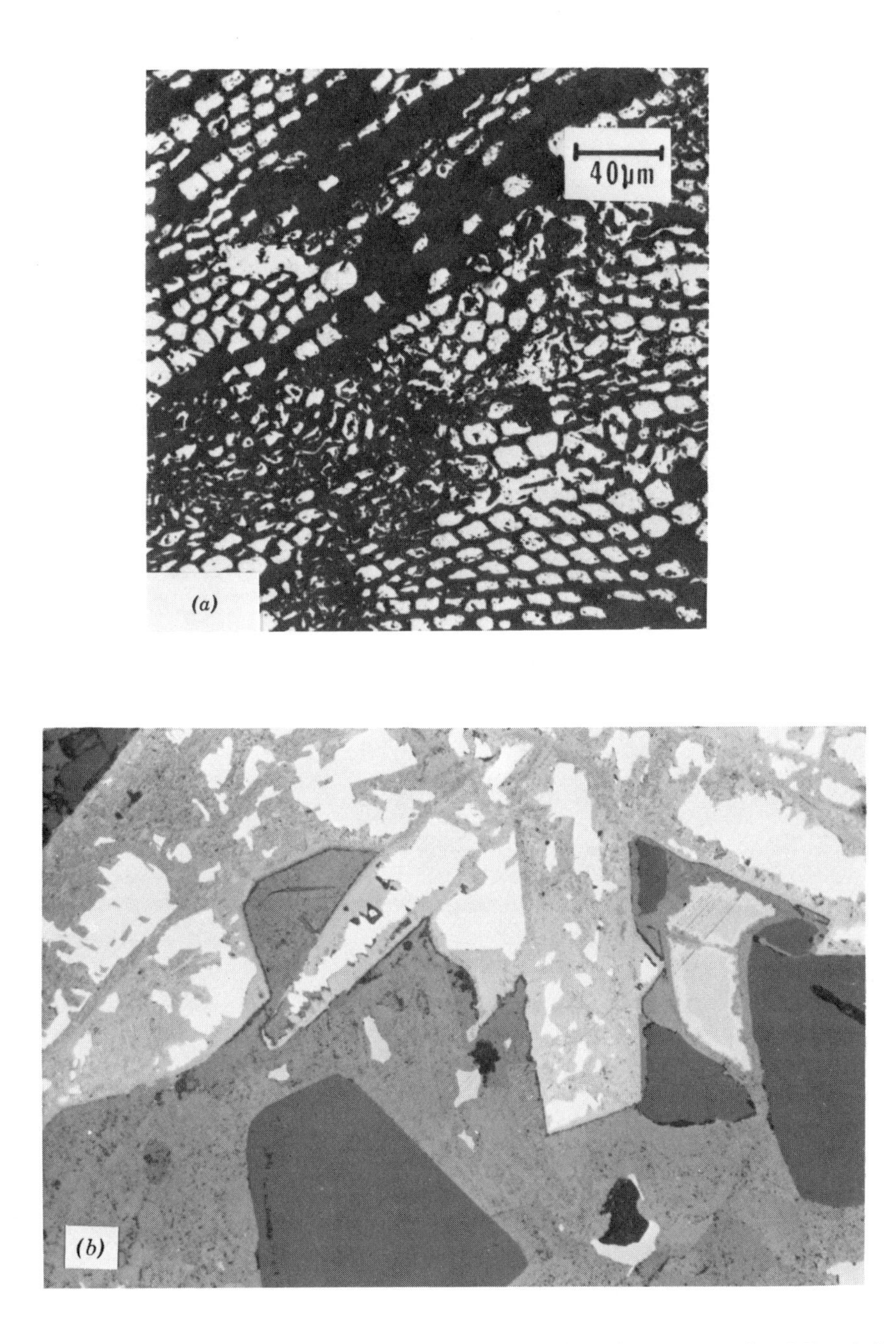

FIGURE 7.9 *(a)* Cellular structure in coal material replaced by pyrite, Minnehaha Mine, Illinois. (Width of field = 270 μm.) (Reproduced from F. T. Price and Y. N. Shieh, *Econ. Geol.* **74**, 1448, 1979, with permission of authors and publisher) *(b)* Pseudomorphous replacement of euhedral marcasite (white) crystals by goethite (light gray), Northern Pennine Orefield, England. (Width of field = 500 μm.)

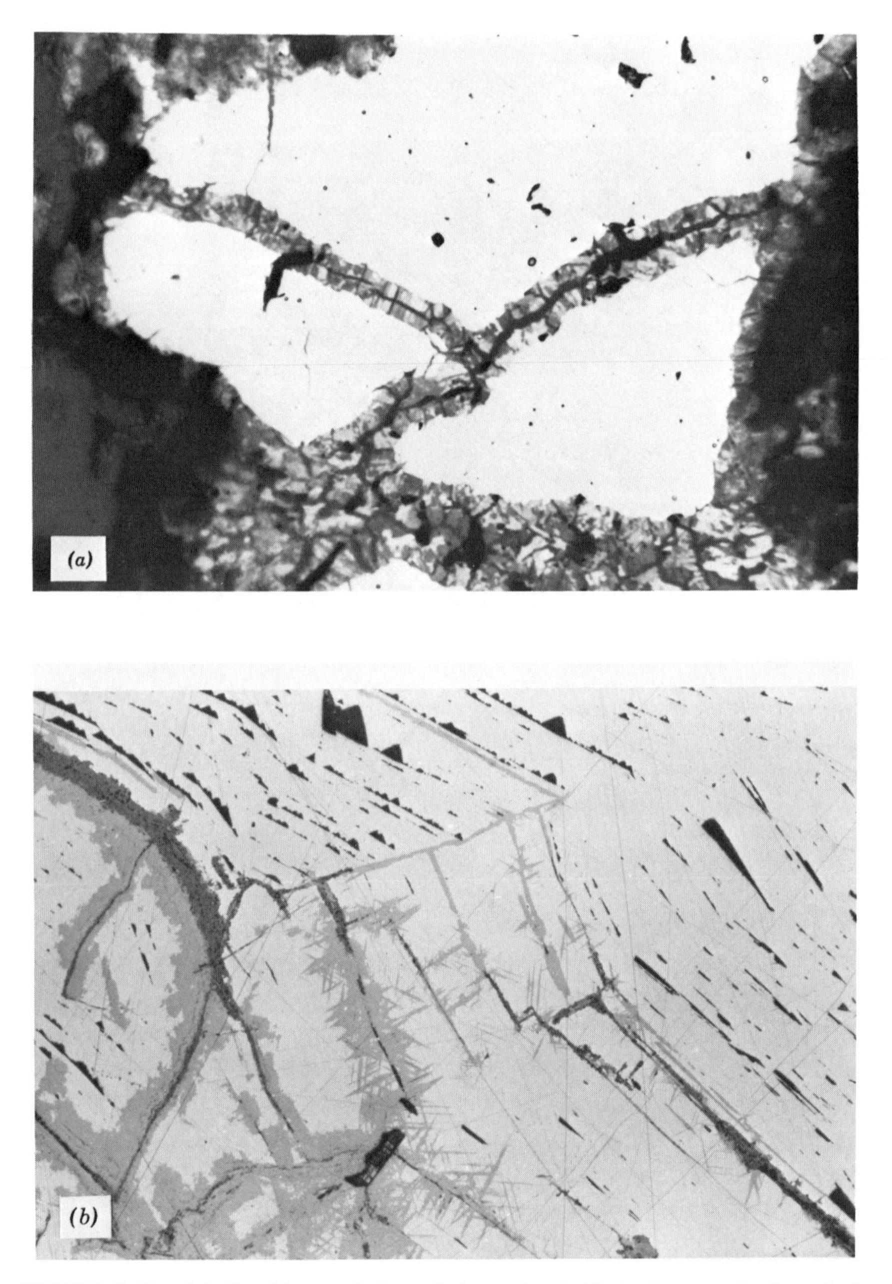

FIGURE 7.10 *(a)* Covellite replacing chalcopyrite (white) along grain boundaries, Great Gossan Lead, Virginia. (Width of field = 520 μm.) *(b)* Chalcocite (medium gray) replacing galena (light gray) along grain boundaries and cleavages, Alderley Edge, Cheshire, England. (Width of field = 500 μm.) *(c)* Galena (white) replaced by cerussite ($PbCO_3$, light gray) from the margins of the original grain, Northern Pennine Orefield,

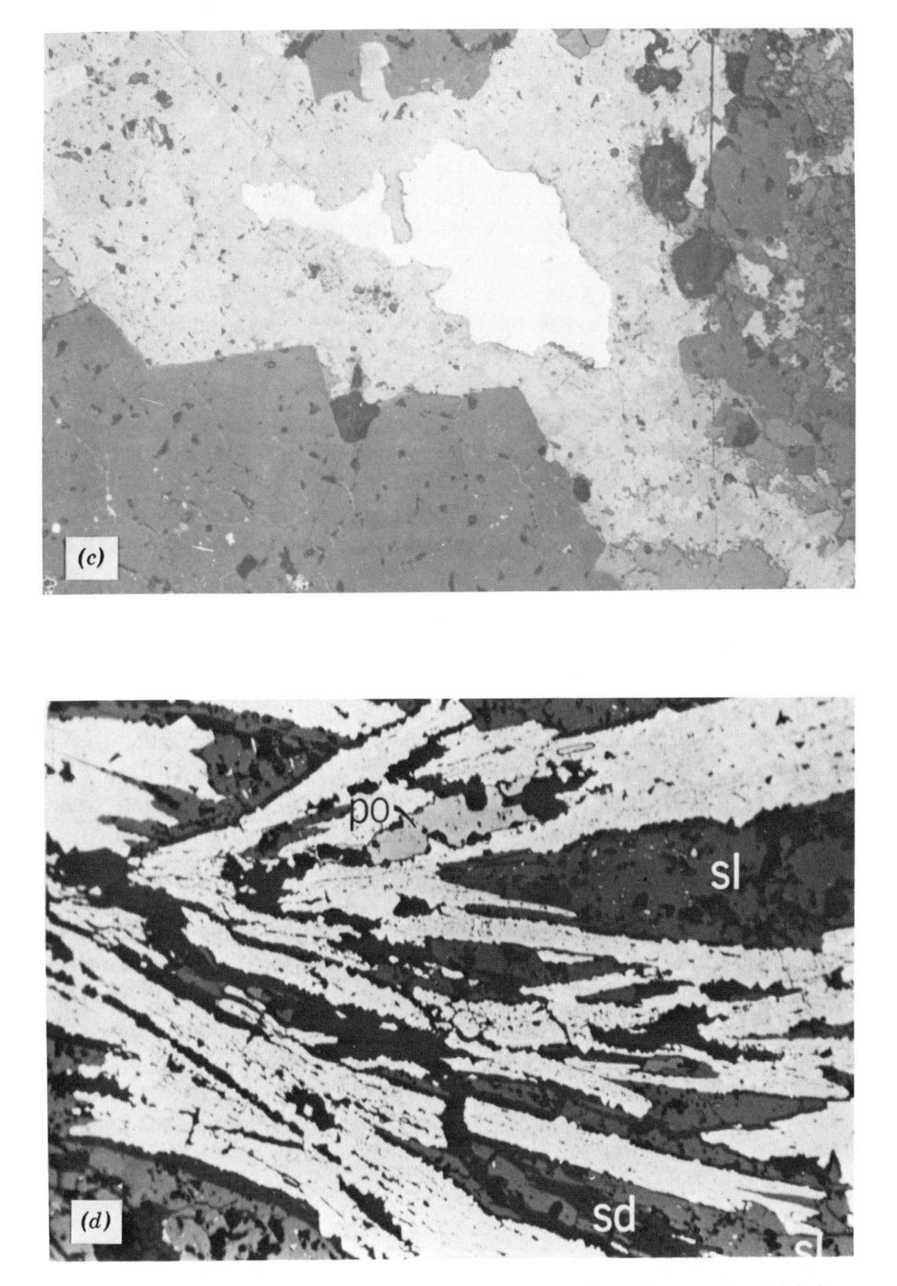

England. (Width of field = 500 μm.) *(d)* Fine-grained pyrite and marcasite (white) pseudomorphously replacing elongate crystals of pyrrhotite. Some remnants of pyrrhotite (po) remain in the cores of these pseudomorphs. Sphalerite (sl) and siderite (sd) comprise the matrix. (Width of field = 3000 μm.) (Reproduced from W. C. Kelly and F. S. Turneaure, *Econ. Geol.* **65**, 620, 1970, with permission of the authors and publisher.)

Fractures, Cleavages, and Grain Boundaries Replacement is the result of a surface chemical reaction, hence any channel between grains or through grains is a prime site for initiation of the replacement process. Replacement along grain boundaries or internal channel ways very often appears in the form of thin laths or equant crystals of the replacing phase projecting into the host. It may also appear as thin concentric coatings developed roughly parallel to the advancing front of replacement. In the early stages of the process, replacement may be readily identified because much of the original phase remains and the original grain boundaries, fractures, or cleavages are still visible (Figures 7.10*a* and 7.10*b*). In more advanced stages, the original phase may be reduced to "islands" now left in a matrix of the secondary minerals (Figure 7.10*c*). If the original material was coarsely crystalline and optically anisotropic, the residual island grains may well show optical continuity (i.e., extinguish simultaneously when viewed under crossed polars). Complete replacement of one mineral by another is often difficult to establish unless vestigial structure, such as the typical morphology of the replaced phase (pyrite cubes now seen as goethite, pyrrhotite laths now seen as pyrite and marcasite, Figure 7.9*b*) are left behind (Figure 7.10*d*).

With careful observation, one can usually distinguish between replacement along fractures and mere fracture-infilling resulting from precipitation of a later phase or its injection during metamorphism. Replacement consumes some of the original phase and tends to produce a rounding off of irregular surfaces, whereas infilling leaves the original fractured surfaces intact. After replacement, the surfaces on either side of a fracture do not "match" (Figure 7.11*a*) whereas after infilling, the surfaces should still "match" (Figure 7.11*b*). It is important to note that textures such as that shown in Figure 7.11*b* have been erroneously interpreted and used as evidence that pyrite is breaking down to form pyrrhotite in these ores.

Replacement along fractures may also resemble some of the exsolution textures discussed later. However, replacement commonly results in an increase in the volume of the secondary replacing phase at the intersections of fractures, whereas this is not common in exsolution. In fact, exsolution often produces the opposite effect in which intersecting exsolution lamellae show depletion of the exsolved phase in the zone of intersection. For the same reasons, exsolution often leaves depleted zones in the host phase (Figure 7.14*d*) adjacent to major concentrations of the exsolved phase. Replacement, however, may result in greater concentration of the secondary phase adjacent to and extending out from major replacement areas.

Crystal Structures The crystal structure of the phase being replaced may control replacement either because this determines cleavage directions or because diffusion can take place more readily along certain crystallographic directions. For example, the oxidation of magnetite commonly results in replacement by hematite along (111) planes (Figure 7.23).

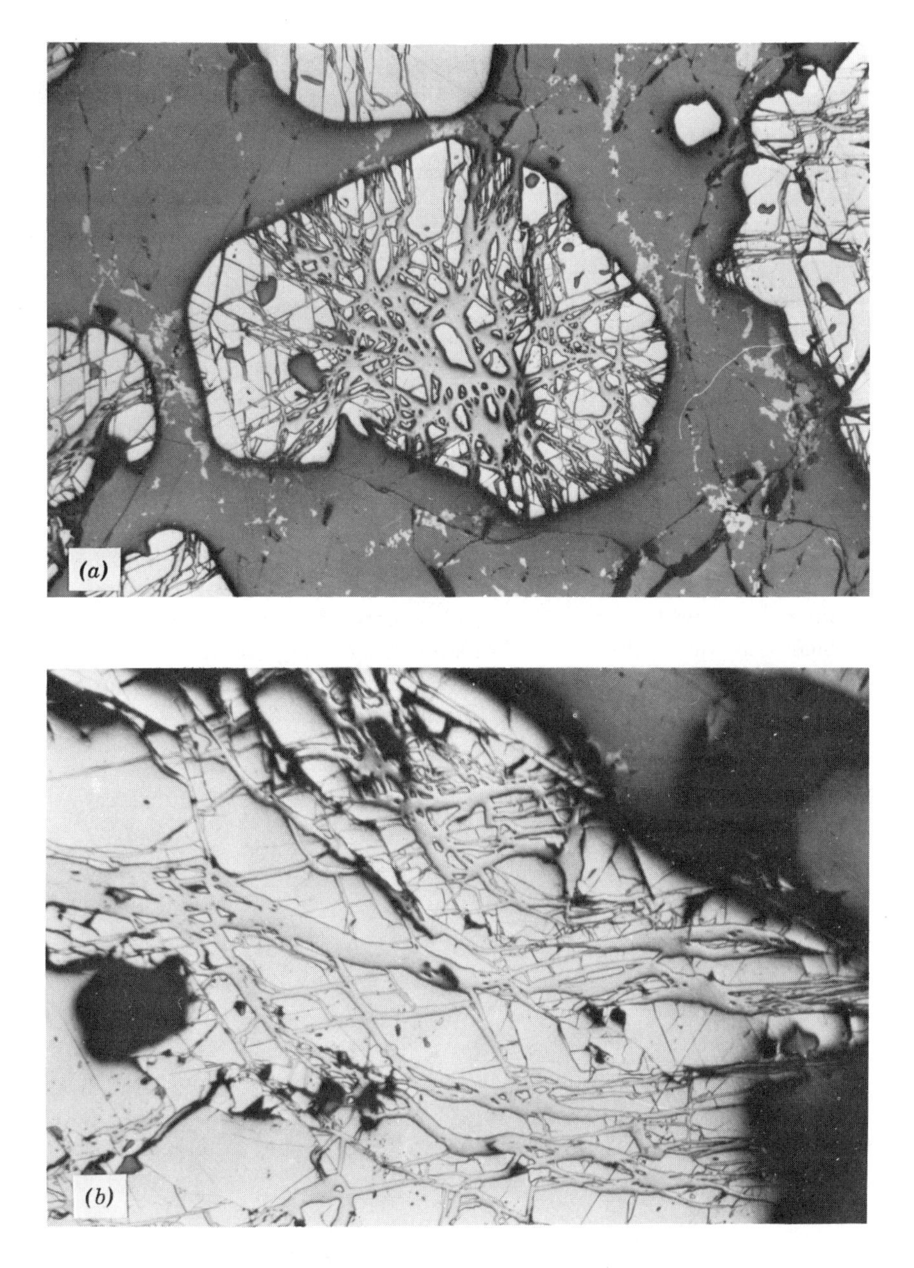

FIGURE 7.11 Subhedral pyrite crystal (white) replaced along fractures by chalcopyrite (light gray) in a matrix of bornite. The chalcopyrite has formed as a reaction product of the pyrite with the bornite deposition fluid. Magma Mine Superior, Arizona. (Width of field = 2000 μm.) *(b)* Fractures in pyrite infilled by injected pyrrhotite with little or no replacement. Great Gossan Lead, Virginia. (Width of field = 520 μm.)

FIGURE 7.12 Differing stages in the development of atoll structure resulting from the selective replacement of intergrown or compositionally zoned crystals.

Chemical Composition The chemical composition of the primary phase may control the composition of the phase that replaces it. During weathering, and often during hydrothermal replacement, the secondary phase retains the same cation composition as the primary phase with merely a change in oxidation state (e.g., hematite forming during oxidation of magnetite or violarite forming on pentlandite, Figures 8.8*d* and 9.8) or a change in anion (e.g., hematite replacing pyrite or anglesite replacing galena). Replacement may also selectively remove one cation while leaving another; this is frequently seen in the replacement of chalcopyrite or bornite by covellite (Figure 7.10*a*). Chemical control of replacement is also demonstrated in Figure 7.11*a* in which a copper-rich fluid has deposited bornite around earlier pyrite. Where the copper-bearing fluid encountered the pyrite and the chemical potential of FeS_2 was highest, the replacement reaction resulted in the formation of chalcopyrite. Replacement may occur selectively, affecting only one phase in an intergrowth or particular zones in a compositionally zoned crystal (producing an *atoll structure,* see Figure

FIGURE 7.13 Boxwork texture of laths of hematite and goethite with residual pyrite in a gossan, Elba, Italy. (Width of field = 500 μm.)

7.12). The reasons for such selective replacement may be extremely subtle and are, in most instances, poorly understood.

A final example of replacement is the open void "boxwork" texture composed of cellular crisscross laths of goethite, hematite, and sometimes pyrite, which is found in gossans (Figure 7.13).

7.5 SECONDARY TEXTURES RESULTING FROM COOLING

Most ores form at elevated temperatures and have undergone cooling over temperature ranges which may be, for example, less than 100°C for many Pb-Zn ores in carbonates but as much as 1000°C for Fe-Ni-Cu ores in ultramafic rocks. As suggested by Figure 7.1 refractory minerals such as magnetite, chromite, pyrite, sphalerite and some arsenides often retain their original compositions and textures through the cooling episode, whereas many sulfides, sulfosalts, and native metals reequilibrate compositionally and texturally during cooling. The textural effects resulting from cooling include those discussed below.

Recrystallization Reequilibration of ores on cooling is usually accompanied to some degree by recrystallization of the primary minerals, an effect that may or may not leave any vestige of the original texture. The resulting textural features are discussed in Section 7.7

Exsolution and Decomposition Many ore minerals undergo compositional or structural adjustments in the form of exsolution or inversion as they cool from the temperatures of initial crystallization or the maximum temperature of metamorphic recrystallization. In exsolution, one phase is expelled from another, often in a characteristic pattern. A list of some commonly observed host and exsolved phases is given in Table 7.1; a more complete list is given by Ramdohr (1969, pp. 190–198).

The form of the exsolved phase varies with the minerals involved, their relative proportions, and the postdepositional cooling history of the ore. The exsolution process results from diffusion (usually of metal atoms through a sulfur or oxygen lattice), the nucleation of crystallites, and the growth of crystallites or crystals. Similarities of crystal structure and chemical bonding between host and exsolved phase, particularly the matching of atomic arrangements in specific layers resulting in a shared plane of atoms, frequently dictate that exsolution is crystallographically controlled (coherent exsolution). For example, pentlandite exsolves such that the (111), (110), and (112) crystallographic planes are parallel to the (001), (110), and (100) planes, respectively, of the host pyrrhotite (Figure 8.8*b*), ulvöspinel exsolves parallel to the (111) planes of host magnetite (Figure 7.14*d*).

If parent and exsolved phases have completely different structures or if there is no crystallographic continuity across the interface between phases, noncoherent exsolution occurs. For example, it is apparent from phase equilibrium studies (Figure 8.15) that if pyrite and pyrrhotite equilibrate at elevated temperature (~400°C or above), pyrite will exsolve on cooling. However, the pyrite exhibits a considerable force of crystallization (i.e., tendency to grow as euhedra at the expense of surrounding phases), which when combined with the dissimilarity of the pyrite and pyrrhotite structures result in the pyrite occurring as individual grains (commonly euhedral cubes) rather than as recognizable exsolution lamellae. The kinetics of the exsolution depend on temperature, degree of supersaturation, and the concentrations of impurities although virtually all pyrrhotites have compositions that have readjusted at low temperatures. A rigorous treatment of the kinetics and mechanisms of exsolution is beyond the scope of this text; for such detail the reader is referred to the work of Yund and McCallister (1970) or Putnis and McConnell (1980). Decrease in the interfacial energy during exsolution is frequently accomplished by the exsolved phase taking on more equant forms. Hence early formed flames of pentlandite in pyrrhotite coalesce

TABLE 7.1 Examples of Ore Minerals Frequently Encountered in Exsolution Textures

Host Phase	Exsolved Phase	Nature of Commonly Observed Intergrowth
Pyrrhotite	Pentlandite	Lamellae or flames
Chalcopyrite	Cubanite	Sharply bounded laths
Chalcopyrite	Sphalerite	Stars, crosses
Chalcopyrite	Mackinawite	Lamellae, irregular wisps
Chalcopyrite	Bornite	Basket weave
Sphalerite*	Chalcopyrite	Rows of blebs
Sphalerite	Pyrrhotite	Rows of blebs
Bornite	Chalcopyrite	Basket weave
Bornite	Chalcocite-digenite	Roughly cubic network
Sphalerite	Stannite	Dispersed blebs
Stannite	Chalcopyrite	Lamellae in triangular pattern
Galena	Matildite	Lamellae
Silver	Dyscrasite	Lamellae
Kamacite	Plessite	Lamellae in triangular pattern
Arsenic or antimony	Stibarsen	Myrmekitic
Stibarsen	Arsenic or antimony	Myrmekitic
Pb-Sb and Pb-Bi sulfosalts	Pb-Sb and Pb-Bi sulfosalts	Lamellar
Magnetite	Ilmenite	Lamellae in triangular pattern
Magnetite	Ulvöspinel	Lamellae in triangular pattern
Ilmenite	Hematite	Lenslike lamellae
Hematite	Ilmenite	Lenslike lamellae

*Usually not the result of exsolution; see text.

into irregular veinlets, and well-defined thin lamellae of chalcopyrite in bornite (or vice versa) retain a basketweave texture but swell into bulbous lenses. A study of the exsolution textures observed in copper-iron sulfides (Brett, 1964) demonstrated that they are not particularly indicative of the rate of cooling or the temperature of initial formation of the ores.

The large variety of exsolution textures observed are difficult to classify using a simple terminology. However, certain terms are widely employed to describe the textures, notably *marginal, lamellar, emulsoid,* and *myrmekitic* exsolution textures (see Figure 7.14). The distinction between crystallographically controlled exsolution and similar replacement textures can often be made because intersecting lamellae show depletion at the junction in the former case and greater concentration in the latter case. Also the depletion of exsolved material around a large bleb known as *seriate distribution* is a distinctive feature well illustrated in Figure 7.14*d*, which shows the exsolution intergrowths in Fe-Ti oxides, among the most important and widely observed of such textures.

An example of the confusion that can arise in the interpretation of ore textures is provided by sphalerite-chalcopyrite intergrowths. Sphalerite in many types of ores contains chalcopyrite in the form of randomly dispersed or crystallographically oriented rows of blebs and rods, each of which may be 1 to 20 μm across (Figure 7.15 and 3.2). This form of chalcopyrite, appropriately termed "chalcopyrite disease," has commonly been ascribed to exsolution on the cooling of the ores after emplacement. Experimental studies (Wiggins and Craig, 1980; Hutchison and Scott, 1980) have demonstrated, however, that chalcopyrite will not dissolve in sphalerite in significant amounts unless temperatures are above 500°C. These data and the observation of chalcopyrite-bearing sphalerites in Zn-Pb ores in carbonates (which formed at 100 to 150°C) and in unmetamorphosed volcanogenic ores (which formed at 200 to 300°C) suggest that temperature-dependent exsolution is not the means by which these intergrowths have formed. Furthermore, detailed studies of doubly polished thin sections of these ores (Barton 1978 and Figure 3.2) reveal that some of the chalcopyrite is actually present as myrmekitic worm- or rod-like bodies which may extend up to several hundred microns. In a detailed study of these features, P. B. Barton and P. M. Bethke (pers. commun., 1979) have concluded that the chalcopyrite results either by epitaxial growth during sphalerite formation or by replacement as copper-rich fluids reacted with the sphalerite after formation. During metamorphism, finely dispersed chalcopyrite may be redistributed when the sphalerite recrystallizes so that it remains concentrated along the sphalerite grain boundaries.

Exsolution itself is a form of decomposition because the original high temperature composition no longer exists as a single homogeneous phase. However, the term "decomposition" is more commonly applied when a phase undergoes an abrupt change into two phases of distinctly different compositions as in *eutectoidal breakdown.* The term decomposition is also applied to the breakdown of the central portion of a complete solid solution series, with the resulting de-

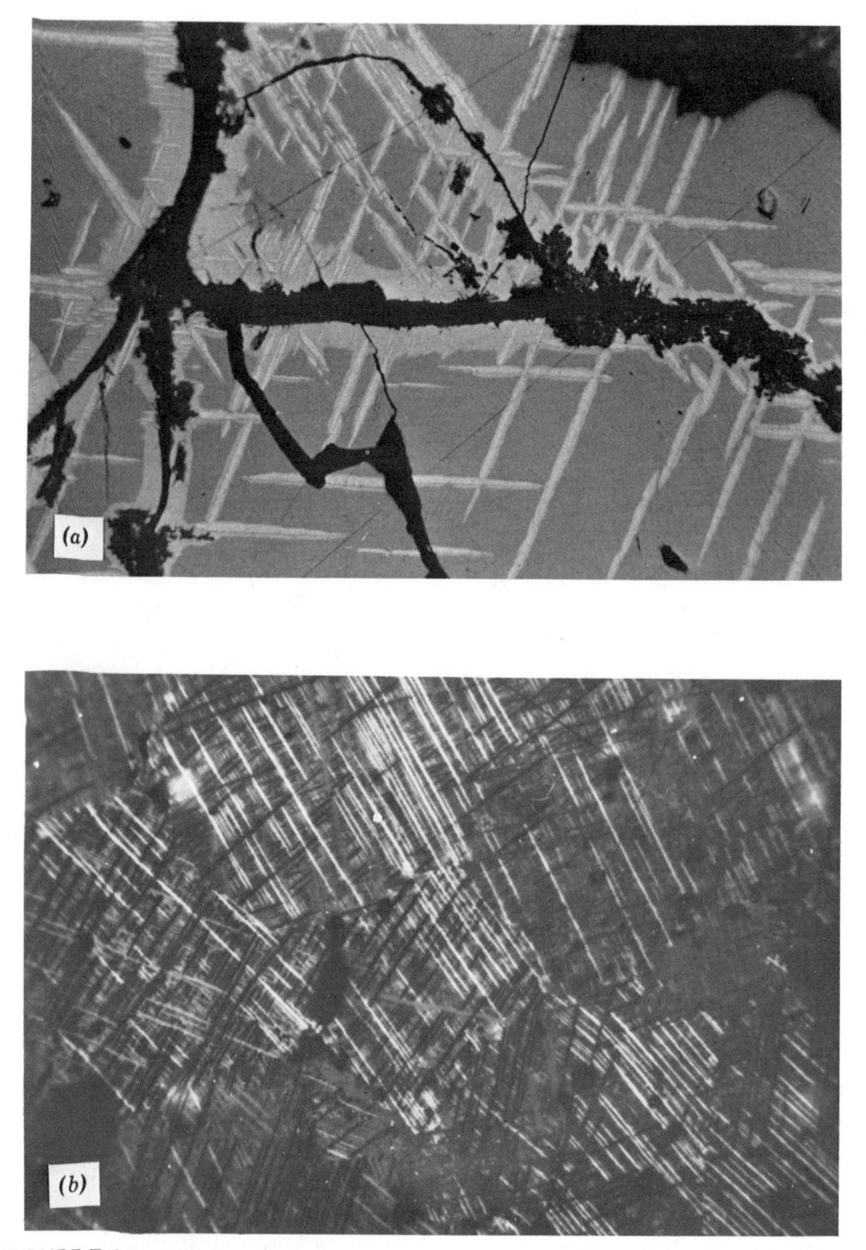

FIGURE 7.14 *(a)* Exsolution of chalcopyrite lamellae (light gray) within bornite (dark gray). Later alteration has resulted in the development of chalcocite (medium gray) along the edges of the fractures and as rims on the chalcopyrite lamellae. Grayson County, Virginia. (Width of field = 520 μm.) *(b)* Fine lamellae of matildite (white and black) within a matrix of galena. This texture has resulted from the decomposition (on cooling) of an initially homogeneous phase. Crossed polars; oil immersion. Leadville, Colorado. (Width

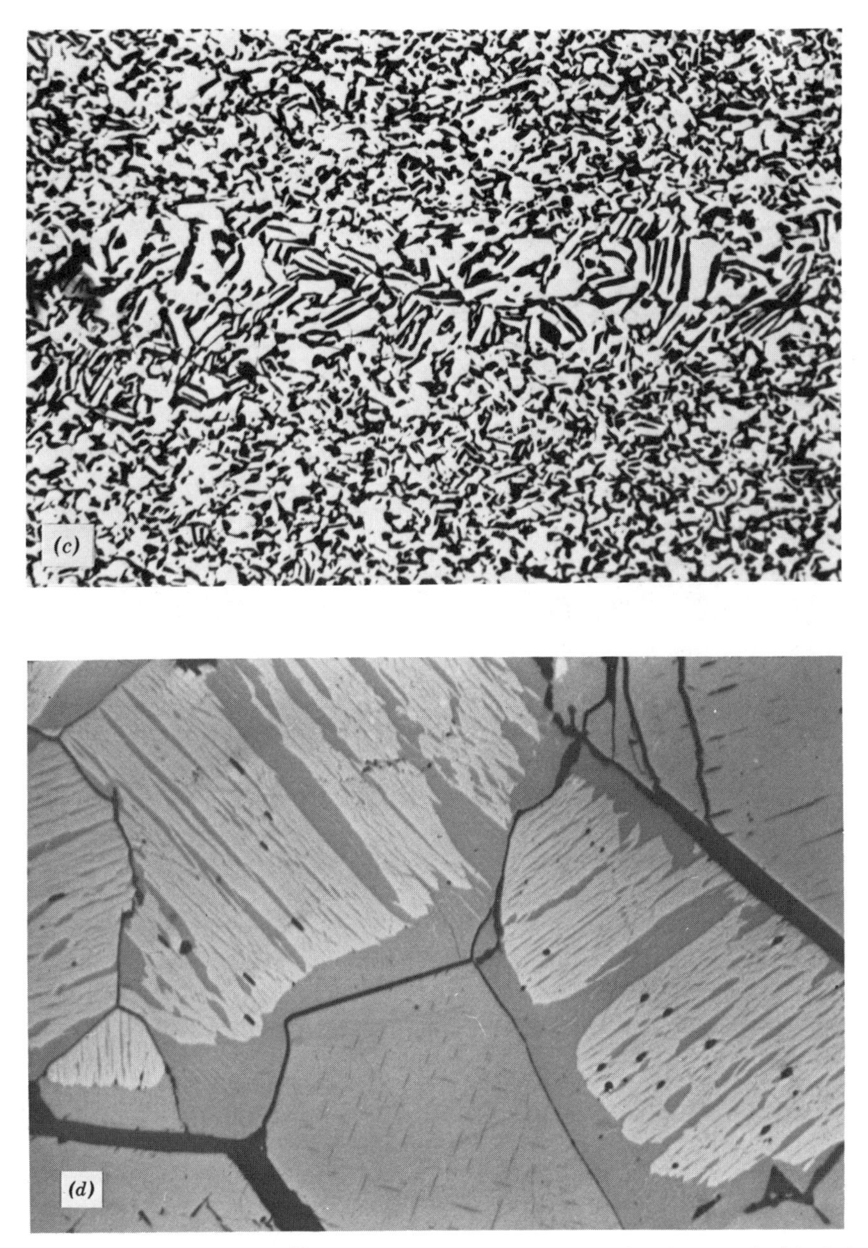

of field = 265 μm.) *(c)* "Allemontite," a myrmekitic texture of native arsenic (black due to oxidation) and stibarson (white), which has resulted from the decomposition of an initially homogeneous phase. (Width of field = 520 μm.) *(d)* Exsolution in coexisting FeTi oxides. Flanking grains of hematite-ilmenite solid solution have separated into hematite (white) and ilmenite (medium gray), and a central grain of Ti-magnetite has exsolved lamellae of ulvöspinel. (Width of field = 2000 μm.)

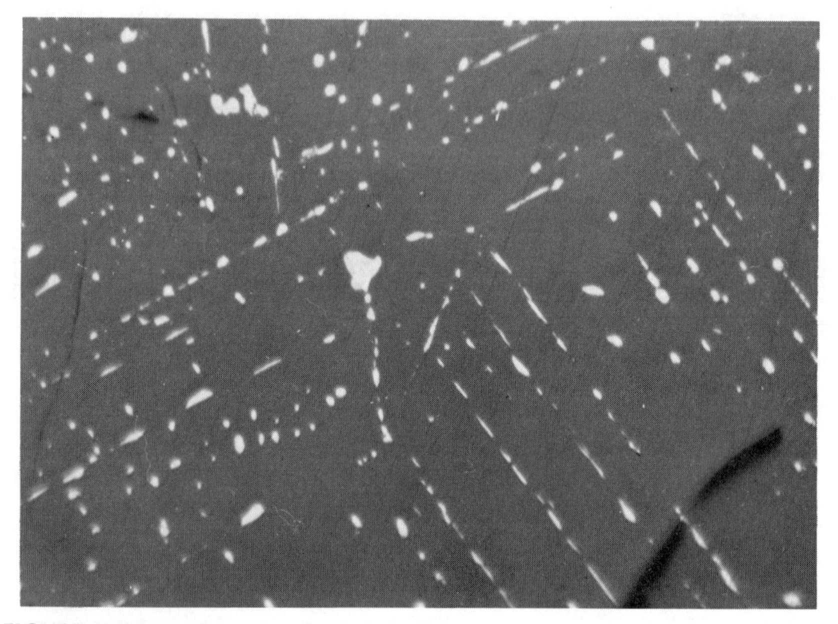

FIGURE 7.15 Grains and rods of chalcopyrite oriented within sphalerite. This assemblage has often been interpreted as the result of exsolution, but experimental studies reveal that sphalerite could not dissolve sufficient copper to form this texture by exsolution. See text for additional discussion. Great Gossan Lead, Virginia. (Width of field = 520 μm.)

velopment of an intimate intergrowth of compositionally distinct phases. Eutectoidal breakdown on cooling is well known in metallurgical studies, but relatively few mineral examples have been verified. Digenite, Cu_9S_5, is not stable below 70°C unless it contains ~1% iron and decomposes on cooling below this temperature to form a complex mixture of anilite and djurleite (see the Cu-S phase diagram in Figure 10.7). If the original composition before cooling is more Cu-rich, the decomposition may result in the formation of a mixture of chalcocite and djurleite. In a detailed study of the Cobalt, Ontario ores, Petruk (1971) has reported a complex intergrowth of galena and chalcocite which has apparently formed as a result of the decomposition of a Cu-Pb sulfide which is only stable at high temperature (Craig and Kullerud, 1968).

Two examples of textures resulting from the decomposition of the central portion of a solid solution series are lathlike matildite-galena intergrowths (Figure 7.14*b*) and the myrmekitic arsenic- (or antimony-) stibarsen intergrowths referred to as "allemontite" (Figure 7.14*c*). The fineness of the intergrowth (sometimes on a scale of a few microns) and the similarily in appearance of constituent phases can easily result in their misidentification as a single phase.

Inversion Inversion of a mineral from one structural form to another of the same composition is not often easily discernible texturally but may produce characteristic twinning. Sometimes, even though inversion has occurred, the crystal morphology of the high temperature phase is retained as a *paramorph.* Some high temperature phases always invert so rapidly on cooling that only the low temperature forms are observed (e.g., troilite, chalcocite, acanthite). Unfortunately, the twinning in acanthite, once thought to be diagnostic of inversion, can form below the inversion temperature of 176°C (Taylor 1969). Certain other phases that are observed as both high and low temperature forms (e.g., cinnabar-metacinnabar; famatinite-luzonite) may not be diagnostic of formation conditions because one of the forms is metastable.

Oxidation-Exsolution and Reduction-Exsolution Exsolution lamellae of ilmenite in magnetite (and less commonly of magnetite in ilmenite) are often present in a relative volume that exceeds the known solubility limits for these minerals. Lindsley (1976) has explained the mechanism by which these lamellae form with reference to the a_{O_2} - T plot for the Fe-Ti oxides given in Figure 9.13. On this plot, the curves for magnetite-ulvöspinel solid solution (Mt-Usp) dip more steeply and those for hematite-ilmenite solid solution (Hem - Ilm) dip less steeply than the curves for buffers such as Ni-NiO, fayalite - magnetite - quartz or most fluids. Consequently, on cooling along a buffer curve or in the presence of a fluid of constant composition, a given Mt-Usp will undergo oxidation and lamellae of ilmenite will form on the (111) planes. Conversely, a Fe_2O_3-rich ilmenite, cooled under similar conditions, will be reduced yielding lamellae of Ti-magnetite parallel to the (0001) planes.

Thermal Stress Most ore minerals have approximately the same coefficients of thermal expansion and thus most mono- or polymineralic masses suffer little induced strain on cooling. One significiant exception is pentlandite, $(Fe,Ni)_9S_8$, which has a coefficient of thermal expansion that is 2 to 10 times larger than sulfides such as pyrrhotite and pyrite with which it is generally associated (Rajamani and Prewitt, 1975). As a result, the chainlike veinlets of pentlandite (Figure 9.5) that form at elevated temperatures (300–600°C) by coalescence of early exsolved lamellae, are typically fractured because they have undergone much greater shrinkage than the host pyrrhotite.

7.6 SECONDARY TEXTURES RESULTING FROM DEFORMATION

Many ores contain textural evidence of deformation. The evidence ranges from minor pressure-induced twinning to complete cataclasis. The degree to which individual mineral grains both respond to and preserve deformational effects ranges widely depending on the mineral, the rate of strain, the nature of the deformation, the associated minerals, the temperature at the time of deformation, and the postdeformational history. The response threshold of minerals appears to be primarily a function of hardness; hence minerals such as many native metals, sulfosalts, copper and silver sulfides deform most readily, copper-iron sulfides and monosulfides less readily, and disulfides, oxides, and arsenides least readily. Accordingly, in polymineralic ores, deformation textures are often evident in only some minerals. The softer minerals deform most readily but they also recrystallize most readily so that the deformational effects are obliterated before those in more refractory minerals. Specific deformation features commonly observed include those discussed in the following.

Twinning, Kinkbanding, Pressure Lamellae These features occur in ores subjected to any degree of deformation and can even be artificially introduced into some of the softer minerals by rough treatment of specimens. Twinning may occur in minerals during initial growth, during structural inversion on cooling (Section 7.5), or as a result of deformation. Although little or no quantitative study has ever been undertaken, Ramdohr (1969) suggests that the three major types of twinning (illustrated Figures 7.16 and 7.17) can be distinguished as follows:

Growth: occurs as lamellar twins of irregular width that are unevenly distributed, present in only some grains and may be strongly interwoven.

Inversion: commonly occurs as spindle-shaped and intergrown networks not parallel throughout grains.

Deformation: occurs as uniformly thick lamellae, commonly associated with bending, cataclasis, and incipient recrystallization (regions of very small equant grains); lamellae often pass through adjacent grains.

These criteria are useful but not infallible guides to the identification of major types. Clark and Kelly (1973), in investigating the strength of some common sulfide minerals as a function of temperature, showed that deformation in pyrrhotite may be as kinkbanding (Figure 7.17c), kinked or bent subparallel lamellae each of which show undulose extinction, or twinning. At less than 2 bar, kinkband deformation predominates below ~300°C, whereas above this temperature both kinking and twinning are common. Pyrrhotite and many other sulfides that are only moderately hard (e.g., stibnite, bismuthinite) also commonly contain "pressure lamellae" (Figure 7.18), slightly offset portions of grains that exhibit either undulatory extinction or slightly different extinction positions. Pressure-induced twins and pressure lamellae often terminate in regions of brittle fracture, crumpling, or very fine-grained regions in which crushed grains have recrystallized.

Breaking of specimens by hammering, or damage caused during grinding or even in careless polishing can induce local pressure twinning in some very soft phases such as native bismuth (Figure 7.17*b*), argentite and molybdenite. The cause of such twinning is usually recognizable because of its local distribution

FIGURE 7.16 *(a)* Growth twin of wolframite. *(b)* Polysynthetic twinning developed in synthetic acanthite that has inverted from argentite on cooling from initial formation at 400°C. Crossed polars; oil immersion. (Width of field = 190 μm.)

or association with scratches; it can often be induced in native bismuth by merely drawing a needle point across a polished surface.

Curvature or Offset of Linear Features Deformation of ores is often evidenced by the curvature or offset of normally linear or planar features such as crystal faces, cleavages, fractures, twins, exsolution lamellae, and primary mineral layering or veining. The triangular cleavage pits in galena, so diagnostic in identification, commonly serve as a measure of deformation. Although the boundary of a single pit may exhibit curvature, the effects are most often seen in the curvature of a row of such cleavage pits (Figure 7.19). Deformation-induced twin lamellae in pyrrhotite, ilmenite, chalcopyrite, and many other minerals frequently exhibit significant curvature and commonly extend across several grains, whereas growth or inversion twins are usually confined within individual grains.

Exsolution intergrowths of cubanite in chalcopyrite, ilmenite in hematite (or vice versa), chalcopyrite in sphalerite, pentlandite in pyrrhotite, and bornite in chalcopyrite (or vice versa) are often linear (or planar) features that are crystallographically controlled. Curvature of the laths, rows of blebs, flames, or rods is indicative of deformation; however, it is not always clear whether the exsolution or the deformation occurred first.

FIGURE 7.17 *(a)* Deformation twins in pyrrhotite. Great Gossan Lead, Virginia. Crossed polars. (Width of field = 520 μm.) *(b)* Polysynthetic twinning developed in native bismuth as a result of deformation. Nipissing Mine, Cobalt, Ontario. Crossed polars; oil immersion. (Width of field = 210 μm.) *(c)* Kinkbanding developed in pyrrhotite as a result of deformation. Great Gossan Lead, Virginia. (Width of field = 700 μm.) (From D. K. Henry et al., *Econ. Geol.* **74**, 648, 1979; used with permission.)

Folding or offset of primary mineral banding is a common feature in many deformed ores on both micro- and macroscales. Folding often results in relatively ductile flow of the softer sulfides such as chalcopyrite, galena and pyrrhotite, but brittle fracture of harder minerals such as pyrite, arsenopyrite, and magnetite. Thus the softer sulfides infill fractures between discontinuous broken portions of harder mineral zones (Figure 10.23). Microscale offsetting of mineral

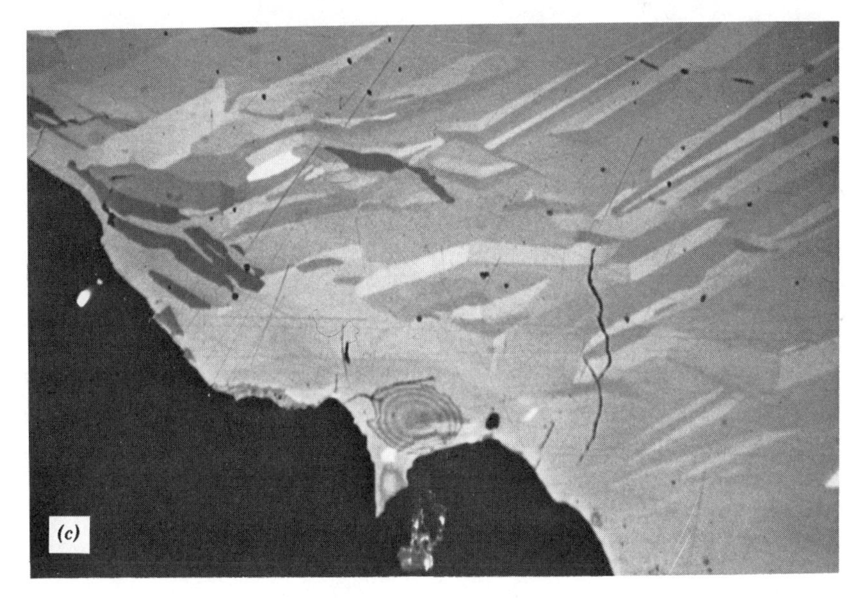

FIGURE 7.17 *(Continued)*

FIGURE 7.18 Complex pressure lamellae developed in stibnite as a result of deformation; the wide variety of colors results from different crystallographic orientations of the lamellae. Crossed polars. (Width of field = 520 μm.)

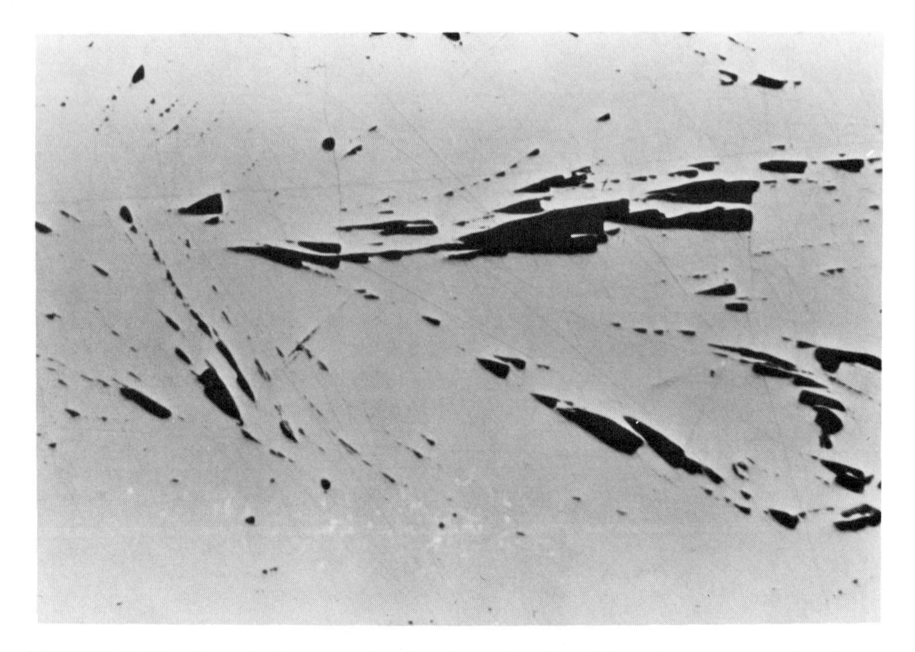

FIGURE 7.19 Curved cleavage pits that have developed in galena as a result of post-depositional deformation. Austinville, Virginia. (Width of field = 520 μm.)

bands or mineralized veins (Figure 8.4), commonly with infilling by later generations of ore or gangue minerals, is frequently seen in ores.

Schlieren Deformed ores often contain zones along which shearing has occurred. In such zones, known as schlieren, the ore minerals may be pulverized and smeared out parallel to the direction of movement. The schlieren are usually planar features in which the ore minerals are very fine grained (sometimes recrystallized) relative to the surrounding rock; typically equant minerals such as galena are frequently present as elongate (often strained and fractured) grains.

Brecciation, Cataclasis, and Durchbewegung Deformation in ores is often evidenced by fracturing or brecciation of ore and gangue minerals, especially but not exclusively, of those that are harder and more brittle such as pyrite, chromite, and magnetite. The amount of brecciation depends on both the degree of deformation and the mineralogy of the ore. Thus moderate deformation will result in considerable brecciation of massive pyrite, magnetite, or chromite ore where all strain is relieved by brittle fracture of these minerals. In contrast, pyrite admixed with pyrrhotite or chalcopyrite usually suffers little even under extreme deformation because the strain is taken up in the softer sulfides. A notable exception to this is the "rolled" pyrite from Sulitjelma, Norway. In this ore, the pyrite cubes have been markedly rounded by being "rolled" in the matrix pyr-

FIGURE 7.20 Durchbewegung texture developed in intensely deformed pyrrhotite. The black silicate inclusions have been stretched and rotated as the pyrrhotite matrix flowed under pressure. Great Gossan Lead, Virginia. The scales are in centimeters and inches.

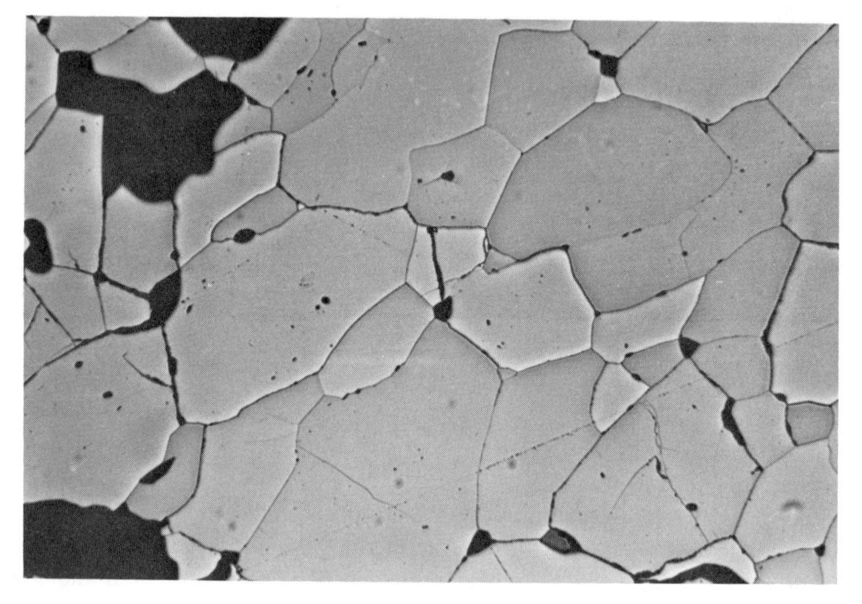

FIGURE 7.21 Annealed texture of recrystallized monmineralic pyrite sample. Note the common development of near 120° triple junctions. Mineral District, Virginia. (Width of field = 520 μm.)

rhotite during severe deformation. Minor brecciation grades into complex cataclasis with an increasing degree of fragmentation and disorientation, eventually involving both ore and gangue minerals (Figure 7.20); this penetrative deformation has been termed *durchbewegung* (literally "move through") by Vokes (1969). In fault zones and in ores that have suffered penetrative high-grade metamorphism, there may be pulverization of ore and gangue minerals, complete randomness of fragment orientation, the development of "ball textures," in which fragments of foliated gangue are rounded into "balls," and extensive development of the deformational features previously described. Injection of softer ore minerals into fractures and cleavages in more brittle ore minerals and gangue minerals is common (Figure 10.23).

7.7 SECONDARY TEXTURES RESULTING FROM ANNEALING

The annealing effects of the slow cooling of ores after deposition or slow heating during metamorphism can significantly alter the original textures. Since cooling and metamorphism are both prolonged annealing processes, the effects discussed here may produce similar textures to those discussed in Section 7.2. The most characteristic feature of annealing is recrystallization to minimize the areas of grain surfaces and interfacial tension through the development of roughly equant grains with 120° interfacial (or *dihedral*) angles (Figure 7.21). The interfacial angles observed at triple junctions of annealed monomineralic aggregates tend toward 120°, whereas those of polymineralic aggregates vary as a function of the mineralogy. The interfacial angles of some equilibrated pairs of common sulfide minerals include galena-sphalerite (103 and 134°), chalcopyrite-sphalerite (106–108°), and pyrrhotite-sphalerite (107–108°) (Stanton, 1972). Since the surface of the polished section, cut at random through the polycrystalline mineral aggregate, yields only apparent angles which can range from 0 to 180°, it is necessary to measure many interfacial angles in a given section in order to statistically determine the true angle. If a large number of angles are measured, that most frequently observed will represent the true angle. During the annealing process, small grains are resorbed at the expense of larger ones; however, small grains of minor phases may remain trapped as lenslike bodies along the grain boundaries of larger grains (Figure 10.23*d*).

The reequilibration that results from annealing can produce either zoned overgrowths on grains or the homogenization of grains containing primary growth-zoning. For example, pyrite overgrowths on primary pyrite or the remains of primary growth-zoning may be visible in normal polished sections, but often require etching to become evident (Figure 2.8*a*). Residual primary growth-zoning in sphalerite or tetrahedrite is rarely evident in polished sections but can

FIGURE 7.22 Annealed texture of recrystallized pyrite euhedra (light gray) within a matrix of sphalerite (dark gray) and minor chalcopyrite (medium gray). Mineral District, Virginia. (Width of field = 520 μm.)

be observed in transmitted light using doubly polished thin sections (Figure 2.7). Recrystallization during annealing may also result in the growth of euhedral, sometimes porphyroblastic crystals, especially of such phases as pyrite (Figure 7.22), arsenopyrite, magnetite, and hematite. The growth of these minerals, like the well-known examples among metamorphic gangue minerals (e.g., garnet and staurolite), depends on the conditions of annealing and the bulk composition of the mineralized zone. Though commonly only a few millimeters in diameter, porphyroblasts may exceed 25 cm across, as observed in the pyrite-pyrrhotite ores of Ducktown, Tennessee. Porphyroblastic growth or overgrowth complicates paragenetic interpretation because there are no unequivocal means of distinguishing porphyroblasts from early formed euhedral crystals. Frequently, however, porphyroblasts contain different amounts and types of inclusions compared to the corresponding primary mineral in the ore.

7.8 SPECIAL TEXTURES

A number of the textures observed in ore minerals are sufficiently distinctive or widely observed to have been given special names. Amongst primary depositional textures are *framboids* (Figures 10.5*b* and 10.16) the aggregates of spherical particles often seen in pyrite and in uraninite. Similar to these primary textures are the *oolitic* textures familiar from carbonate rocks but also found in iron and manganese ores (see Section 10.2). Among replacement textures, the replace-

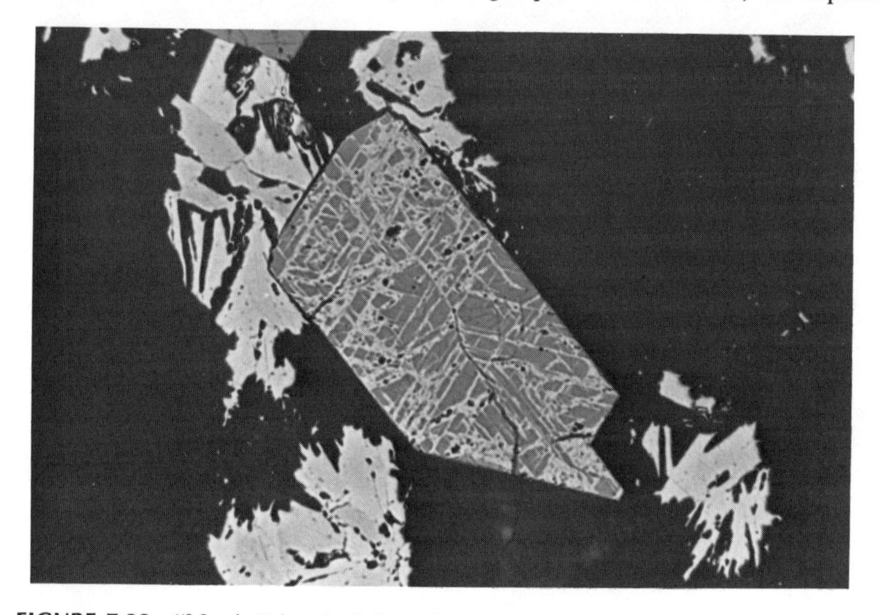

FIGURE 7.23 "Martite" in which hematite (white) has developed along crystallographically preferred planes in magnetite as a result of oxidation. (Width of field = 240 μm.)

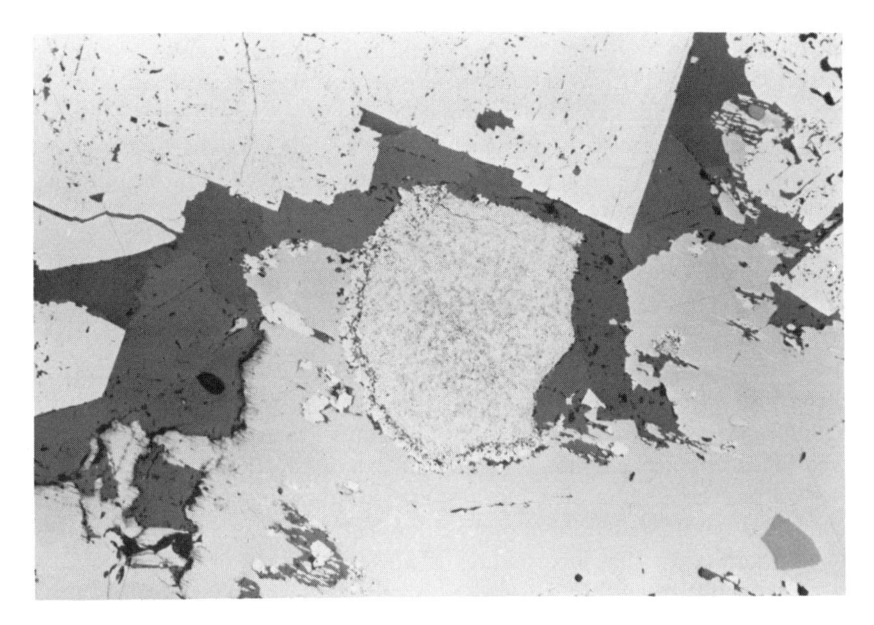

FIGURE 7.24 Bird's-eye texture of pyrite and marcasite developed during the alteration of pyrrhotite. (Width of field = 520 μm.)

FIGURE 7.25 Sphalerite "star" in chalcopyrite. Mineral District, Virginia. (Width of field = 330 μm.)

ment of magnetite by hematite along cleavage (111) directions is termed *martitization* (Figure 7.23) and the characteristic alteration of pyrrhotite to a fine mixture of pyrite and marcasite results in *birds eye* texture (Figure 7.24) Some exsolution textures are particularly characteristic, such as the *flames* of pentlandite in pyrrhotite (Figure 8.8*b*) and the *stars* of exsolved sphalerite found in some chalcopyrite (Figure 7.25).

7.9 CONCLUDING SUMMARY STATEMENT

Rarely does a single texture provide unequivocal evidence regarding the origin or history of an ore deposit. Commonly, a variety of textures representing different episodes in the development and subsequent history of a deposit are observed. This chapter has not provided an exhaustive discussion of the great variety of textures seen in ores, but has introduced some of the most commonly encountered types. With careful observation, common sense, and a little imaginative interpretation that incorporates whatever is known of the geological setting of a deposit much can be learned about the origin and postdepositional history of an ore from the study of ore textures.

BIBLIOGRAPHY

Barton, P. B. (1978) Some ore textures involving sphalerite from the Furutobe Mine, Akita Prefecture, Japan. *Mining Geol.* **28**, 293–300.

Bastin, E. S. (1950) *Interpretation of Ore Textures.* Geol. Soc. Am. Mem. No. 45.

Brett, P. R. (1964) Experimental data from the system Cu-Fe-S and its bearing on exsolution textures in ores. *Econ. Geol.* **59**, 1241–1269.

Clark, B. R. and Kelly, W. C. (1973) Sulfide Deformation Studies: I. Experimental deformation of pyrrhotite and sphalerite to 2,000 bars and 500°C. *Econ. Geol.* **68**, 332–352.

Craig, J. R. and Kullerud, G. (1968) Phase relations and mineral assemblages in the copper-lead-sulfur system. *Am. Mineral.* **53**, 145–161.

Edwards, A. B. (1947) *Textures of the Ore Minerals.* Australian Institute of Mining and Metallurgy, Melbourne.

Hutchison, M. N. and Scott, S. D. (1980) Sphalerite geobarometry in the Cu-Fe-Zn-S system, *Econ. Geol.* (in press).

Lindsley, D. H. (1976) Experimental studies of oxide minerals. In D. Rumble ed., *Oxide Minerals.* Mineral. Soc. Am. Short Course Notes 3, L61–L88.

Petruk, W. and Staff (1971) Characteristics of the Sulphides. In *The Silver-arsenide deposits of the Cobalt-Gowganda Region Ontario. Can. Mineral* **11**, 196–221.

Putnis, A. and McConnell, J. D. C. (1980) *Principles of Mineral Behavior,* Elsevier, New York.

Rajamani, V. and Prewitt, C. T. (1975) Thermal expansion of the pentlandite structure. *Am. Mineral.* **60**, 39–48.

Ramdohr, P. (1969) *The Ore Minerals and Their Intergrowths.* Pergamon, New York.

Roedder, E. (1968) The noncolloidal origin of "colloform" textures in sphalerite ores. *Econ. Geol.* **63**, 451–471.

Stanton, R. L. (1972) *Ore Petrology.* McGraw-Hill, New York.

Taylor, L. A. (1969) The significance of twinning in Ag_2S. *Am. Mineral.* **54**, 961–963.

Vokes, F. M. (1969) A review of the metamorphism of sulphide deposits. *Earth. Sci. Rev.* **5**, 99–143.

Wiggins, L. B. and Craig, J. R. (1980) Reconnaissance of the Cu-Fe-Zn-S system: Sphalerite phase relationships. *Econ. Geol.* **75**, 742–752.

Yund, R. A. and McCallister, R. H. (1970) Kinetics and mechanisms of exsolution. *Chem. Geol.* 5–30.

8

Paragenesis, Formation Conditions, and Fluid Inclusion Geothermometry of Ores

8.1 INTRODUCTION

Following mineral identification and textural characterization, two major objectives in ore microscopy are the determination of the order of formation of associated minerals in time succession or *paragenesis* and the estimation of the conditions under which the minerals formed or have reequilibrated.* Such determinations, although not vital to the extraction or exploitation of the ore, are important in deciphering the geological history of the ore and may be of value in exploration, in correlating various parts of ore bodies, and in the correlation of specific trace metals (e.g., gold) with certain episodes or types of mineralization. Paragenetic determination requires the detailed examination of polished sections to identify phases, recognize diagnostic textures (as described in Chapter 7), and decipher "time diagnostic" features. This in turn requires well-prepared, representative ore samples, the application of relevant phase equilibria data, and the integration of all geological and mineralogical data available for the deposit. One of the most powerful tools in determining temperatures of formation and the nature of the ore-forming fluids is the study of fluid inclusions, discussed in the latter part of the chapter.

The sampling and sample examination procedures useful in deciphering paragenesis are also discussed, although not all of the points considered are applicable to all ores; indeed, some ores, especially those that have been intensely metamorphosed, are not amenable to paragenetic studies either because the original record is insufficiently distinctive or because it has been subsequently altered beyond recognition.

*"Paragenesis" has also been used, particularly in the European literature, to refer to characteristic ore mineral assemblages but is used in this text only in reference to the sequential formation of minerals.

8.2 PARAGENETIC STUDIES

There is no "standard method" for carrying out paragenetic studies because each ore deposit is unique. However, as the goal of all such studies is to decipher the sequence of mineral formation, certain principles outlined in the following can be applied to most examples.

8.2.1 Sample Selection and Preparation

The samples available for study rarely comprise more than an infinitesimally small fraction of the total of the deposit; hence samples must be representative of the whole deposit if they are to be useful in paragenetic studies. Of course, the larger and more complex an orebody, the greater the number of samples needed to study it adequately. However, more important than the number of samples is their quality and, in many cases, their orientation—a factor especially critical in ores that possess planar or linear features (graded bedding, mineral bands parallel to vein walls, cross-cutting mineralized veins, etc.). It should be noted that samples of the unmineralized or only slightly mineralized host rocks of a deposit are often as useful as samples of massive ore in deciphering the paragenesis. Such samples can reveal the opaque minerals present before mineralization or those introduced early in the paragenesis. Conventional polished sections may be too small to display textural and paragenetic relationships in very coarse-grained ores, complex veins, or bedded ores; this problem can be overcome by combining hand samples or oriented slabs of ore with polished and thin sections and by the use of both high- and low-power objectives in microscopy. In some ores, the doubly polished thin section provides information superior to that provided by the conventional polished section for paragenetic studies. It allows observation of gangue and ore minerals in the *same* sample and of internal structure in some ore minerals (e.g., sphalerite, tetrahedrite, pyrargyrite), which is visible neither in the standard thin section nor the polished section.

It is also necessary to reemphasize the importance of using well-polished sections. Many subtle features are missed in sections that are poorly polished or that show too much relief.

8.2.2 Crystal Morphology and Mutual Grain Boundary Relationships

The shapes of individual crystals and the nature of the contacts between adjacent grains have often been used as criteria for determining paragenesis. In general, euhedral crystals have been interpreted as forming early and growing unobstructed; grains with convex faces have been interpreted as forming earlier than those with concave faces. Such simplistic interpretations are often correct but must be used with caution. Indeed, for many minerals euhedral crystal morphology is an indication of growth into open space, especially in vein deposits. For example, calcite, quartz, fluorite, sphalerite, cassiterite, galena, covellite, and sulfosalts usually form well-developed euhedral crystals only in directions where growth is unobstructed (Figure 8.1). The existence of such crystals, mixed with or overgrown by other minerals, indicates that the euhedra were first formed; furthermore, it usually indicates the direction of general growth (i.e., in the direction of the euhedral crystal faces). However, certain minerals (e.g., pyrite, arsenopyrite) tend through their force of crystallization to form well-developed crystals regardless of their position in the paragenetic sequence. For example, early pyrite in Cu-Pb-Zn veins occurs as isolated euhedra or intergrown subhedra with many well-developed faces; secondary pyrite resulting from exsolution of primary pyrrhotite in Fe-Cu-Ni ores often occurs as well-formed cubes; pyrite that forms as a result of metamorphic recrystallization commonly occurs as perfectly developed cubic or pyritohedral porphyroblasts up to several centimeters in diameter (Figure 8.2).

Sometimes the evidence of crystal morphology rather than the crystal itself aids in paragenetic interpretation. Thus in some Pb-Zn ores of the Mississippi Valley, dissolution has removed euhedral 1 cm galena crystals that had grown on the surfaces of open fractures. The evidence of the galena is preserved because, prior to its dissolution, fine-grained pyrite and marcasite was precipitated on top of it in a 2–3 mm thick band. Now the ore specimens reveal the following sequence of events: (1) fracturing of wall rock, (2) formation of sphalerite and euhedral galena, (3) formation of colloform overgrowths of pyrite and marcasite

FIGURE 8.1 Euhedral crystals of covellite formed through unobstructed growth into open space. These crystals are embedded in clear epoxy, with the edges of some exposed at the surface of the polished section. Creede, Colorado. (Width of field = 1 cm.)

FIGURE 8.2 Porphyroblasts of pyrite grown in a matrix of pyrrhotite during regional metamorphism. Great Grossan head, Virginia. (Width of field = 5 cm.)

FIGURE 8.3 Mutually interpenetrating grains of sphalerite (medium gray) and galena (white). Montezuma, Colorado. (Width of field = 520 μm.)

that faithfully record the galena morphology on their undersurface, and (4) leaching, leaving euhedral voids once occupied by the galena.

Mutual grain boundaries (equal degrees of penetration) (Figure 8.3) must be interpreted with care and with the recognition that the ore microscopist has only a two-dimensional view of a three-dimensional material. The equal interpenetration of minerals, the absence of characteristic first-formed crystals, and the absence of replacement features usually prevent determination of any paragenetic sequence and may indicate simultaneous crystallization of the minerals.

8.2.3 Colloform Banding and Growth Zoning

Colloform banding, a concentric botryoidal overgrowth of fine radiating crystals, (Figure 8.4*a*) is a texture commonly encountered in open-space filling ores. It is especially common in iron and manganese oxides (Figure 8.4*b*), including manganese nodules, uranium minerals, arsenides, and in pyrite and sphalerite. Although colloform banding has often been attributed to gel formation, Roedder (1968) has shown that typical colloform sphalerites owe their origin to direct crystallization of fine fibrous crystals from a fluid. The colloform bands are actually composed of radiating masses of crystals growing from many adjacent sites along a vein wall, the surface of a wall rock fragment, or previously formed ore minerals. Although generally forming a smooth or undulating surface, colloform growth may occur locally as stalactites. Several Mississippi Valley and vein-type Zn-Pb deposits have also yielded well-developed sphalerite stalactites complete with the hollow central tube. The colloform structures grow from some substrate outward, with sequential growth periods evidenced by overlying bands. Individual bands are often readily distinguished by interlayering of other minerals, by change in the size, shape, or orientation of crystals, or by color zoning, each of these representing some change in the ore fluid or the conditions of precipitation. Minor chemical changes, as evidenced by micron-scale color-banding in colloform sphalerites (only seen in doubly polished thin sections in transmitted light) do not disturb crystal growth since individual crystals exhibit continuity for 1 to 2 cm and may contain hundreds of growth bands.

Growth zoning in individual crystals is a common feature in many types of ore minerals and in a wide variety of deposits. Magmatic precipitates such as chromites and magnetites may display zonal compositional and color variations reflecting changes in the magma from which they precipitated. Several hydrothermally deposited vein minerals may contain distinct color bands (Figure 7.3), which also record a changing environment of formation. Such bands often contain fluid or solid inclusions trapped at the time of precipitation and thus can yield considerable paragenetic information.

The existence of the same, or at least portions of the same, color or compositional zonal sequence in adjacent sphalerite crystals or in crystals growing simultaneously along a fracture provides the basis for *sphalerite stratigraphy* (a term coined by Barton et al., 1977). McLimans et al. (1980) have applied spha-

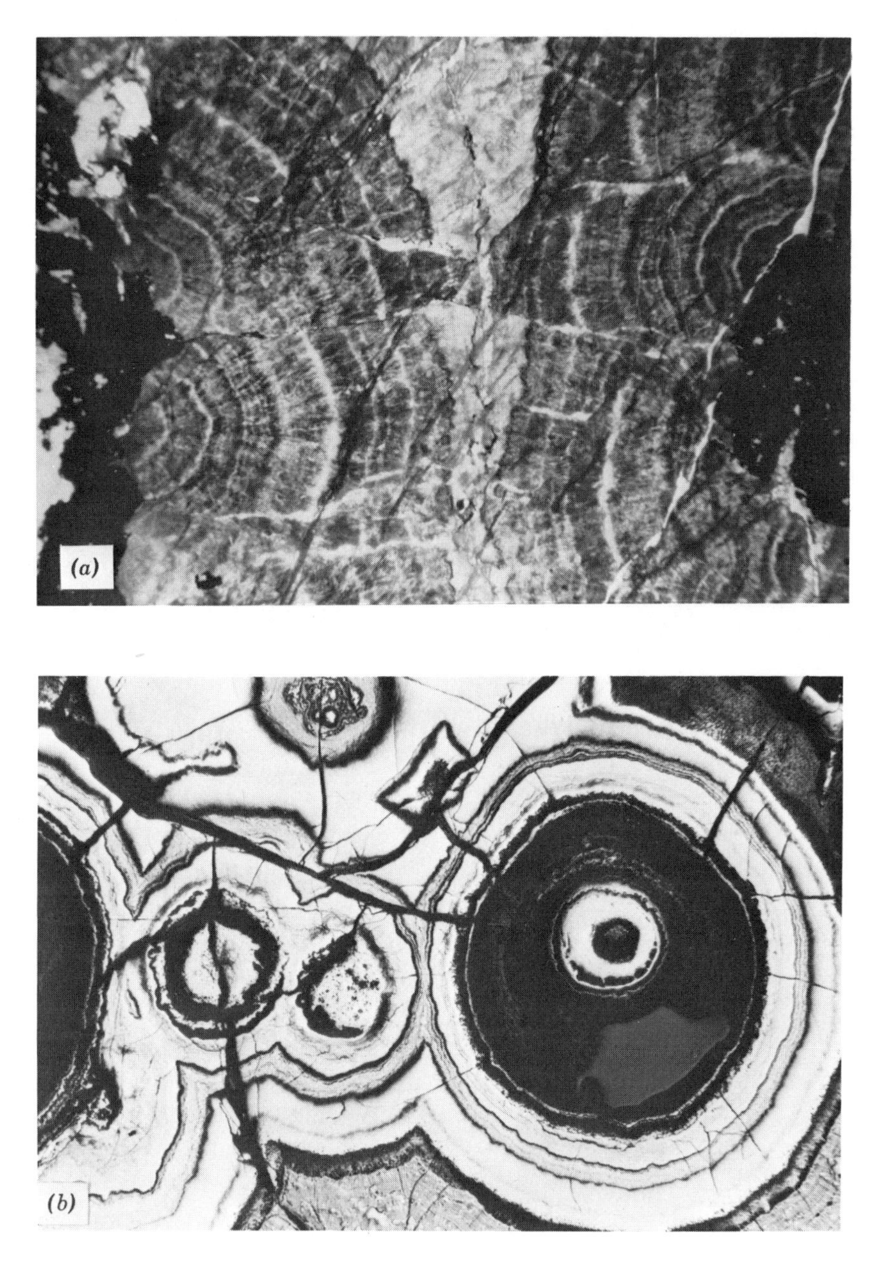

FIGURE 8.4 *(a)* Colloform banding illustrating sequential growth of sphalerite inwards from the walls of a fracture. Transmitted light photomicrograph of doubly polished thin section. Early pyrite (black) has been successively overgrown by banded sphalerite and dolomite (white). Austinville, Virginia. (Width of field = 2000 μm.) *(b)* Concentric growth banding, showing sequential development of hematite and goethite in pisolitic iron ore. Schefferville, Quebec. (Width of field = 2000 μm.)

lerite stratigraphy in the Upper Mississippi Valley District by correlating individual color bands in colloform sphalerite over a few hundred meters and certain bands over several kilometers (Figure 8.5). Sphalerite stratigraphy represents a powerful technique in paragenetic studies but requires carefully collected, preferably oriented, specimens and their examination in doubly polished thin sections.

8.2.4 Cathodoluminescence and Fluorescence

In recent years, cathodoluminescence has become a useful ancillary technique in paragenetic studies of certain minerals. In this technique, a 1 cm diameter beam of electrons accelerated at a potential of 10–15 kv strikes a sample (thin section or polished section) contained in an evacuated viewing chamber on a microscope stage. The sample may be viewed in transmitted or reflected light or only by the luminescence excited by the electron beam. Although most ore minerals exhibit no visible response to the electron beam, some ore minerals such as cassiterite (Hall and Ribbe, 1971), sphalerite, scheelite, powellite, and willemite and some common gangue minerals such as fluorite, calcite, dolomite, feldspar, and quartz (Smith and Stenstrom, 1964) fluoresce visibly. The luminescence of these min-

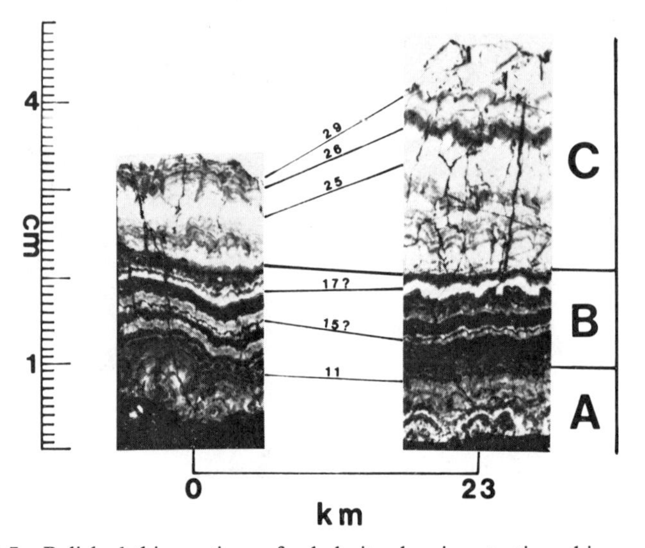

FIGURE 8.5 Polished thin sections of sphalerite showing stratigraphic zones *A* (early), *B* (middle), and *C* (late) and correlation of stratigraphy over 23 km northeast from the Amelia orebody, Illinois (left), to the Hendrickson orebody, Wisconsin (right). (Reproduced from R. K. McLimans et al., *Econ. Geol.* 75, 354, 1980, with permission of the authors and publisher.)

erals depends on the presence of trace to minor amounts of an activator element (e.g., Mn, Dy, Cr) incorporated at the time of initial crystallization.

Since the fluids depositing both ore and gangue minerals commonly change as a function of time, sequentially deposited ore and gangue minerals sometimes possess luminescent growth zones that are not visible either using transmitted- or reflected-light microscopy (Figure 8.6). These growth zones may be useful in sample-to-sample correlation and in the interpretation of paragenesis. The technique has been extensively used in the study of sedimentary cements (Meyers, 1978; Nickel, 1978) and has proved effective in the correlation of carbonate gangue associated with sphalerite ores in the East Tennessee zinc district (Ebers and Kopp, 1979).

Fluorescence, the emission of visible light in response to exposure to ultraviolet (U.V.) light, is very similar to cathodoluminescence and thus provides another means by which paragenetic information may be derived from polished or thin sections. U.V. light sources, usually mercury arc lamps ("black lamps"), are generally divided into long wave (300–400 nm) and short wave (< 300 nm) varieties and are available as inexpensive hand-held and more elaborate microscope-mounted models. Most common ore and gangue minerals exhibit no visible response to U.V. light but many of the same minerals noted above as responsive to cathodoluminescence also fluoresce visibly, especially under short wave U.V. light. The name, in fact, is derived from the fluorescence of fluorite. As with cathodoluminescence, the fluorescence depends upon the presence of certain activator elements and varies widely in intensity and color.

Routine examination of samples with a U.V. lamp is very simple and often leads to the immediate recognition of some minerals (e.g. powellite, cassiterite, scheelite) which do not have other unique visible distinguishing characteristics. Furthermore, characteristic broad zones (e.g. fluorescent bands in calcite, fluorite, dolomite) and sometimes delicate growth zoning, invisible under normal lighting conditions, may become visible. Such zoning may, of course, be very useful in the correlation of one sample with another. The simplicity of use and the widespread availability of U.V. lamps combined with the ease of obtaining valuable information has made this technique a routine tool in many laboratories.

8.2.5 Cross-Cutting Relationships

In mineralogical examination, just as in geological field studies, cross-cutting relationships are a key to paragenetic interpretation. The *veinlet or other feature that cross-cuts another is younger than that which it cuts across,* except where the older phase has been replaced, or where both features result from metamorphic remobilization. Therefore, the veinlet that cuts across another veinlet (Figure 8.7) or crystal is later in the paragenetic sequence, whether it represents simple open-space filling or replacement. Deformational episodes are often indicated by the presence of microfaults (Figure 8.4*a*) which offset bands or veins of earlier-

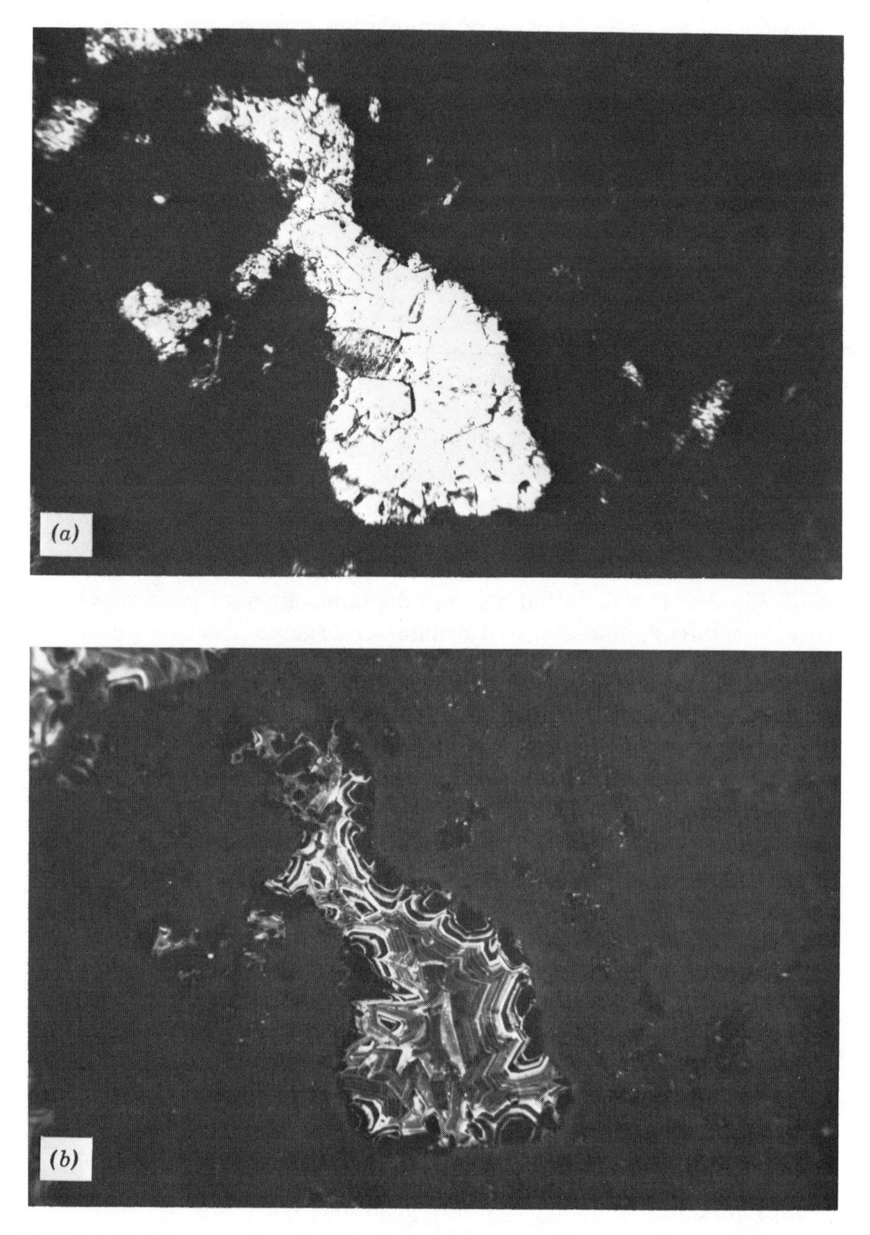

FIGURE 8.6 The use of cathodoluminescence in defining growth banding in calcite: *(a)* a thin section with a calcite filled vug as viewed in transmitted plane polarized light; *(b)* the same area under cathodoluminescence reveals the delicate growth zoning as defined by differing brightness of the bands. (Width of field = 4400 μm.)

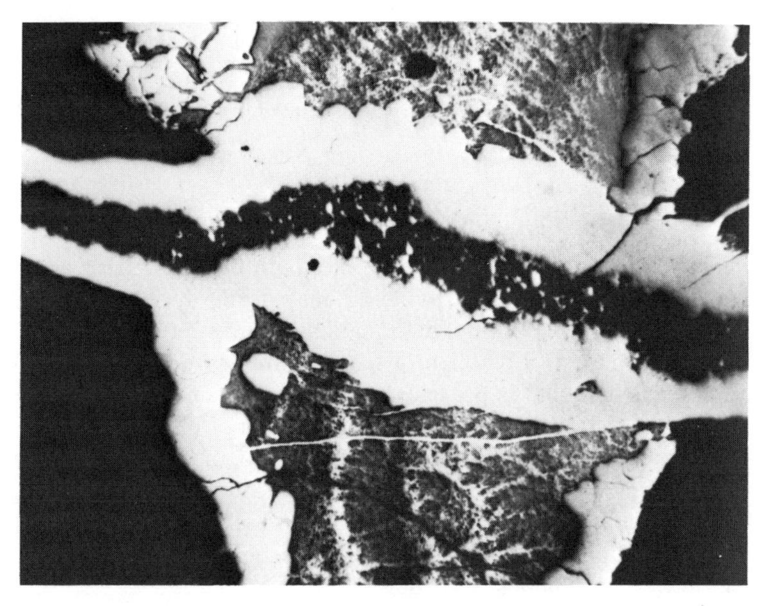

FIGURE 8.7 Cross-cutting relationships shown in a manganese oxide ore in which early chalcophanite is cut by a later veinlet of the same mineral. Red Brush Mine, Virginia. (Width of field = 2000 μm.)

formed minerals or by crushing of earlier grains that may have been subsequently infilled by later minerals. Detailed studies of some syngenetic ores have revealed cross-cutting relationships in the form of scour marks, channels, soft sediment slumping, and cross bedding, some of which may be observed on the microscale.

8.2.6 Replacement

Replacement features are very useful in the determination of paragenesis; clearly the mineral being replaced predates the one replacing it. Since replacement is generally a surface chemical reaction, it usually proceeds inward from crystal boundaries or along fractures. In general, during advanced replacement, the replacing phase possesses convex boundaries, whereas the replaced phase possesses concave boundaries and may remain as residual "islands" within a matrix of the later phase (Figure 7.10*c*). Replacement and weathering processes often lead to one mineral occupying the site another originally possessed in the paragenesis; the difficulty often lies in the unequivocal identification of the now absent mineral. Sometimes, replacement has been incomplete and a few fragments of the original phase remain. A good example is illustrated in Figure 9.8 where some original pentlandite remains within the replacing violarite; most pentlandite grains of this ore are completely replaced but the grain shown in Figure 9.8 reveals the true paragenesis. The cubic morphology of pyrite is probably the most readily recognized of replaced minerals; thus cubes of chalcopyrite, covellite, or goethite are usually regarded as pseudomorphs after pyrite. Similarly in Figure 7.9*b* characteristic bladed crystals of marcasite have been partly replaced by goethite.

8.2.7 Twinning

Twinning can be useful in the interpretation of both the paragenesis and the deformational history of an ore. Twinning may form during initial growth, through inversion, or as a result of deformation (Figures 7.16, 7.17*a*, and 7.17*b*). Since growth twinning is a function of temperature and degree of ore fluid supersaturation and kinetics, its presence in only some grains of a specific mineral may be useful in distinguishing different generations of that mineral. Inversion twinning, if unequivocally identified, is indicative of the initial formation of a higher temperature phase and of at least partial reequilibration on cooling. Deformational twinning can serve as an indicator of deformation during the ore-forming episode (if present only in early formed minerals) or after ore deposition (if present in the ore minerals of all stages).

8.2.8 Exsolution

Exsolution is common in some ore types and may be useful in deciphering certain stages of the paragenesis. In the Fe-Cu-Ni (-Pt) ores associated with ultramafic rocks (Section 9.3), virtually all of the nickel remains incorporated within the $(Fe,Ni)_{1-x}S$ monosulfide solid solution from the time of initial formation at 900–1100°C until the ores cool below 400°C (Figures 9.6 and 9.7). Laboratory phase equilibria studies demonstrate that much of the nickel then exsolves as oriented lamellae of pentlandite. The origin of the earliest exsolved pentlandite is not immediately obvious as it coalesces to form chainlike veinlets (Figures 8.8 and 9.5) at the margins of the host pyrrhotite grains; however, the later (and lower temperature) exsolved pentlandite is retained within the pyrrhotite as crystallographically oriented lamellae and "flames" (Figure 8.8). These characteristic exsolution lamellae indicate that the pentlandite is a secondary phase formed later than the pyrrhotite; although the phase equilibria indicate that this is also true of the granular vein pentlandite, it is not obvious from the texture.

8.3 EXAMPLES OF PARAGENETIC STUDIES

Although it is difficult to generalize, the opaque minerals in many ores can be associated with one of four major divisions:

1 The host rock (or wallrock) materials, which, if igneous, may contain primary oxides, if sedimentary, may contain detrital or authigenic opaques (e.g., framboidal pyrite, titanium oxides).

2 The main mineralization episode, which, although often multiphase, is usually one major introduction of fluids, volatiles, or magma that undergo cooling.

3 A phase of secondary enrichment (in the zone of supergene alteration) resulting in overgrowths and replacement textures.

4 A phase of oxidation and weathering again resulting in replacement textures and the formation of oxides, hydroxides, sulfates, carbonates, and so on.

Normally the sequence of mineral formation (paragenesis) would follow (1)–(4) although many deposits contain evidence of only divisions (1) and (2). It is also important to note that many minerals may have more than one paragenetic position, although different generations may have different habits (e.g., very early pyrite framboids → pyrite cubes → late colloform pyrite) or chemical compositions.

The mineralogical literature contains many paragenetic studies undertaken in varying amounts of detail. Here, three recent reports are used as examples of what can be done through careful observation and application of available data.

8.3.1 The Nickel-Copper Ores of Sudbury District, Ontario

The paragenetic sequence of the minerals in the massive nickel-copper ores of the Sudbury basin has become apparent through the combination of field, microscopic, and phase equilibrium studies. The setting of these ores at the base of a mafic intrusive body led to the view that these ores are a product of the intrusive episode. The common trend of massive ore grading upward into disseminated ore in which isolated "droplets" of sulfide are suspended in a silicate matrix suggested that the sulfides—pyrrhotite with lesser amounts of pentlandite, chalcopyrite, pyrite, and magnetite—had their origin as immiscible sulfide melts, which after separation from the parent silicate magma, coalesced at the base of the intrusion through gravitational setting. Subsequent discovery of similar sulfide droplets (Figure 7.2*b*) in crusts on basaltic lava lakes on Hawaii and in pillow lavas on mid-ocean ridges and laboratory studies on the solubility of sulfur in ultramafic melts certainly confirm the possibility of this general mode of origin. Although direct separation and segregation of a sulfide melt is no doubt the means by which many Sudbury-like ores formed, current theories for the origin of the Sudbury ores involve a more complex history with several phases of injection of sulfide- and silicate-rich magmas that had already undergone some differentiation.

The first mineral to crystallize from the Sudbury sulfide-oxide melts was magnetite, which formed as isolated skeletal to euhedral or subhedral grains (Figure 8.8*a*). Subsequent to the onset of magnetite crystallization, all or most of the sulfide mass crystallized as a nickel- and copper-bearing, high-temperature, pyrrhotitelike phase (the mss, monosulfide solid solution, as shown in Figures 9.6 and 9.7). The phase equilibrium studies of Yund and Kullerud (1966) and Nal-

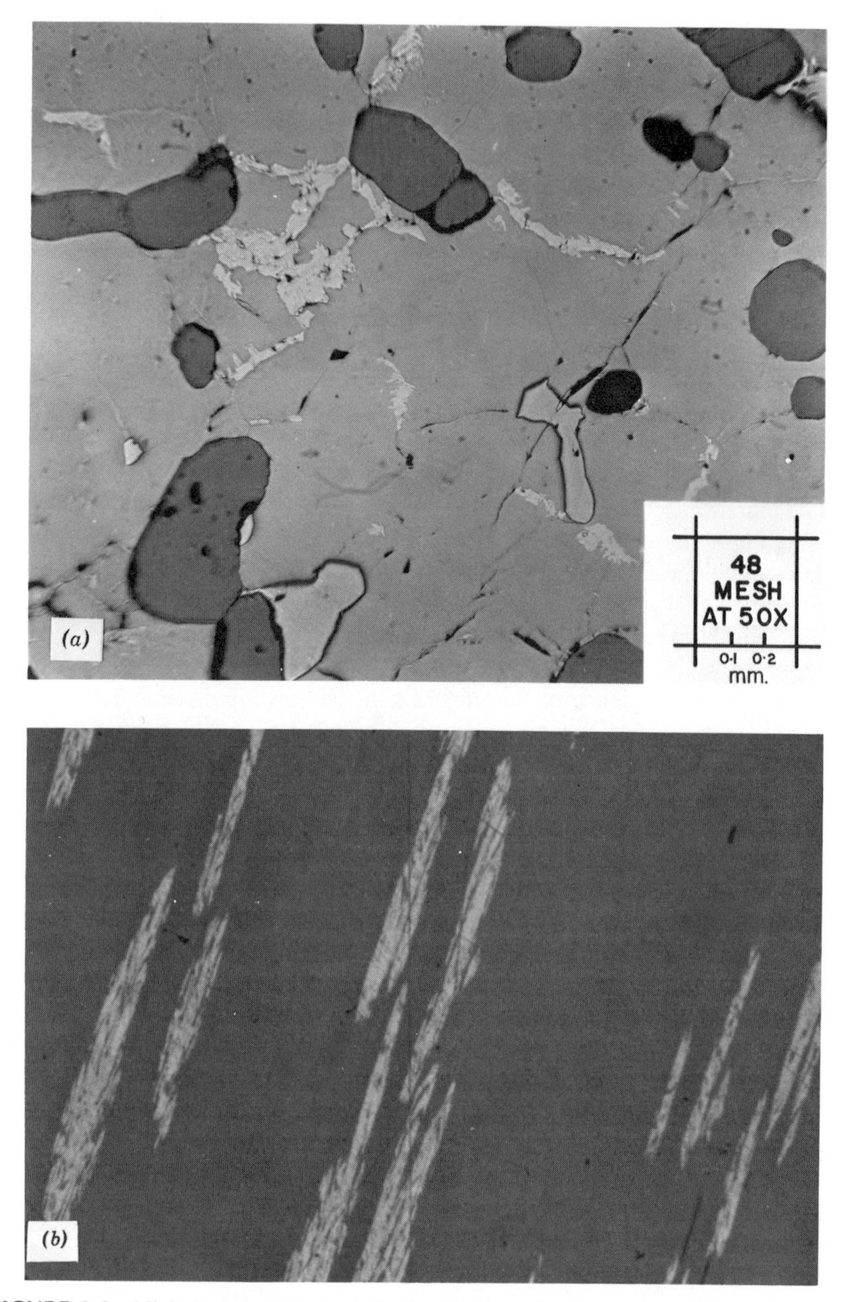

FIGURE 8.8 Nickel-copper ore from Sudbury, Ontario, Canada, illustrating the paragenesis of the ore. *(a)* Early-formed subhedral grains of magnetite (dark gray) within coarse granular pyrrhotite (medium gray) rimmed by granular pentlandite that has coalesced after exsolution. Also present are two anhedral grains of chalcopyrite. (Width of field = 1700 μm.) *(b)* Exsolution "flames" of pentlandite (light gray) in a pyrrhotite ma-

trix. (Width of field = 330 μm.) *(c)* Irregular areas and lamellae of cubanite (dark gray) within chalcopyrite produced by exsolution on breakdown of iss (crossed polars). (Width of field = 330 μm.) *(d)* A veinlet of violarite retaining the blocky fracture pattern of the pentlandite that it has replaced. It is surrounded by altered pyrrhotite that exhibits a single cleavage direction. (Width of field = 520 μm.)

drett et al. (1967) have demonstrated that during subsequent cooling the mss could not have continued to accomodate the copper and nickel in solid solution. Expulsion of the copper as a high-temperature chalcopyrite-like phase (the iss, intermediate solid solution) probably would have begun when temperatures cooled to 400–500°C; most of the copper would have been expelled before cooling reached 300°C, but small amounts would have continued forming from the mss as cooling reached 100°C or less. Exsolution of nickel in the form of pentlandite would have begun as soon as the sulfur-poor boundary of the shrinking mss (Figure 9.7) reached the bulk composition of the ore in any local region. Pentlandite is not stable above about 610°C (pentlandite decomposes to mss + heazlewoodite, Ni_3S_2, above this temperature) but a plot of the bulk compositions of most nickel-copper ores, including those at Sudbury, indicates that pentlandite formation would not have occurred above about 400°C. In laboratory studies (Francis et al., 1976), the expulsion of nickel as pentlandite occurs as crystallographically oriented lamellae in the mss. However, during the slow cooling from 400° down to ~100°C, the previously expelled chalcopyritelike phase would have tended to coalesce into anhedral masses (Figure 8.8*a*) and the pentlandite to diffuse and recrystallize into chainlike veinlets interstitial to mss grains. During continued cooling at temperatures below 100 to 200°C, the diffusion rates of nickel would be much reduced and the last pentlandite exsolved would be trapped as fine oriented "flames" (Figure 8.8*b*). Locally, the cooling iss phase would exsolve cubanite (Figure 8.8*c*) and recrystallize as chalcopyrite.

The paragenetic position of pyrite, which is irregularly distributed in the Sudbury ores, would have depended on the local bulk sulfur content of the sulfide mass (see Figures 9.6 and 9.7); however, its tendency to form euhedral crystals now masks its position in the sequence. If the local sulfur content were less than about 38 wt. %, pyrite would not have formed from the mss until the ores cooled below 215°C and much pentlandite had been exsolved. However, if the local sulfur content were more than about 39 wt. %, pyrite would have begun forming when the sulfur-rich boundary of the mss retreated to the local bulk composition. In either situation, the pyrite and pentlandite would not have coexisted until the temperature had cooled below about 215°C (Craig, 1973).

The pyrrhotite of the Sudbury ores consists of a mixture of hexagonal and monoclinic forms and represents the low temperature remnant of the mss after the copper and nickel have been exsolved as iss (or chalcopyrite) and pentlandite. As is evident in Figure 8.9 these two forms of pyrrhotite formed only after cooling of the ores was nearly complete.

The last stage in the paragenesis of the Sudbury nickel-copper ores was the local development of violarite as an alteration product of pentlandite (and sometimes pyrrhotite) (Figure 8.8*d*). The formation of violarite from pentlandite probably does not reflect an equilibrium state but the ease with which the pentlandite structure is converted to the violarite structure as iron and nickel are removed during weathering.

The paragenesis of the Sudbury ores is summarized in Figure 8.9; the mss and

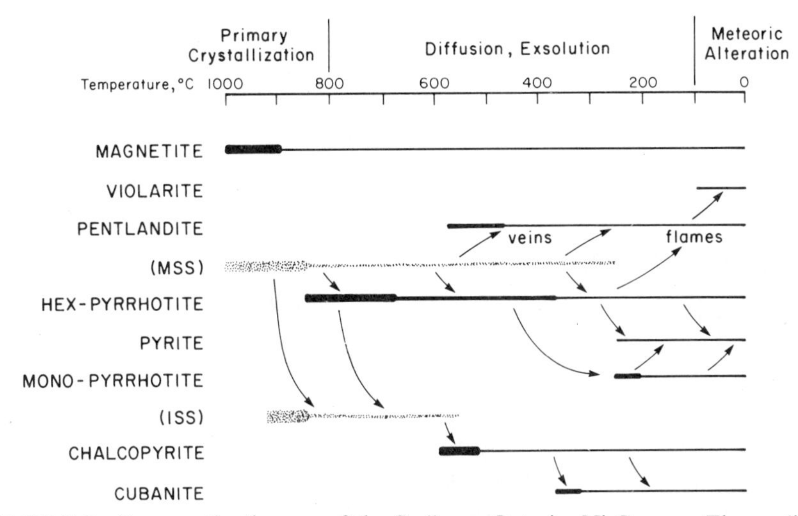

FIGURE 8.9 Paragenetic diagram of the Sudbury, Ontario, Ni-Cu ores. The medium- or heavyweight block line indicates the period of formation; the lightweight line indicates persistence. The stippled lines indicate phases that exist only at elevated temperatures.

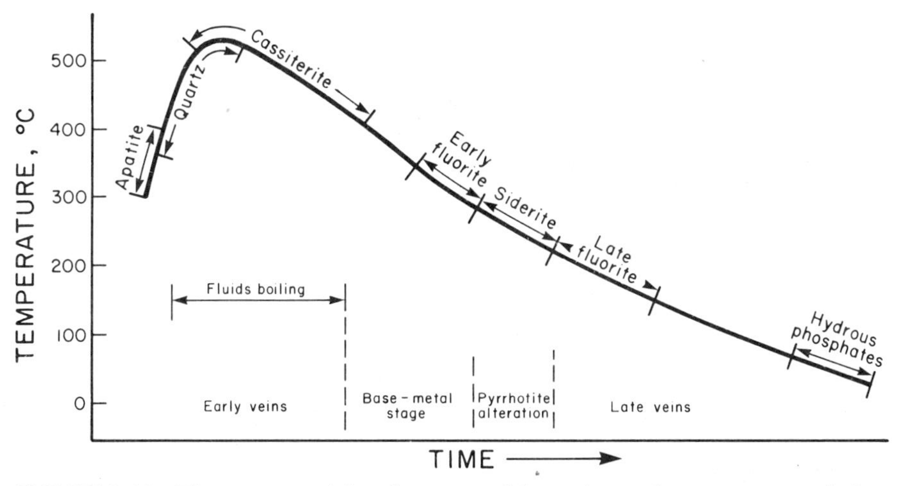

FIGURE 8.10 Figure summarizing the stages of formation and temperature variation during deposition of the tin-tungsten ores of Bolivia. (After Kelly and Turneaure, *Econ. Geol.* **65**, 673, 1970; used with permission.)

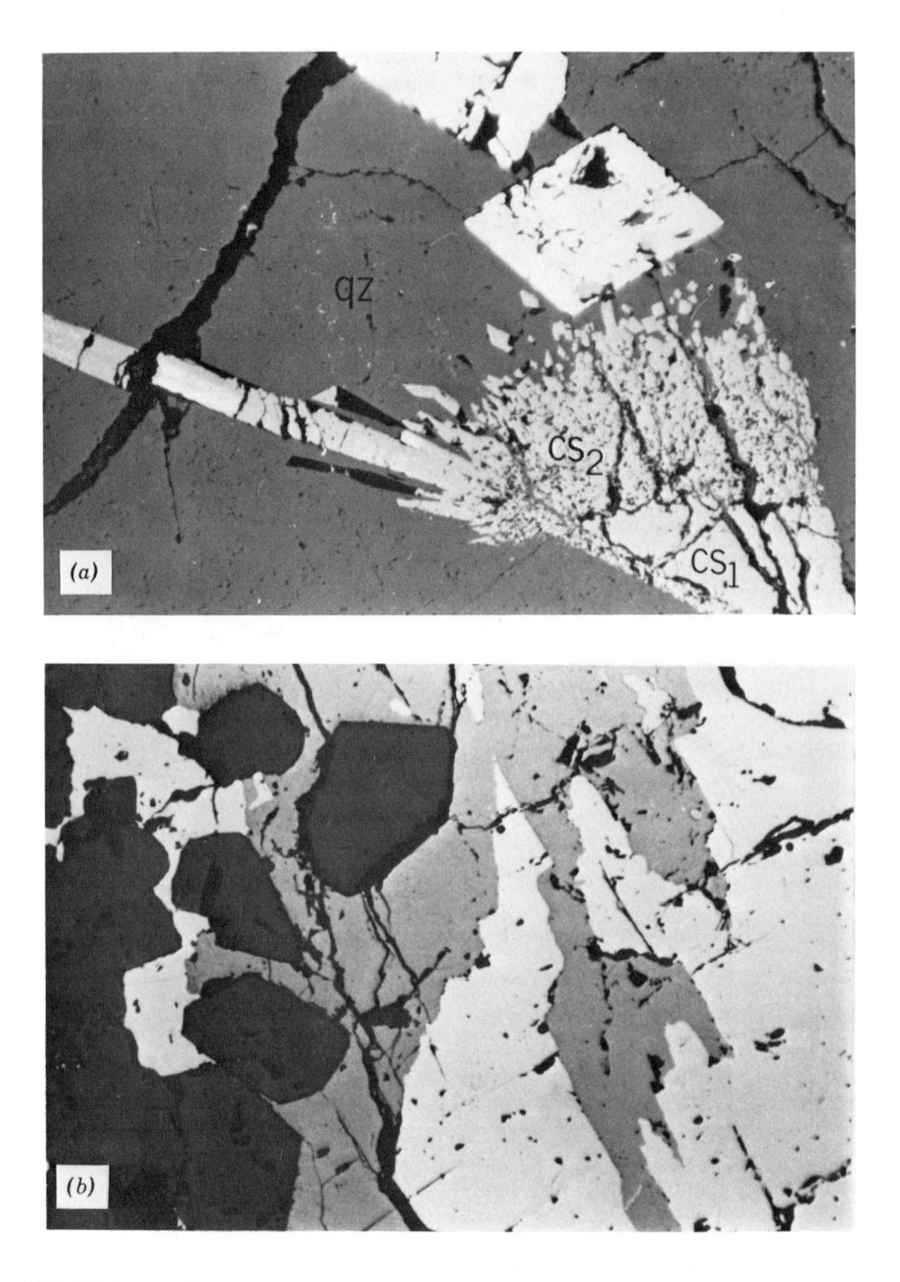

FIGURE 8.11 Tin-tungsten ores from Bolivia illustrating paragenesis of the ore. *(a)* Early cassiterite (cs_1) overgrown by needlelike cassiterite crystals (cs_2). Both generations of cassiterite are enclosed in and veined by quartz (qz), which also surrounds euhedral pyrite crystals (white). Milluni Mine. (Width of field = 2260 μm.) *(b)* Early stannite concentrated along contacts of pyrrhotite (white) with older cassiterite (black). The distribution of stannite (gray) is controlled by the basal parting in pyrrhotite. Huanuni Mine. (Oil immersion, width of field = 1600 μm.) *(c)* Later sulfides (py = pyrite, po = pyr-

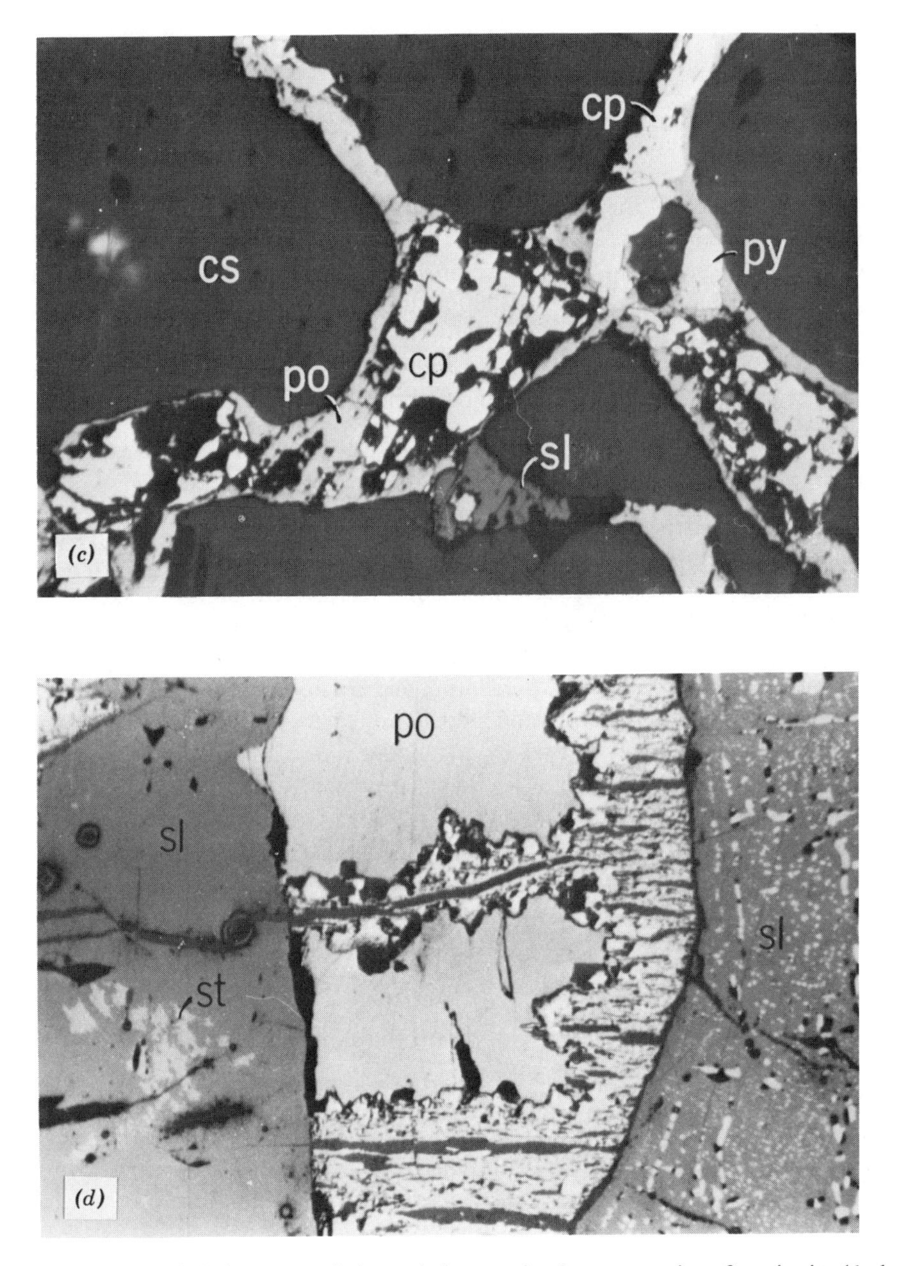

rhotite, sl = sphalerite, cp = chalcopyrite) occurring between grains of cassiterite (dark gray). Araca Mine. (Oil immersion, width of field = 960 μm.) *(d)* Pyrrhotite (po) containing lamellar intergrowths of pyrite, marcasite, and siderite developed along the contact with early sphalerite (sl) to the right, which contains pyrrhotite and chalcopyrite blebs. Sphalerite to the left contains exsolved stannite (st). Colquiri Mine. (Width of field = 1450 μm.) (Reproduced from W. C. Kelly and F. S. Turneaure, *Econ. Geol.* **65**, 616, 620, 1970, with permission of authors and publisher.)

iss are noted in parentheses because their role is reconstructed from studies of phase equilibria.

8.3.2 Tin-Tungsten Ores in Bolivia

In a detailed study of the mineralogy, paragenesis, and geothermometry of Bolivian tin-tungsten vein ores, Kelly and Turneaure (1970) have unraveled a complex history of ore formation. These ores are interpreted as subvolcanic vein deposits formed at depths of 350–2000 m and in a temperature range of 530 to 70°C. The ore solution, as evidenced by fluid inclusions, were NaC1-rich brines of low CO_2 content that were boiling during at least part of the ore forming period. The paragenetic sequence, noted only briefly later, is summarized in Figure 8.10.

In the earliest vein stage, quartz with apatite intergrown as a coarse band along the vein walls is observed, quartz being overgrown by cassiterite which makes up the central part of the vein and which lines cavities with well-developed crystals (Figure 8.11*a*). Local bismuthinite is interpreted as having crystallized before cassiterite because it is both surrounded and replaced by the cassiterite. In the base metal sulfide stage, pyrrhotite and sphalerite fill space between and partly replace quartz and cassiterite (Figure 8.11*b*). The sphalerite contains exsolved chalcopyrite, pyrrhotite, and stannite. Early fluorite is mutually intergrown with sphalerite and pyrrhotite (Figure 8.11*c*). The pyrite-marcasite-siderite stage is evidenced by pyrrhotite alteration (Figure 8.11*d*); the hypogene nature of the alteration is demonstrated by high filling temperatures of fluid inclusions in siderite. Pyrrhotite alteration begins along veins and fractures but may proceed to perfect pseudomorphs of pyrite-marcasite after pyrrhotite crystals. The final stages evident in the paragenesis arc veinlets and crustifications of (1) siderite, sphalerite, and late fluorite, and (2) hydrous phosphates. In Figure 8.10, Kelly and Turneaure (1970) trace the general paragenesis and the thermal history of the ores.

8.3.3 The Lead-Zinc Ores of the North Pennines, England

The North Pennine Orefield contains lead-zinc-barite-fluorite mineralization, chiefly as fissure infilling veins in mainly Lower Carboniferous strata. Fluid inclusion studies suggest that the solutions responsible for the precipitation of these minerals were brines and that deposition occurred at temperatures below 250°C.

The sulfide mineral assemblages of the North Pennines were studied by Vaughan and Ixer (1980) who noted distinct differences in the assemblages to the north in the Alston Block area and to the south of the orefield in the Askrigg Block. In the latter area a consistent generalized paragenesis is observed, as shown in Figure 8.12. An early diagenetic phase of framboidal pyrite is found in the host rock limestone and intimately associated with small carbonaceous

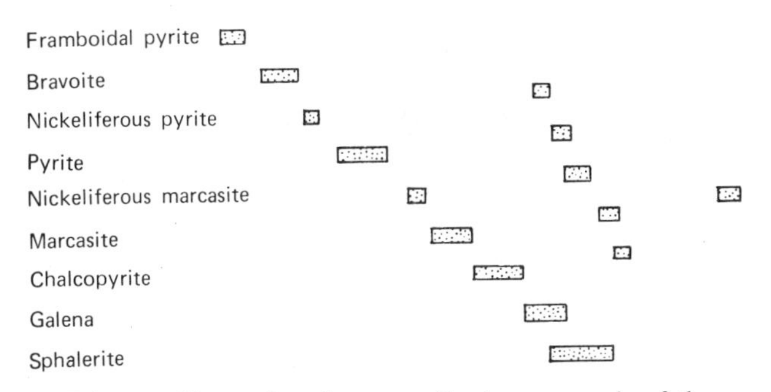

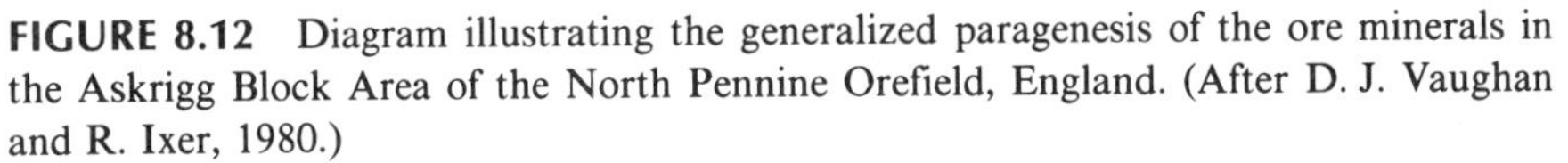
FIGURE 8.12 Diagram illustrating the generalized paragenesis of the ore minerals in the Askrigg Block Area of the North Pennine Orefield, England. (After D. J. Vaughan and R. Ixer, 1980.)

laths. These pyrite framboids may act as nuclei to later pyrite cubes or radiating marcasite crystals (Figure 8.13*a*). Bravoite, the earliest epigenetic sulfide, may show multiple zoning and occurs as inclusions in fluorite, calcite, or barite (Figure 8.13*b*). Bravoite is commonly succeeded by nickeliferous pyrite and the first of several generations of normal pyrite that is intergrown with marcasite and nickeliferous marcasite (Figure 8.13*c*). Minor chalcopyrite (showing late supergene alteration to covellite and limonite) is followed by galena, which can enclose all the earlier sulfides (Figure 8.13*d*) and commonly alters in weathering to cerussite or anglesite. Sphalerite is the last primary sulfide to form.

This overall paragenesis is very similar to that observed in the South Pennine Orefield, whereas in the northernmost Alston Block area much greater diversity is found. The authors discuss the observed paragenesis in terms of the conditions of ore formation, which suggest higher temperatures and more diverse fluid compositions in this northernmost area.

8.4 ORE FORMATION CONDITIONS AND THE APPLICATION OF PHASE EQUILIBRIA DATA

Reference to relevant phase diagrams can help in: (1) anticipation and recognition of phases; (2) recognition of trends in ore chemistry (i.e., the character of the ore fluid and its variation in time or space); (3) understanding of reactions and some textural features (e.g., exsolution); (4) understanding of the correlation or antipathetic relationships between phases; (5) recognition of equilibrium or disequilibrium mineral assemblages; (6) interpretation of the nature of the ore-forming fluid and the mechanisms that were operable during mineralization; and (6) estimation of the temperature and pressure during ore formation or subsequent metamorphism.

FIGURE 8.13 Sulfide mineral assemblages in the lead-zinc ores of the North Pennines, England, illustrating the paragenesis of the ores. *(a)* Framboidal pyrite overgrown by later euhedral pyrite (Oil immersion, width of field = 500 μm.) *(b)* Bravoite with overgrowths of later pyrite in a veinlet surrounded by carbonates. (Oil immersion, width of

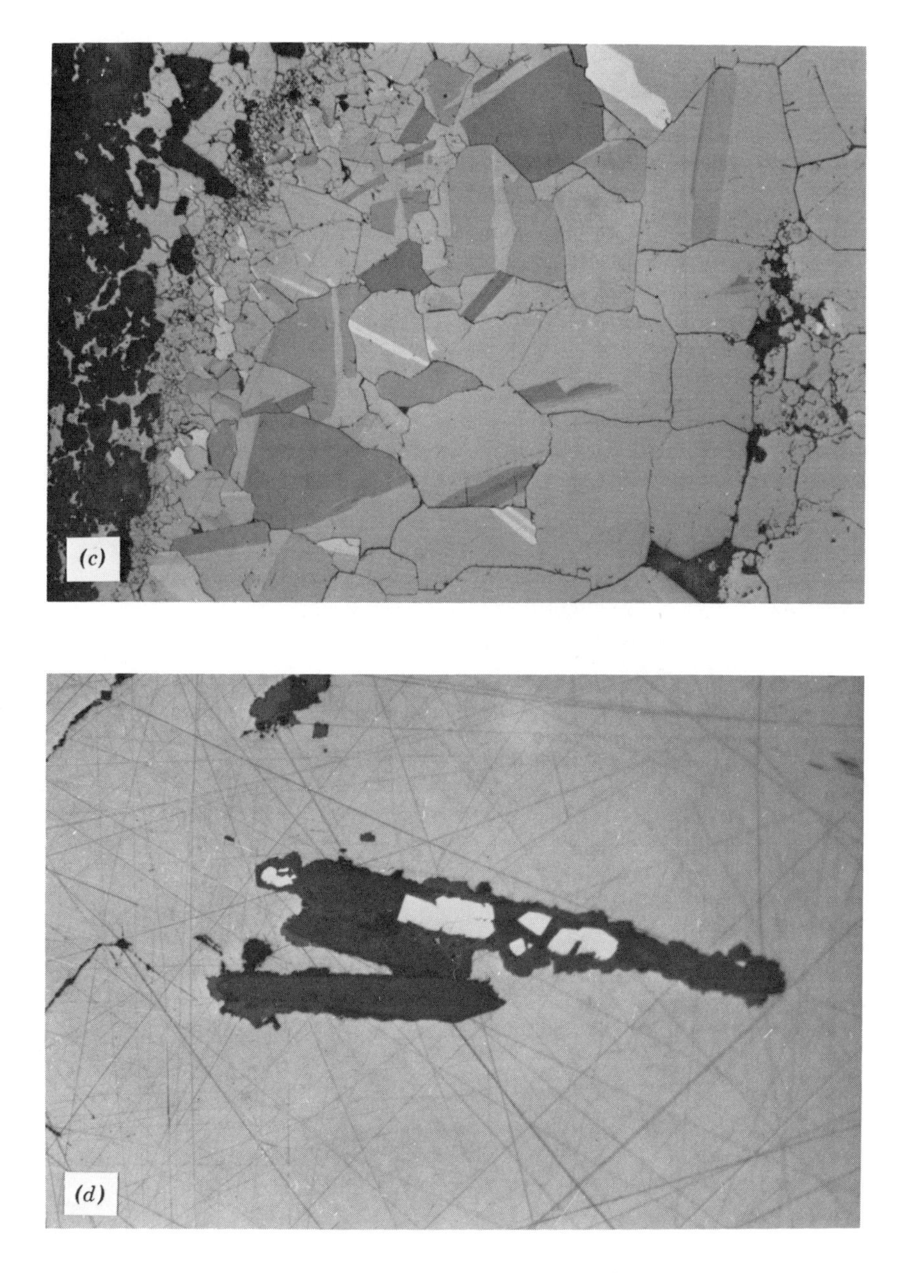

field = 500 μm.) *(c)* Pyrite with associated marcasite. (Oil immersion and partly crossed polars, width of field = 500 μm.) *(d)* Subhedral pyrite grains associated with carbonates and bladed crystals of barite enclosed in later galena. (Width of field = 500 μm.)

It is impossible to present more than a few of the many relevant phase diagrams but the following systems are discussed, at least in part, in this book:

Fe-S	Figure 8.15
Cu-S	Figure 10.7
Cu-Fe-S	Figure 8.14
Fe-Ni-S	Figures 9.6 and 9.7
Fe-Zn-S	Figure 10.24
Fe-As-S	Figure 8.16
FeO-Fe_2O_3-TiO_2	Figures 9.12 and 9.13
Cr_2O_3-(Mg,Fe)O-SiO_2	Figure 9.4
Au-Ag-Te	Figure 9.25
Ca-Fe-Si-C-O	Figure 10.2
Cu-O-H-S-Cl	Figure 10.10
U-O_2-CO_2-H_2O	Figure 10.8
NaCl-H_2O	Figure 8.21
H_2O (P-T)	Figure 8.22
Iron Minerals (Eh-pH)	Figure 10.3

For additional information on these and other systems the reader is referred to *Mineral Chemistry of Metal Sulfides* (Vaughan and Craig, 1978), "Sulfide Phase Equilibria" (Barton and Skinner) in *Geochemistry of Hydrothermal Ore Deposits* (1979), *Sulfide Minerals* (P. H. Ribbe, ed., 1974), and *Oxide Minerals* (D. Rumble, ed., 1976). Most diagrams in the literature are "equilibrium" diagrams and may be cautiously applied to the natural ores, which also represent, at least locally, conditions of equilibrium. For example, in Cu-Fe-sulfide ores, assemblages such as pyrite-pyrrhotite-chalcopyrite, pyrite-chalcopyrite-bornite, or even pyrite-digenite-bornite are common because they are stable (see Figure 8.14) whereas pyrrhotite-covellite or cubanite-chalcocite assemblages are unknown and are not expected, because they are not stable. It is important to note that although most ore mineral assemblages do represent equilibrium, disequilibrium assemblages are not uncommon. This is especially true in weathering zones in which reaction kinetics are slow because of the low temperatures involved.

Phase diagrams are important in providing geothermometric and geobarometric data. The geothermometers are of two types—"sliding scale" and "fixed point." The "sliding-scale" type is based on the temperature dependence of the composition of a mineral or pair of minerals when it is part of a specified assemblage (e.g., the composition of pyrrhotite in equilibrium with pyrite in Figure 8.15*a*). Ideally, determination of the composition of pyrrhotite coexisting with pyrite would uniquely define the temperature of equilibration; in practice, however, it has been found that nearly all pyrrhotites have reequilibrated to near

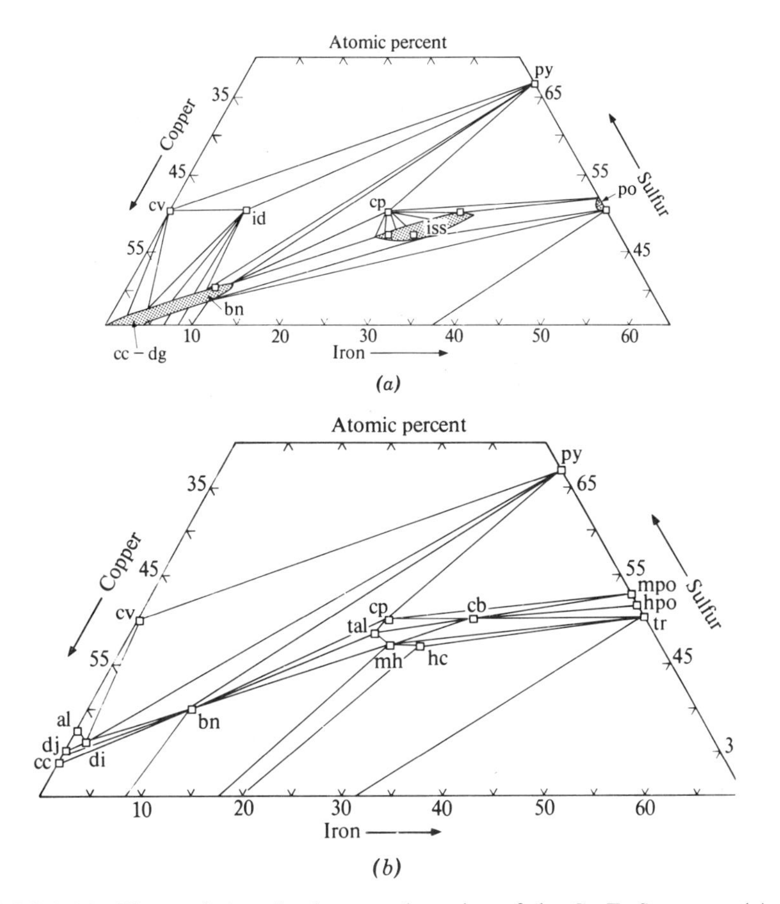

FIGURE 8.14 Phase relations in the central portion of the Cu-Fe-S system; *(a)* schematic relations at 300°C; *(b)* possible phase relations at 25°C. Abbreviations: cc, chalcocite; dj, djurleite; di, digenite; al, anilite; cv, covellite; bn, bornite; id, idaite; cp, chalcopyrite; tal, talnakhite; mh, mooihoekite; hc, haycockite; cb, cubanite; mpo, monoclinic pyrrhotite; hpo, hexagonal pyrrhotite; tr, troilite; py, pyrite; iss, intermediate solid solution. (From D. J. Vaughan and J. R. Craig, 1978; used with permission.)

room temperature conditions (see Figure 8.15*b*). Unfortunately, the rapid rates of sulfide reequilibration processes have limited the usefulness of many sliding scale geothermometers. Two notable exceptions involve refractory minerals that retain their high temperature compositions during cooling and thus are applicable as sliding scale geothermometers, namely:

1 Arsenopyrite when equilibrated with pyrite and pyrrhotite (and some other less common assemblages) as shown in Figure 8.16.

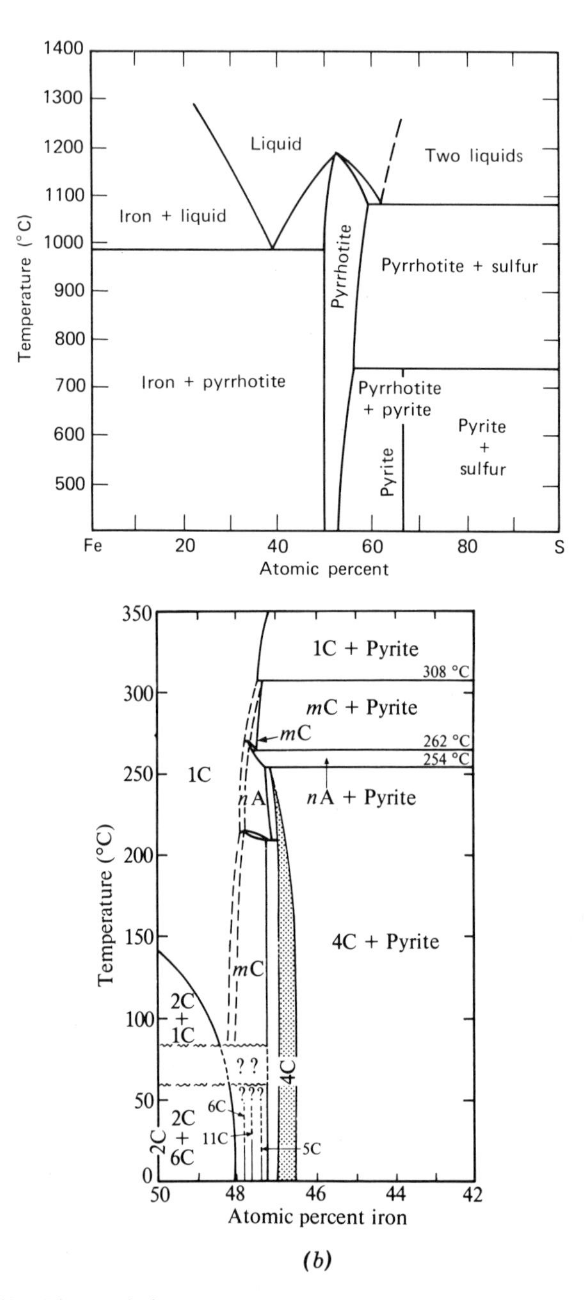

FIGURE 8.15 Phase relations among condensed phases in the Fe-S system: *(a)* above 400°C; *(b)* in the central portion of the system below 350°C. The notations 2C, 1C, 6C, 4C, 11C, and *m*C refer to *c*-axis superstructure dimensions, and *n*A to *a*-axis superstructure dimensions in pyrrhotites. (From D. J. Vaughan and J. R. Craig 1978; used with permission.)

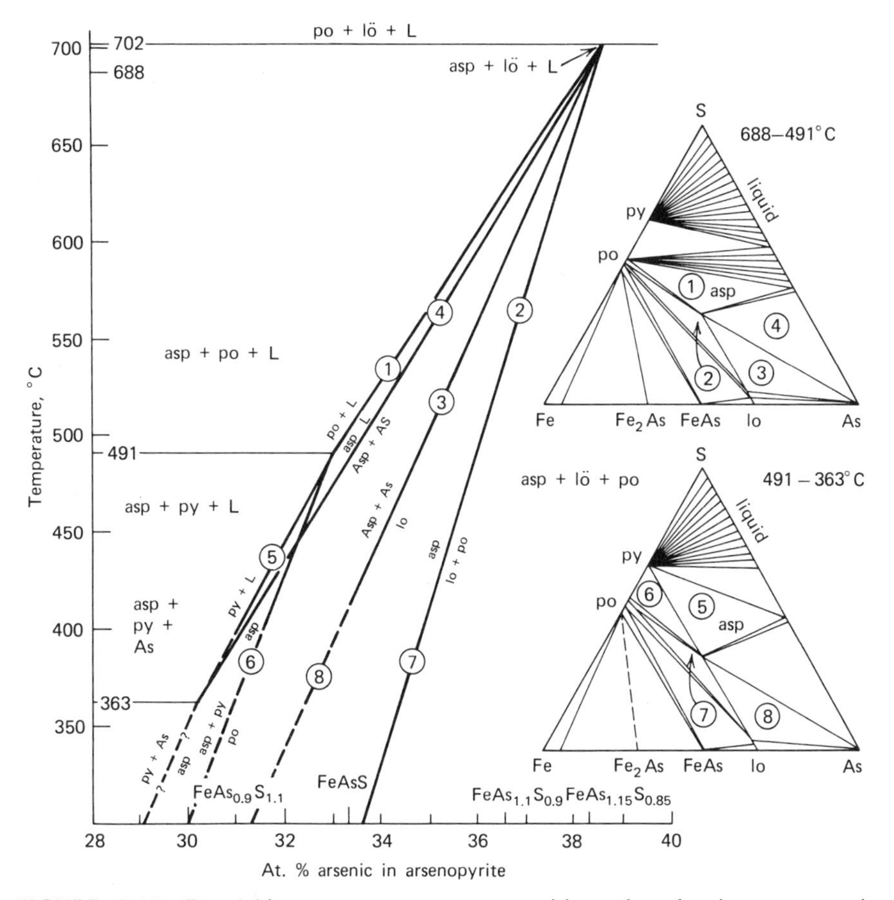

FIGURE 8.16 Pseudobinary temperature—composition plot showing arsenopyrite composition as a function of temperature and equilibrium mineral assemblage. The assemblages numbered 1–8 in the Fe-As-S phase diagrams on the right correspond to the labeled curves in the left-hand diagram. Abbreviations: asp, arsenopyrite; py, pyrite; po, pyrrhotite; lö, löllingite; *L*, liquid. (After U. Kretschmar and S. D. Scott, *Can. Mineral.* **14**, 366–372, 1976, used with permission.)

2 Coexisting magnetite-ulvöspinel and ilmenite-hematite. The compositions of these minerals, if equilibrated together, uniquely define both the temperature and the oxygen activity of equilibration as shown in Figure 9.13.

"Fixed point" geothermometers are minerals or mineral assemblages that undergo a reaction (e.g., melting, inversion, reaction to form a different assemblage) at a defined temperature. For example, crystals of stibnite must have formed below its melting point (556°C) and the mineral pair pyrite + arsenopyrite must have formed below 491°C. The fixed points thus do not sharply define the temperature of equilibration but rather set upper and lower limits. Barton and Skinner (1979) and Vaughan and Craig (1978) have prepared extensive lists of reaction points that serve as potentially useful fixed point geothermometers. Most fixed points are known only for very low pressures; those whose pressure dependence has been determined generally rise at a rate of about 10°C kbar.

Phase equilibria studies have revealed that the iron content of sphalerite equilibrated with pyrite and pyrrhotite, although temperature independent between about 300 and 550°, is pressure dependent. This relationship has been defined (Scott and Barnes 1971; Scott 1973) and thus allows the sphalerite composition in this assemblage to serve as a geobarometer (Figure 10.24). The equation below relates iron content (as FeS) to the pressure of equilibration (Hutchison and Scott 1980): P bar $= 42.30–32.10 \log$ mole % FeS. Although sphalerite is among the most refractory of ore minerals and thus may preserve a composition indicative of the original pressure of equilibration with pyrite and pyrrhotite, sphalerite often undergoes at least partial reequilibration (Skinner and Barton, 1979). Such reequilibration, which is most evident where the sphalerite is in contact with pyrrhotite or chalcopyrite, results in a decrease in the FeS content of the sphalerite. Consequently, Hutchison and Scott (1979) found that sphalerite inclusions that equilibrated with pyrite and pyrrhotite in metamorphosed ores and were trapped within recrystallizing pyrite seemed to preserve compositions that best reflected the metamorphic pressures. When partial reequilibration of sphalerite has occurred, those with highest FeS contents are probably most indicative of the pressures of original equilibration because lower temperature reequilibration reduces the FeS content.

8.5 FLUID INCLUSION STUDIES

The study of fluid inclusions, although commonly carried out on nonopaque minerals and using a transmitted-light microscope has become a fruitful and expanding field of investigation that supplements conventional ore microscopy and is therefore worthy of brief discussion in this text. In particular, it has provided valuable data on the temperatures of ore formation or subsequent metamorphism and on the chemistry of the ore fluids. Fluid inclusions are abundant in many common ore and gangue minerals and can be observed with a standard petrographic microscope. Accordingly, a brief discussion of the nature and significance of fluid inclusions and the means of preparing samples for their study is presented in the following; for more detailed descriptions the reader is referred to the bibliography presented at the end of the chapter.

8.5.1 The Nature and Location of Fluid Inclusions

Fluid inclusions are small amounts of fluid that are trapped within crystals during initial growth from solution or during total recrystallization *(primary inclusions)* or during isolated recrystallization along fractures at some later time *(secondary inclusions).* Fluid inclusions are very abundant in common ore and gangue minerals, sometimes occurring as a billion or more per cubic centimeter. Their volumes are often less than 10 μm^3 but may reach as much as a cubic millimeter or more. The inclusions are, however, seldom recognized in conventional polished sections because so little light actually enters most ore minerals; even among the most transparent of ore minerals, inclusions are rarely recognized as anything except internal reflections. However, when transparent ore and gangue minerals are properly prepared and observed in transmitted light as described later, they are often found to contain abundant tiny inclusions, commonly oriented along well-defined crystallographic planes (Figure 8.17) and having a wide variety of shapes. Some inclusions (Figure 8.18) may contain visible bubbles or mineral grains that precipitated at the time of trapping or that precipitated from the fluid after trapping (daughter minerals).

Primary inclusions, those trapped during growth of the host mineral, may be samples of the ore-forming fluid and may reveal important information regarding the conditions of ore transport and deposition. Roedder (pers. commun., 1980) has pointed out that there has been alternation of ore and gangue mineral deposition in many ores without simultaneous deposition. If such has occurred, fluid inclusions in gangue minerals may not represent the fluids from which the ore minerals formed.

Secondary inclusions do not reliably give information on the nature of ore-forming fluids or conditions but may provide some insight into postdepositional conditions. The distinction between primary and secondary inclusions is often not unequivocal; Roedder (1979) has offered the criteria listed on Table 8.1 to help in interpretation. The reader is directed to Roedder's paper for additional discussion.

Commonly, the fluids trapped along growing crystal faces are homogeneous; however, sometimes two or more immiscible liquids (i.e., water and oil or water and CO_2), liquids and gases (i.e., boiling water and steam), or liquids plus solids (i.e., water plus salts or other minerals) may be trapped together. Such inclusions (termed multiphase inclusions) are difficult to interpret geothermometrically but may provide considerable data on the nature of the ore-forming fluid. Typical host minerals in which fluid inclusions are observed include sphalerite, cassiterite, quartz, calcite, dolomite, fluorite, and barite, but nearly any transparent mineral may contain visible inclusions. Roedder (1979) even notes that some granite feldspars contain so many fluid inclusions with daughter NaCl crystals that NaCl diffraction lines appear in single-crystal X-ray photographs of the feldspars. The opaque ore minerals such as galena and pyrite contain inclusions, the forms of which may be seen on some fractured or cleaved surfaces, but present techniques do not permit their undisturbed *in situ* observation.

8.5.2 Changes in Fluid Inclusions Since Trapping

Most fluid inclusions were trapped as a homogeneous fluid at elevated temperatures and pressures. During the subsequent cooling the fluid may separate into

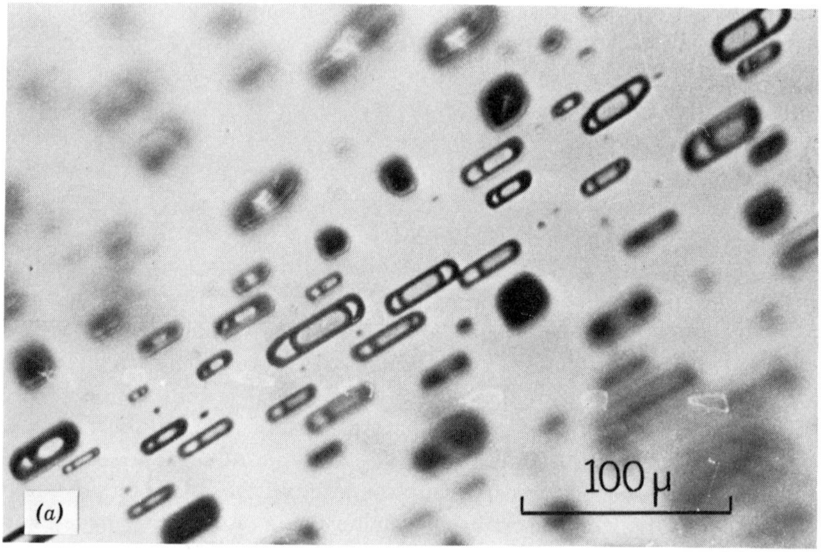

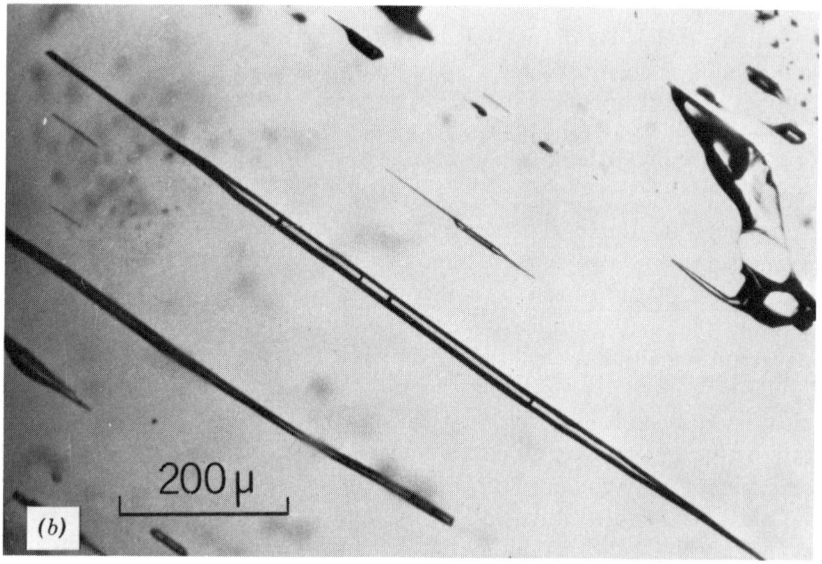

FIGURE 8.17 Fluid inclusions in cassiterite, Oruro district, Bolivia. *(a)* Inclusions lying along a healed cleavage plane. The gas phase fills the inclusions at 424–434°C. *(b)* Needlelike inclusions, some with double bubbles due to constriction of the chamber. (Reproduced from W. C. Kelly and F. S. Turneaure, *Econ. Geol.* **65**, 649, 1970, with permission of the authors and publishers.)

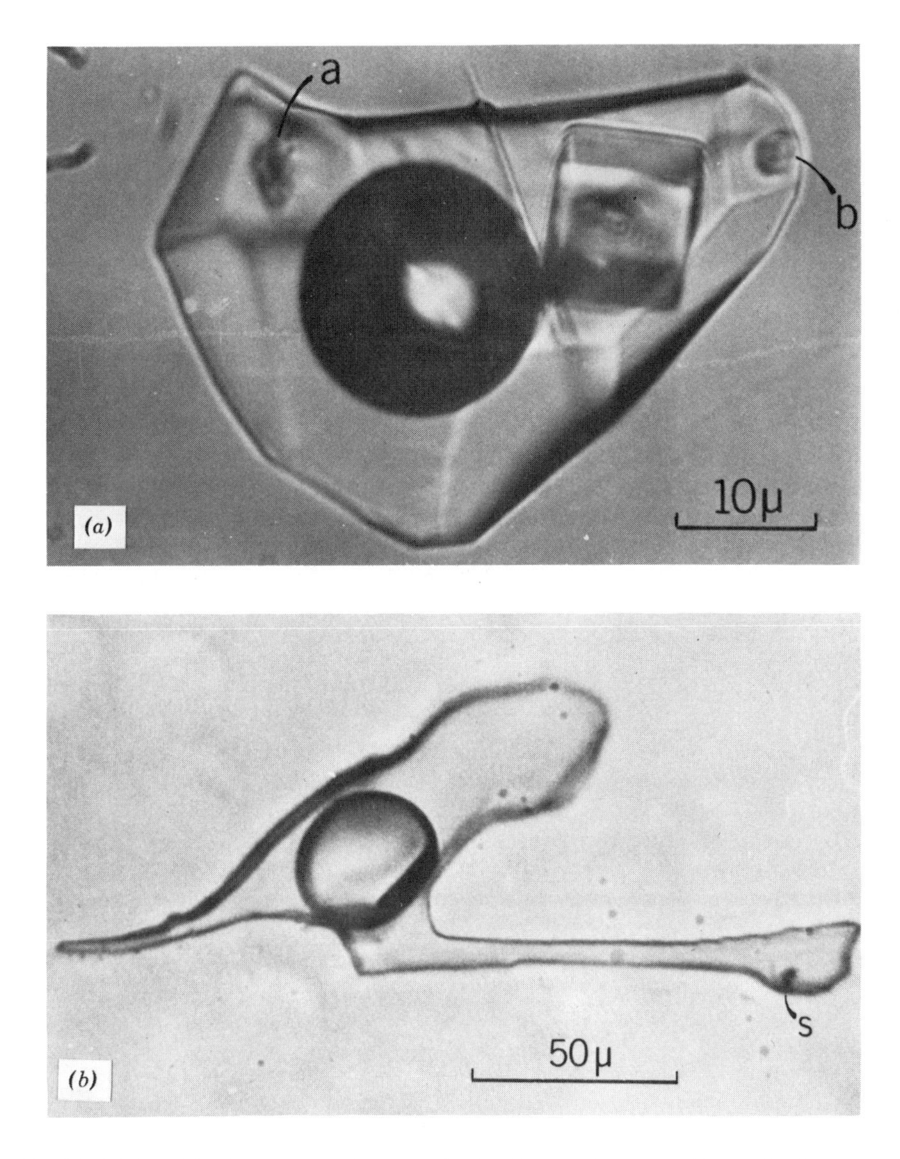

FIGURE 8.18 Fluid inclusions that contain daughter inclusions. *(a)* Inclusion in quartz with a large halite cube and unidentified daughter salts at *a* and *b*. The total salinity is approximately 47%, and the fluid fills the inclusion at 430°C. Gigante Chica, Laramcota Mine, Bolivia. *(b)* Inclusion in apatite having an irregular form suggestive of necking down. A grain of an opaque inclusion at *s* lies in front of a small halite cube. The inclusion fills with liquid at 350°C. Lallaqua Mine, Bolivia. (Reproduced from W. C. Kelly and F. C. Turneaure, *Econ. Geol.* **65**, 651, 1970, with permission of the authors and publishers.)

TABLE 8.1 Criteria for Recognition of the Origin of Fluid Inclusions (revised from Roedder 1976, 1979)

- Criteria for primary origin
  - Single crystals with or without evidence of direction of growth or growth zonation.
    - Occurrence as a single inclusion (or isolated group) in an otherwise inclusion-free crystal
    - Large size of inclusion(s) relative to enclosing crystal (e.g., $^{1}/_{10}$ of crystal) and/or equant shape
    - Isolated occurrence of inclusion away from other inclusions (e.g., $\geq$ inclusion diameters)
    - Random three dimensional occurrence of inclusions in crystal
    - Occurrence of daughter minerals of the same type as occur as solid inclusions in the host crystal or contemporaneous phases
  - Single crystals showing evidence of directional growth
    - Occurrence of inclusion along boundary between two different stages of growth (e.g., contact between zone of unimpeded growth and zone containing extraneous solid inclusions)
    - Occurrence of inclusion in a growth zone beyond a visibly healed crack in earlier growth stages
    - Occurrence of inclusion at boundary between subparallel growth zones
    - Occurrence of inclusion at intersection of growth spirals
    - Occurrence of relatively large flat inclusions in the core or parallel to external crystal faces
    - Occurrence of inclusion(s) at the intersection of two crystal faces
  - Single crystals showing evidence of growth zonation (on the basis of color, solid inclusions, clarity, etc.)
    - Occurrence of different frequencies or morphologies of fluid inclusions in adjacent growth zones
    - Occurrence of planar arrays outlining growth zones (unless parallel to cleavage directions)
  - Crystals evidencing growth from heterogeneous (i.e., two phase) or changing fluid
    - Occurrence of inclusions with differing contents in adjacent growth layers (e.g., gas inclusions in one layer, liquid in another layer, or oil and water in another layer, etc.)
    - Occurrence of inclusions containing some growth medium at points where host crystal has overgrown and surrounded adhering globules of an immiscible phase (e.g., oil droplets)
    - Occurrence of primary-appearing inclusions with "unlikely" growth medium (e.g., mercury in calcite, oil in fluorite or calcite, etc.)
  - Hosts other than single crystals
    - Occurrence of inclusions at growth surfaces of nonparallel crystals (these have often leaked and could be secondary)
    - Occurrence of inclusions in polycrystalline hosts (e.g., vesicles in basalt, fine-grained dolomite, vugs in pegmatites—these have usually leaked)
    - Occurrence in noncrystalline hosts (e.g., bubbles in amber, vesicles in pumice)

TABLE 8.1 Criteria for Recognition of the Origin of Fluid Inclusions (revised from Roedder 1976, 1979) ***(Continued)***

Criteria for secondary origin
- Occurrence of inclusions in planar groups along planes that cross-cut crystals or that parallel cleavages
- Occurrence of very thin, flat or obviously "necking-down" inclusions
- Occurrence of primary inclusions with filling representative of secondary conditions
- Occurrence of inclusions along a healed fracture
- Occurrence of empty inclusions in portions of crystals where all other inclusions are filled
- Occurrence of inclusions that exhibit much lower (or more rarely higher) filling temperatures than adjacent inclusions

Criteria for pseudosecondary origin
- Occurrence of secondary-like inclusions with a fracture visibly terminating within a crystal
- Occurrence of equant and negative crystal-shaped inclusions
- Occurrence of inclusions in etch pits cross-cutting growth zones

liquid and vapor because the fluid contracts much more than the solid host mineral. Immiscible fluids may separate on cooling, and daughter crystals, usually halite or sylvite, may precipitate as saturation of the fluid occurs. Many inclusions do not now have the shape they originally possessed because of solution and deposition in different parts of the inclusion cavity. In general, inclusions will tend, by solution and redeposition, to reduce surface area and to become more equant. Through this process, elongate inclusions may separate into several more equant inclusions as a result of "necking down" as shown in Figure 8.19. If the necking down occurs after phase separation, the process may isolate the vapor bubble in one of the newer inclusions while leaving another newer inclusion completely fluid-filled. As a result, neither inclusion would be representative of the originally trapped fluid and the information that could be derived would be limited. Larger flat primary inclusions or secondary cracks may also undergo considerable recrystallization (Figure 8.20) in which one large inclusion is reduced to many small ones occupying the same region within the crystal. Roedder (1977) has cautioned the student of fluid inclusions as follows: "It is important to remember always that the fluid inclusions in a mineral provide information only on the fluids present *at the time of sealing* of the inclusion, whether that be during the growth of the host crystal or during rehealing of a later fracture."

Leakage, the movement of material into or out of the original inclusion, can occur but is not common. It is evident, however, when one observes planes containing large numbers of inclusions, all of which are empty. In general, quartz, fluorite, calcite, and sphalerite are free from leakage problems; barite and gypsum are more prone to such problems.

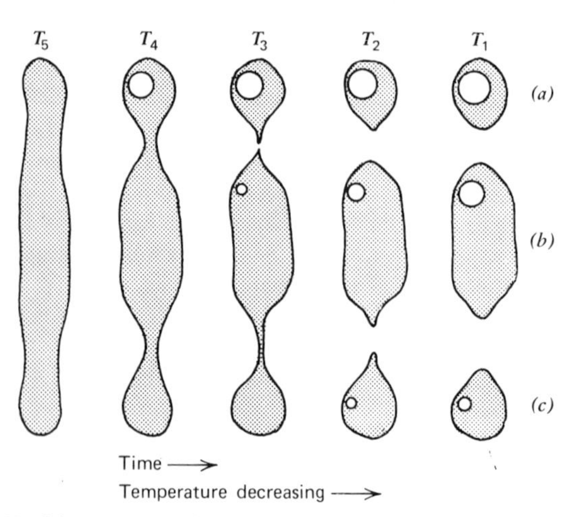

FIGURE 8.19 Necking down of a long tubular inclusion. The original inclusion, trapped at temperature T_5, breaks up during slow cooling to form three separate inclusions, *a*, *b*, and *c*. Upon reheating in the laboratory, inclusion *a* would homogenize above the true trapping temperature T_5; inclusion *b* would homogenize above T_3, inclusion *c* would homogenize between T_2 and T_3 (Reproduced from E. Roedder, in *Geochemistry of Hydrothermal Ore Deposits,* 2nd ed., 1979, p. 903, with permission of the publisher.)

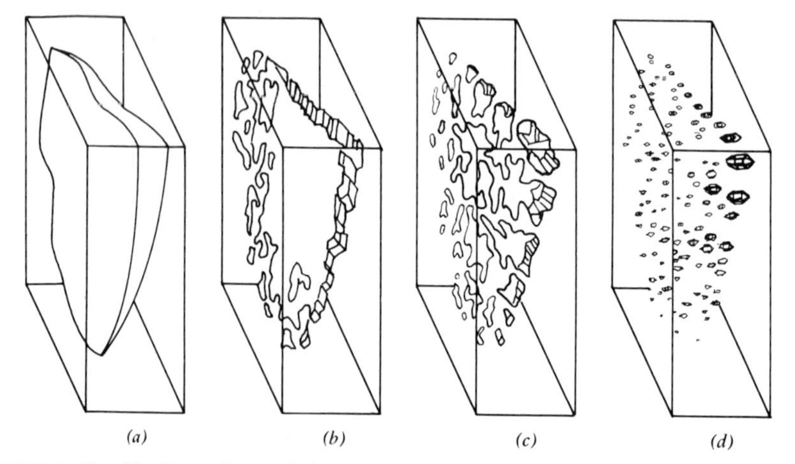

FIGURE 8.20 Healing of a crack in a quartz crystal, resulting in secondary inclusions. Solution of some of the curved surfaces having nonrational indices and redeposition as dendrite crystal growth on others eventually result in the formation of sharpy faceted negative crystal inclusions. If this process occurs with falling temperature, the individual inclusions will have a variety of gas-liquid ratios. (Reproduced from E. Roedder in *Geochemistry of Hydrothermal Ore Deposits*, 2nd ed., 1979, p. 702, with permission of the publisher.)

8.5.3 The Preparation of Samples and the Observation of Fluid Inclusions

Fluid inclusions commonly go unnoticed both because the observer is not looking for them and because conventional polished sections and thin sections are poorly suited for their observation. Fluid inclusions are best seen and studied in small single crystals, cleavage fragments, or cut mineral plates that are thick enough to contain the undamaged inclusion, but thin enough to readily transmit light and that are doubly polished to minimize the interferences of surface imperfections and excessive diffuse light scattering. The ideal sample thickness varies from one specimen to another depending on the transparency, the grain size, and the size of the inclusions; for most samples 1.0–1.5 mm is quite satisfactory. Some crystals or granular aggregates can be cut directly but many need support and are best cut after having been cast in a polyester resin. The polyester is readily dissolved in chloroform; thus the sample plate can be removed after cutting and polishing has been completed. The polyester block with enclosed sample is cut into one or more 1.0–1.5 mm thick plates. High-speed diamond saws may induce considerable fracturing in specimens and should not be used to cut samples for fluid inclusion studies; slow-speed thin-blade diamond saws give clean cuts with minimal sample damage. Polishing of the plates is often facilitated by bonding them to a supporting aluminum, brass, or glass disk with a low melting point resin (such as Lakeside Type 30C). After one side is polished, the plate is released from the supporting metal disk, turned over and readhered so that the other side may be similarly polished. When polishing is completed, the plate is released from the support disk, the polyester removed, and the plate is ready for examination.

Fluid inclusions range in size from rare megascopically visible examples that are greater than 1 cm in length to submicroscopic examples; most work has, however, been carried out on those which range upward from 10 μm. Inclusions of 10 μm or larger are readily observed within small crystals, cleavage fragments, or polished plates by examination using a standard microscope. Such specimens may be placed on a glass microscope slide and viewed with transmitted light. Care must be taken, however, not to allow high magnification lenses, which have short free-working distances, to strike the specimen. Since the depth of field is very limited on such lenses, it is easy for the observer to strike the lens on the sample when adjusting focus to follow a plane of inclusions into the sample. Lenses with long free-working distances, such as those designed for use with the Universal Stage, are very useful and are necessary when using heating and cooling stages. Because bubbles seem frequently to be located in the least visible corner of an inclusion, auxiliary oblique lighting systems are useful. In addition, a colored filter or monochromatic light source may prove useful in seeing the inclusions and daughter phases.

It is obviously important to study inclusions in samples that have been well documented in terms of their mineralogy, location, and paragenesis; measurements of freezing point depression and inclusion homogenization on heating should be made on the same inclusions.

8.5.4 The Compositions of Fluid Inclusions

Fluid inclusions are extremely important in the study of ore deposits because they often represent unaltered, or at least minimally altered, samples of the ore-forming fluid. Most workers do not have facilities to determine the actual chemical composition of the inclusions they observe, but can determine the salinity of the trapped solution by measuring the freezing temperature as described in Section 8.5.5.

The most comprehensive listing of compositional data for fluid inclusions is that by Roedder (1972). By far the most abundant type of inclusion is that which contains a low-viscosity liquid and a smaller volume gas or vapor bubble. The liquid is generally aqueous, has a pH within one unit of neutral and contains a total salt concentration between 0 and 40 wt. %. The salts consist of major amounts of Na^+, K^+, Ca^{2+}, Mg^{2+}, Cl^-, and SO_4^{2-} with minor amounts of Li^+, Al^{3+}, BO_3^{3-}, H_4SiO_4, HCO_3^-, and CO_3^{2-}. Na^+ and Cl^- are usually dominant; carbon dioxide, both liquid and gas forms and liquid hydrocarbons, are fairly common. Liquid hydrogen sulfide has also been observed, but it is rare. Fluid carbon dioxide will never be observed above 31°C, its critical point; hence the fluid inclusion observer must be careful of sample heating by the light source and even of working in a hot room. Daughter minerals, usually cubes of halite (NaCl) or sylvite (KCl), form when nearly saturated fluids cool from the initial temperature of entrapment. The presence of such crystals obviously indicates that the fluid is salt-saturated. Other crystals that are observed in fluid inclusions but that are not simple precipitates of a supersaturated solution include sulfides, quartz, anhydrite, calcite, hematite, and gypsum. Such crystals probably either formed before the inclusion was finally sealed, as a result of secondarily introduced fluids, or even through oxidation resulting from hydrogen diffusion. For example, $2Fe^{2+}$ (soln.) + $3H_2O$ = Fe_2O_3 (hematite) + $4H^+$ + H_2 (lost through diffusion).

The total NaCl-equivalent salinity of fluid inclusions can be determined by the freezing-point depression method. In practice, this is achieved by freezing the sample, then observing it through the microscope as it is warmed and measuring the temperature at which the last ice melts. This temperature is then used to read off the solution composition from curve *A-B* on a diagram such as Figure 8.21 or calculated from equations such as those prepared by Potter et al. (1978). Curve *A-B* represents the freezing-point depression of water as a function of salt content.

Fluid inclusions are cooled or heated by means of stages that mount on a conventional microscope. Commercial designs such as the Leitz 350, the R. Chaix M.E.C.A. or the Linkham are in common use, as are a wide variety of "home-made" models. In all such fluid inclusion stages, the samples are held and heated

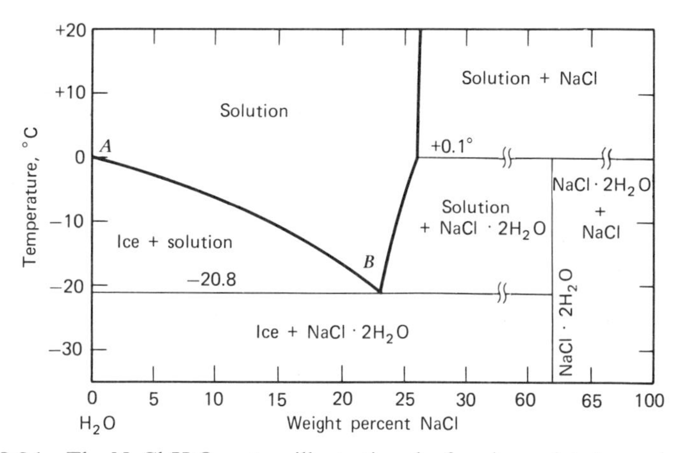

FIGURE 8.21 The NaCl-H_2O system illustrating the freezing-point depression of water as a function of the NaCl content (curve *A-B*).

or cooled within a chamber equipped with a viewing window. Roedder (1976) has pointed out that "the operation of any heating stage must be done with care and constant consideration of the possible sources of error, as it is surprisingly easy to get beautifully consistent, reproducible, but incorrect numbers." Accordingly, prior to use, the stage should be carefully tested for thermal gradients and calibrated with standards. The problems are especially acute in freezing-point determinations because an error of 1°C is equivalent to an error of about 1 wt. % NaCl equivalent.

8.5.5 Fluid Inclusion Geothermometry

Fluid inclusion geothermometry, now recognized as one of the most accurate and widely applicable techniques for determining the temperatures at which a crystal formed or recrystallized, consists of determining the temperature at which a heterogeneous fluid inclusion homogenizes. In practice, a sample is heated while being viewed on a microscope stage until the liquid and a coexisting bubble that occupy the inclusion at room temperature homogenize and fill the inclusion as a single fluid. Filling is usually accomplished by disappearance of the bubble, but may also occur by conversion of the liquid phase to vapor. The actual filling temperature is in practice often reproducible to ~ 1°C, but it represents a minimum value for the temperature of formation because, in general, an appropriate pressure correction is necessary. In low-temperature deposits formed from dense, high-salinity fluids at shallow depths (e.g., many Pb-Zn ores in carbonate rocks), corrections are usually $<$ 25°C, but in high-temperature ores formed from low salinity fluids at depths $>$ 10 km, corrections may exceed 300°C. The correction procedure is illustrated in the temperature-density diagram for H_2O shown in Figure 8.22. The heavy curve extending from *G* (gas)

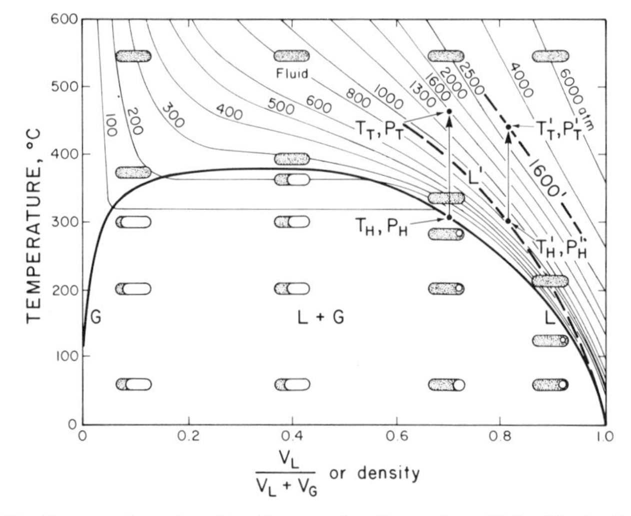

FIGURE 8.22 Temperature-density diagram for the system H_2O, illustrating the method of applying pressure corrections in fluid inclusion studies. The heavy solid line is the boundary of the two-phase field for pure H_2O and represents the homogenization temperature for inclusions filled with water. The light solid curves are isobars for pure H_2O. The heavy dashed line *L'* illustrates the position of the two-phase field for H_2O with 20 wt. % NaCl and the dashed line labeled 1600 represents the 1600 atm isobar for a 20 wt. % NaCl solution. The pressure correction procedure is described in the text.

to *L* (liquid) represents the boundary of the two-phase field for pure H_2O and thus defines the filling temperatures for H_2O inclusions of various densities. Thus an inclusion with a 70% fluid filling (density = 0.7 if pure H_2O) would homogenize at 300°C (T_H, P_H). If, however, independent geologic information indicated that the actual pressure at the time of trapping (P_T) was 1600 bars, then the true trapping temperature (T_T) was actually 470°C (i.e., the temperature at which the 0.7 density coordinate cuts the 1600 isobar). If the inclusion were found to contain a 20% NaCl solution (density = 1.15 at 20°C) by a freezing point determination, the pressure correction would have to be made from another family of curves (examples of which are shown in the heavy dashed curves on Figure 8.21) representing the two-phase boundary and isobars for a 20% NaCl solution. The two-phase boundary for this liquid is shown by the heavy dashed curve (*L'*). The isobars for salt solutions are not well-known, but the 1600 isobar for a 20% NaCl solution is shown schematically as the heavy dashed 1600' curve. Accordingly, the corrected trapping temperature (T'_T, P'_T) of an inclusion that homogenized at 300°C (T'_H, P'_H) and that contains a 20% NaCl solution is 450°C.

The data employed in making pressure corrections were originally derived by

Lemmlein and Klevtsov (1961) and recently summarized by Potter (1977). Since halite solubility in water is temperature dependent, determination of the temperature at which any daughter halite crystals dissolve establishes a minimum for the initial temperature of trapping.

8.5.6 Applications of Fluid Inclusion Studies

Fluid inclusion geothermometry has been extensively employed in determining the temperatures of ore mineral formation. However, Roedder (1977, 1979) has pointed out that there are several other uses for fluid inclusion studies, including mineral exploration and even the determination of geologic age relations. For additional information on the examples noted in the following, the reader is directed to Roedder's papers and the references at the end of this chapter.

The two most obvious applications of fluid inclusion studies are the determination of the temperature of ore formation or recrystallization and the determination of the salinity of the fluid entrapped. The study of fluid inclusions in epigenetic ores has often revealed that the temperatures at which the ores were emplaced were different from the temperatures recorded in fluid inclusions in the enclosing host rocks. Furthermore, temperature variations both temporal and spatial in origin, have been observed within single ore deposits (e.g., Creede, Colorado; Panasqueira, Portugal; Casapalca, Peru) and even within different growth zones of single crystals. In such cases the temperature differences observed may be employed either to locate "blind" ore bodies or to extend known ones. Variations within a mineralized zone may also serve to define directions of ore fluid movement, to aid in the interpretation of parageneses, and as records of the changing nature of the ore fluid as a function of time. Since ore-forming brines are often more concentrated than fluids not associated with ores, trends in salinity obtained from freezing point measurements may supplement temperature data in the exploration or extension of ore deposits.

In some districts, veins formed only during one episode, or of only one affinity, carry mineralization whereas other veins of identical gangue mineralogy are barren. In some instances (e.g., Sadonsk, Soviet Union; Cobalt, Ontario) the fluid inclusion data are significantly different for the two types of veins and thus can be used to aid in exploration.

In structurally complex areas cut by several generations of veins of similar mineralogy, fluid inclusion data may help identify segments of individual veins and aid in clarifying the chronological relationships between veins. Similarly the chronology of crosscutting veins of similar mineralogy may be clarified because natural decrepitation (i.e., bursting due to heating) or halo effects may be observed in the fluid inclusions of the older vein.

During weathering and erosion of ore deposits, resistant gangue minerals such as quartz are preserved with fluid inclusions intact in the gossan, the residual soils, and in stream sediments derived during erosion. The fluid inclusions in such quartz may be used as an aid in deciphering the nature of the original deposit and even its location.

Clearly, the many potential uses of fluid inclusion studies will result in their increasing application to the study of ore deposits.

BIBLIOGRAPHY

Barton, P. B. and Skinner, B. J. (1979) Sulfide mineral stabilities. In H. L. Barnes, ed., *Geochemistry of Hydrothermal Ore Deposits*, 2nd ed. Wiley-Interscience, New York.

Brumby, G. R. and Shepherd, T. J. (1978) Improved sample preparation for fluid inclusion studies. *Mineral. Mag.* **42**, 297–298.

Craig, J. R. (1973) Pyrite-pentlandite assemblages and other low temperature relations in the Fe-Ni-S system. *Am. J. Sci.* **273A**, 496–510.

Cunningham, C. F. (1976) Fluid Inclusion Geothermometry. *Geol. Rundsch.* **66**, 1–9.

Ebers, M. L. and Kopp, O. C. (1979) Cathodoluminescence microstratigraphy in gangue dolomite, the Mascot-Jefferson City District, Tennessee, *Econ. Geol.* **74**, 908–918.

Francis, C. A., Fleet, M. E., Misra, K. C., and Craig, J. R. (1976) Orientation of exsolved pentlandite in natural and synthetic nickeliferous pyrrhotite. *Am. Mineral.* **61**, 913–920.

Hall, M. R. and Ribbe, P. H. (1971) An electron microprobe study of luminescence centers in cassiterite. *Am. Mineral.* **56, 31–45**.

Hutchison, M. N. and Scott, S. D. (1979) Application of the sphalerite geobarometer to Swedish Caledonide and U.S. Appalachian metamorphosed massive sulfide ores (abs.). *Symposium Volume on Caledonian-Appalachian Stratabound Sulphides* Trondheim, Norway, pp. 14–15.

Hutchison, M. N. and Scott, S. D. (1980) Sphalerite geobarometry in the Cu-Fe-Zn-S system. *Econ. Geol.* (in press).

Kellanch, Y. A. and Badham, J. P. N. (1978) Mineralization and paragenesis at the Mount Wellington Mine, Cornwall. *Econ. Geol.* **73**, 486–495.

Kelly, W. C. and Turneaure, F. S. (1970) Mineralogy, paragenesis, and geothermometry of the tin and tungsten deposits of the Eastern Andes, Bolivia. *Econ. Geol.* **65**, 609–680.

Kretschmar, U. and Scott, S. D. (1978) Phase relations involving arsenopyrite in the system Fe-As-S and their application. *Can. Mineral.* **14**, 364–386.

Lemmlein, G. G. and Klevtsov, P. V. (1961) Relations among the principal thermodynamic parameters in a part of the system H_2O-NaC1. *Geokhimiya* No. 2, 133–142 (in Russia); trans. in *Geochemistry* **6**, 148–158.

Lusk, J. and Ford, C. E. (1978) Experimental extension of the sphalerite geobarometer to 10 bar. *Am. Mineral.* **63**, 516–519.

McLimans, R. K., Barnes, H. L., and Ohmoto, H. (1980) Sphalerite stratigraphy of the Upper Mississippi Valley zinc-lead district, Southwest Wisconsin. *Econ. Geol.* **75**, 351–361.

Meyers, W. J. (1978) Carbonate cements: their regional distribution and interpretation in Mississippian limestones of southwestern New Mexico. *Sedimentology* **25**, 371–400.

Naldrett, A. J., Craig, J. R., and Kullerud, G. (1967) The central portion of the Fe-Ni-S system and its bearing on pentlandite exsolution in iron-nickel sulfide ores. *Econ. Geol.* **62**, 826–847.

Nickel, E. (1978) The present status of cathode luminescence as a tool in sedimentology. *Min. Sci. Eng.* **10**, 73–100.

Potter, R. W. (1977) Pressure corrections for fluid-inclusions homogenization temperatures based on the volumetric properties of the system NaC1-H_2O. Journ. Res. U. S. Geol. Surv. **5**, 603–607.

Potter, R. W., Clynne, M. A., and Brown, D. L. (1978) Freezing point depression of aqueous sodium chloride solutions. *Econ. Geol.* **73**, 284–285.

Ribbe, P. H., ed. (1974) *Sulfide Mineralogy.* Mineral. Soc. Short Course Notes, Vol. 1.

Roedder, E. (1962) Ancient Fluids in Crystals, *Sci. Am.* Oct. 1962, 38–47.

Roedder, E. (1968) The noncolloidal origin of "colloform" textures in sphalerite ores. *Econ. Geol.* **63**, 451–471.

Roedder, E. (1972) Composition of fluid inclusions. In Fleischer, M., ed. *Data of Geochemistry.* 6th ed. *U.S. Geological Survey Prof. Paper 440-JJ.*

Roedder, E. (1976) Fluid-inclusion evidence on the genesis of ores in sedimentary and volcanic rocks. In Wolf, K. H., ed., *Handbook of Stratabound and Stratiform Ore Deposits.* Elsevier, Amsterdam, Vol. 2, pp. 67–110.

Roedder, E. (1977) Fluid inclusions as tools in mineral exploration, *Econ. Geol.* **72**, 503–525.

Roedder, E. (1979) Fluid inclusions as samples of ore fluids. In Barnes, H. L., ed., *Geochemistry of Hydrothermal Ore Deposits*, 2nd ed. Wiley-Interscience, New York, pp. 684–737.

Rumble, D., ed., (1976) *Oxide Minerals.* Min. Soc. Am. Short Course Notes, Vol. 3.

Scott, S. D. and Barnes, H. L. (1971) Sphalerite geobarometry and geothermometry. *Econ. Geol.* **66**, 653–669.

Smith, J. V. and Stenstrom, R. C. (1964) Electron-excited luminescence as a petrographic tool. *J. Geol.* **73**, 627–635.

Vaughan, D. J. and Craig, J. R. (1978) *Mineral Chemistry of Metal Sulfides.* Cambridge University Press, Cambridge, England.

Vaughan, D. J. and Ixer, R. A. (1980) Studies of the sulfide mineralogy of North Pennine ores and its contribution to genetic models. *Trans. Inst. Min. Metall.* **89**, B99–B109.

Yund, R. A. and Kullerud, G. (1966) Thermal stability of assemblages in the Cu-Fe-S system. *J. Petrol.* **7**, 454–488.

9

Ore Mineral Assemblages Occurring in Igneous Rocks and Vein Deposits

9.1 INTRODUCTION

Ore minerals are not uniformly distributed in the earth's crust, but generally occur in associations that are characteristic in their mineralogy, textures, and relationships to specific rock types. The existence of these characteristic associations, each containing its own typical suite of ore minerals, considerably simplifies the task of the ore microscopist because it permits him or her to anticipate the minerals likely to be encountered once the general association has been recognized. The grouping of ores into characteristic associations is, of course, a useful empirical rule of thumb but is not intended as a rigid scientific classification. Such an empirical division also has the advantage that no genetic models are implied, even though a similar mode of origin is likely. Indeed, these associations largely result from the formation of the ores under characteristically limited physico-chemical conditions, the nature of which may often be inferred from detailed study of the ores. Chapters 9 and 10 briefly discuss the most commonly encountered ore mineral assemblages and their textures, with the ores categorized according to their most widely recognized types. The various sections are not intended as exhaustive discussions of ore petrology but do include the currently accepted theories on ore genesis because the authors believe that an understanding of ore mineralogy and textures is enhanced by some knowledge of the ore forming process. The sections included should prepare the student for most of the ore mineral assemblages encountered in an introductory course and in most ore samples he might examine; however, it is important to be ever watchful for the unexpected and unusual minerals. In the identification of minerals other than those included in Appendix I, we recommend reference to encyclopedic works such as Ramdohr's *The Ore Minerals and Their Intergrowths,* Uytenbogaardt and Burke's *Tables for the Microscopic Identification of Ore Minerals,* and Picot and Johan's *Atlas des Mineraux Metalliques.*

9.2 CHROMIUM ORES ASSOCIATED WITH MAFIC AND ULTRAMAFIC ROCKS

Mineralogy

Major: chromite (ideally $FeCr_2O_4$ although always containing significant MgO, Al_2O_3, Fe_2O_3).

Minor: sulfides of nickel, copper and iron (pentlandite, pyrrhotite, chalcopyrite, gersdorffite, bornite, valleriite).

Trace: platinum group minerals (ferroplatinum, cooperite, laurite, stibiopalladinite, sperrylite, nickeliferous braggite) and rutile.

Mode of Occurrence

There are two distinct modes of occurrence:

1 In layered basic intrusions as magmatic sediment layers.

2 In peridotite or serpentinized peridotite masses associated with orogenic belts (sometimes termed "podiform" or "Alpine-type" chromites).

Examples

The classic example of a chromite deposit associated with a layered basic intrusion is the Bushveld Complex, South Africa; other examples include the Stillwater Complex, Montana (United States), and the Great Dyke (Rhodesia).

"Podiform" chromites occur in many orogenic belts and are generally much smaller deposits; important examples include deposits in Turkey, the Ural Mountains (U.S.S.R.), the Phillipines, and Cuba.

Mineralogy and Textures

The few, isolated (though economically important) layered intrusions that can be regarded as chromium deposits occur in tectonically stable environments. The layers of the intrusive complexes can be regarded as magmatic "strata," which may be of considerable lateral extent. Within these, the chromite bands may range from a few millimeters to over 20 m in thickness and show many features analogous to those shown in sedimentary rocks (lensing or wedging out, intraformational contortion, scour and fill structures). Pure chromite rocks (chromitites) may grade through various amounts of chromite + silicate (olivine, pyroxene) to normal dunites, peridotites, and so on. While commonly occurring in the olivine-rich layers, the chromite may occur in significant amounts in any association that is *basic* in terms of overall composition. Although chromite itself is virtually the only ore mineral, it may show considerable differences in composition within deposits and between deposits of the "layered" and "podiform" types:

1. MgO/FeO ratios tend to be greater in podiform chromites (1 to 2.3) than in layered chromites (0.6 to 1).
2. Fe_2O_3 contents tend to be lower ($<$ 8 wt. %) and Cr/Fe ratios higher ($\sim$1.5 to 4.5) in podiform chromites than in layered chromites ($\sim$10 to 24 wt. % Fe_2O_3 and 0.75 to 1.75 Cr/Fe ratios, respectively).
3. Al_2O_3 and Cr_2O_3 have reciprocal relations (Cr_2O_3 being $\sim$6.5 to 16 wt. %, Al_2O_3 $\sim$6 to 52 wt. %) in podiform chromites and vary widely in layered chromites. The overall Al_2O_3/Cr_2O_3, ratio tends to be higher in podiform chromites.

In the layered intrusives, chromite commonly occurs as well developed octahedral crystals (Figure 9.1), particularly when associated with larger amounts of interstitial material. In the cases where there is less interstitial material, the crystals develop polygonal interference boundaries (see Section 7.2).

The "podiform" or "Alpine-type" chromite ores occur in highly unstable tectonic environments, so that in addition to the compositional differences noted previously there are marked textural differences. Although polygonal interfer-

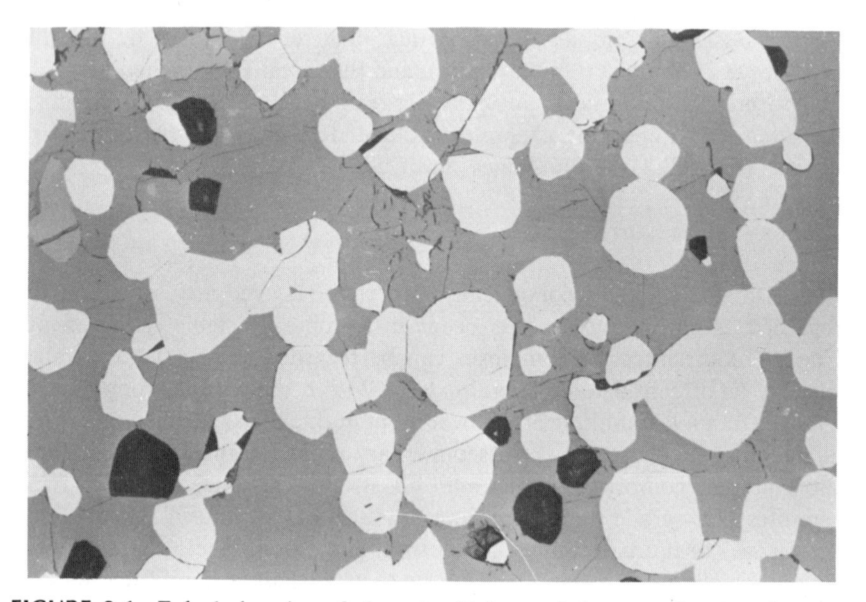

FIGURE 9.1 Euhedral grains of chromite (light gray) in a matrix of mafic silicate, Bushveld Complex, South Africa. (Width of field = 2000 μm.)

ence boundaries develop when there is very little interstitial material, the chromite grains are nearly always rounded when surrounded by silicate (Figure 9.2). These chromite grains can range from fine disseminations (< 1.0 mm diameter) to the coarse textures (~ 1.5 cm diameter) of "leopard" or "grape" ore. Sometimes concentric shells of chromite and serpentine produce *orbicular* ores, and in the chromites of both layered and podiform types, concentric compositional zoning may be developed with outer zones exhibiting relative enrichment in iron (often observable under the microscope as a lighter peripheral zone). Such textures are often the result of hydrothermal alteration during serpentinization. Textures caused by deformation are also characteristic.

Both of the chromite associations described above may contain nickel concentrations of minor importance and concentrations of the platinum group metals that may be of considerable economic significance. Nickel in the layered intrusions occurs as sulfides and arsenides (assemblages of pyrrhotite-pentlandite-chalcopyrite with very minor gersdorffite, bornite, valleriite) in mafic horizons. These sulfide assemblages are similar to those discussed in Section 9.3 and are the result of a complex series of exsolution and inversion reactions. In the podiform chromites, most of the nickel occurs in solid solution in olivine and may be concentrated during weathering processes, although small amounts may occur as disseminated sulfides (pentlandite, heazlewoodite). The classic example of the concentration of platinum group metals in layered intrusives is the *Merensky*

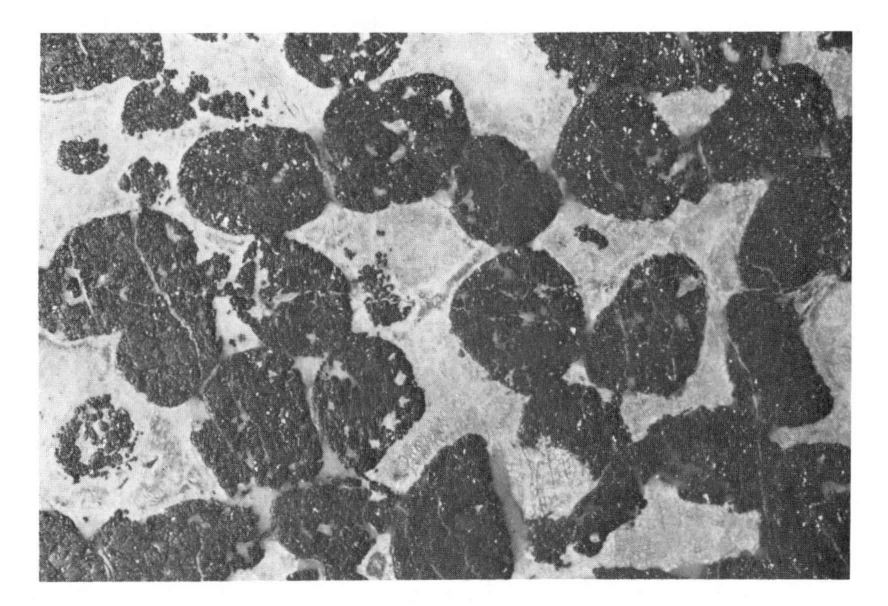

FIGURE 9.2 Rounded chromite aggregates in mafic silicate matrix, Greece. Note that in the megascopic view the chromite appears black, whereas in the microscopic view in Figure 9.1 the chromite appeared white. (Width of field = 3.5 cm.)

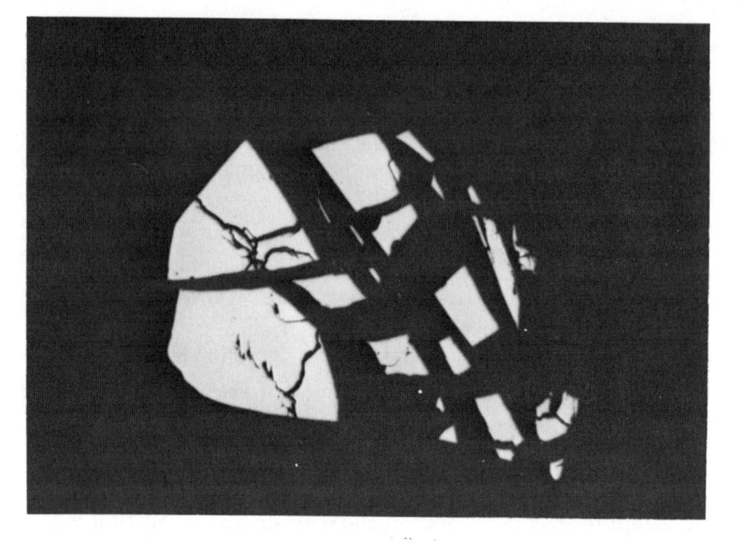

FIGURE 9.3 Fractured grain of sperrylite in a matrix of altered mafic silicates, Potgieterust, South Africa. (Width of field = 2000 μm.)

"*Reef*" which is a persistent (~300 km) but thin (<1 m) layer of the Bushveld Complex. Between the top and bottom bounding chromite-rich bands of the so-called reef the precious metal concentrations (as ferroplatinum, cooperite, laurite, stibiopalladinite, sperrylite, nickeliferous braggite, and native gold, Figure 9.3) reach a maximum and are associated with base metal sulfides of iron, nickel, and copper (pyrrhotite, pentlandite, chalcopyrite, valleriite).

Origin of the Ores

It is universally accepted that the chromite ores of layered intrusives are magmatic in origin and related to processes of fractional crystallization and gravitative settling of layers of crystals on the floor of the intrusive sheets. The textures of the ores are wholly in accordance with such an origin. The major problems in the origin of these orebodies are concerned with the mechanisms for producing essentially monomineralic chromite layers. Suggestions have included concentration by current sorting or preferential precipitation in response to changes in pressure, water content, oxygen fugacity or through multiple injections of magma. A mechanism recently proposed (Irvine, 1974, 1977) is of precipitation resulting from sudden extensive contamination of the parental basic magma with more acid liquid that has differentiated to a relatively siliceous composition. Here, the addition of the silica-rich material forces the composition of the crystallizing melt (Figure 9.4) from the olivine + chromite cotectic curve (along which disseminated chromite admixed with olivine is forming) into the field of primary chromite crystallization (in which only chromite forms).

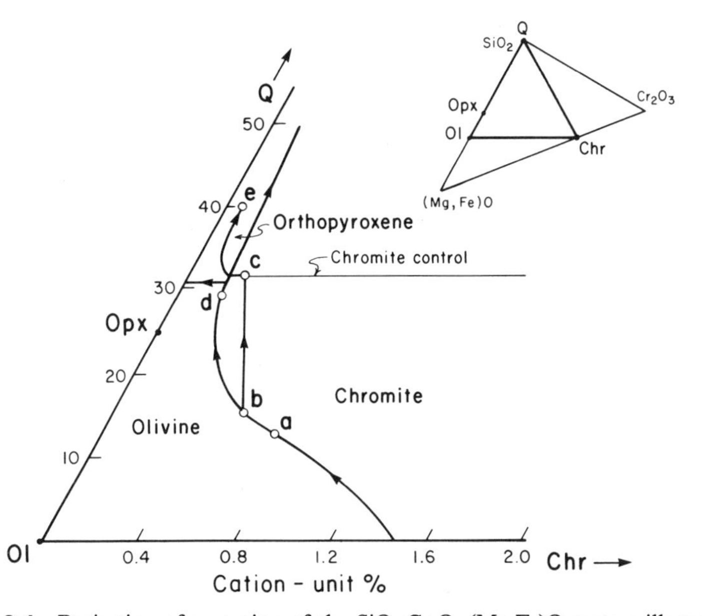

FIGURE 9.4 Projection of a portion of the SiO_2-Cr_2O_3-(Mg,Fe)O system illustrating a possible mode of origin of chromite layers in stratiform intrusions. Coprecipitation of chromite and olivine from a magma beginning at point *a* would normally occur as the melt cooled along the cotectic *a b d* to *e* and would result in the formation of small and decreasing amounts of chromite. In contrast, if after the melt had progressed from *a* to *b*, the melt were blended with another more silica-rich melt, the bulk composition of the liquid would shift into the primary crystallization field of chromite (point c). This would cause precipitation of only chromite, which would settle as a monomineralic layer, until the crystallizing liquid compsition returned to the chromite-olivine or chromite-orthopyroxene cotectic. (After T. N. Irvine, 1977; used with permission.)

The origin of the podiform chromites is clearly very different, and it is closely related to the problem of the origin of the ultramafic rocks of Alpine type, which form part of so-called ophiolite complexes. Current theories relate the creation of ophiolites to processes along spreading boundaries between lithospheric plates. It has been suggested (see, e.g., Dickey, 1974) that the podiform chromites form first as magmatic cumulates (much like the chromites of layered intrusives) in magma pockets along these plate boundaries. Subsequent segregation with episodic mechanical disruption both during crystallization and in lateral transport away from the spreading zone results in "snowball" aggregation, rounding, and deformation of individual chromite blebs.

Selected References

Dickey, J. S., Jr. (1975) A hypothesis of origin of podiform chromite deposits. *Geochim. Cosmochim. Acta* **39**, 1061–1074.

Irvine, T. N. (1974) Crystallization sequences in the Muskox Intrusion and other layered intrusions II. Origin of chromitite layers and similar deposits of other magmatic ores. *Geochim. Cosmochim. Acta* **39**, 991–1020.

Irvine, T. N. (1977) Origin of chromitite layers in the Muskox and other stratiform intrusions: A new interpretation. *Geology* **5**, 273–277.

Wilson, H. D. B., ed. (1969) *Magmatic Ore Deposits.* Econ. Geol. Monograph No. 4 (a collection of articles including several on chromite deposits).

9.3 IRON-NICKEL-COPPER SULFIDE ORES ASSOCIATED WITH MAFIC AND ULTRAMAFIC IGNEOUS ROCKS

Mineralogy

Major: pyrrhotite (both monoclinic and hexagonal varieties), pentlandite, pyrite, magnetite, chalcopyrite.

Minor: cubanite, mackinawite, platinum-group metal minerals, argentian pentlandite.

Secondary: millerite, violarite.

Mode of Occurrence

Massive to disseminated in, or immediately associated with, mafic to ultramafic intrusive or extrusive rocks (gabbro, basalt, peridotite, norite) or metamorphosed mafic to ultramafic rocks.

Examples

Sudbury, Ontario; Thompson and Lynn Lake, Manitoba; Pechenga, Monchegorsk and Noril'sk, Soviet Union; Kambalda, W. Australia.

Mineral Associations and Textures

The iron-nickel-copper ores of this association occur as massive to disseminated sulfides (or sulfides plus oxides) in close association with mafic or ultramafic igneous rocks, notably norites and basalts. In most deposits of this type, pyrrhotite (commonly in large part the magnetic monoclinic variety) constitutes the principal ore mineral making up 80% or more of the ore. Chalcopyrite, the primary source of copper in these ores, is present as irregularly dispersed anhedral polycrystalline aggregates and veinlets. Pentlandite, the primary nickel- (and sometimes cobalt-) bearing phase, is usually not visible in rough hand specimens. It is, however, readily observed megascopically and microscopically in polished surfaces by its lighter color and higher reflectance relative to pyrrhotite. Under the microscope, pentlandite is commonly observed in two distinct textures: (1) as granular polycrystalline veinlets (see Figure 9.5 and 8.8*a*); (2) as oriented lamellae and "flames" (see Figure 8.8*b*). The granular veinlets occur as irregular chainlike structures interstitial to pyrrhotite, chalcopyrite, and magnetite. The pentlandite, in addition to being slightly lighter in color than the pyrrhotite is usually intensely fractured, a feature attributed to its having undergone a much greater volume reduction on cooling than the host pyrrhotite. The "flame" structure of pentlandite is one of the most diagnostic textures among ore minerals and results from crystallographically oriented exsolution of pentlandite from an originally formed nickel-bearing pyrrhotite.

Magnetite is present in these ores in rather variable quantities as euhedral, rounded, or even skeletal crystals dispersed within the pyrrhotite (see Figure 9.5). The magnetite may be titaniferous and contain oriented exsolution lamellae of ilmenite or ulvöspinel. Pyrite, although nearly always present, ranges from rare in the ores of the Sudbury district to a major constituent at Kambalda. Cubanite is generally present only in very minor proportions and nearly always as sharply defined laths within the chalcopyrite. The cubanite may be overlooked

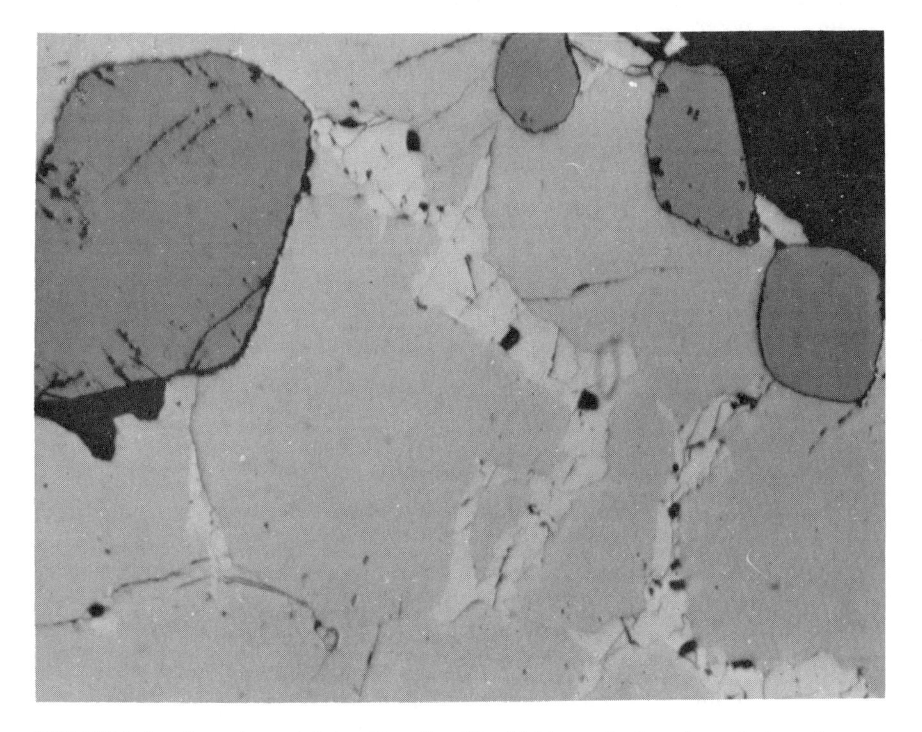

FIGURE 9.5 Granular veinlets of pentlandite (light gray) along boundaries of coarse pyrrhotite grains with associated subhedral grains of magnetite (dark gray). Sudbury, Ontario. (Width of field = 360 μm.)

in a cursory examination because of its similarity to chalcopyrite, but it is readily noted under crossed nicols due to its strong anisotropism. Mackinawite is common in small amounts as irregular grains and "wormlike" features within the chalcopyrite and is recognized by its strong anisotropism and bireflectance.

In addition to copper and nickel, these ores constitute major sources of the world's platinum group metals in the form of arsenides (sperrylite, $PtAs_2$), sulfarsenides (hollingworthite, RhAsS); bismuth- and antimony-bearing phases (froodite, $PdBi_2$; insizwaite, $PtBi_2$; sudburyite, PdSb), and tellurides (moncheite, $PtTe_2$; michenerite, PdBiTe), which may be present in trace quantities. These minerals are all characterized by high reflectances and occur as small grains, which are rarely encountered in polished sections. Silver also occurs in some ores as solid solution in the pentlandite structure mineral argentian pentlandite [$(Fe,Ni)_8AgS_8$] and significant cobalt and gold are recovered from the ores.

Origin of the Ores and Textures

These iron-nickel-copper sulfide ores are generally considered to have formed as a result of the separation of an immiscible sulfide-oxide melt from a sulfur-saturated silicate melt shortly before, during, or shortly after, emplacement at temperatures of 900°C or above. The sulfide-oxide melt may have settled through the partly crystalline silicate magma or, if segregated early, may have been intruded separately; the resulting ores range from massive to disseminated, or even brecciated. When emplacement involves subaerial or submarine basaltic extrusions, the separated sulfides are commonly present as millimeter or smaller-sized rounded droplets dispersed within, or interstitial to, the silicates.

Studies of phase equilibria in the Cu-Fe-Ni-S and Fe-O-S systems reveal that the earliest formed sulfide phase, which may be accompanied by the formation of magnetite, is a nickeliferous and cupriferous pyrrhotite phase commonly referred to as the monosulfide solid solution (mss). The compositional limits of this phase at 600°C and above (Figure 9.6*a*) include the bulk compositions of many of the ores; hence it appears that all or much of the entire sulfide mass formed initially as mss. During the subsequent cooling, as the compositional limits of the phase were considerably reduced (Figures 9.6*b* and 9.7), exsolution of chalcopyrite and then pentlandite or pyrite (depending on the bulk composition) occurred. The mss does not decompose entirely until temperatures of about 200°C are reached and, even then, pentlandite exsolution from the remaining nickeliferous pyrrhotite apparently continues to temperatures as low as 100°C. The mineral paragenesis is summarized in Figure 8.9.

The early formed magnetite crystallized either as euhedral to subhedral or skeletal grains that appear to have changed very little except for internal exsolution of titanium as oriented blades of ilmenite or ulvöspinel. The major primary phase, the mss, has undergone considerable recrystallization and compositional change as copper and nickel have exsolved to form chalcopyrite and pentlandite, respectively. On cooling, the capability of the mss to retain copper

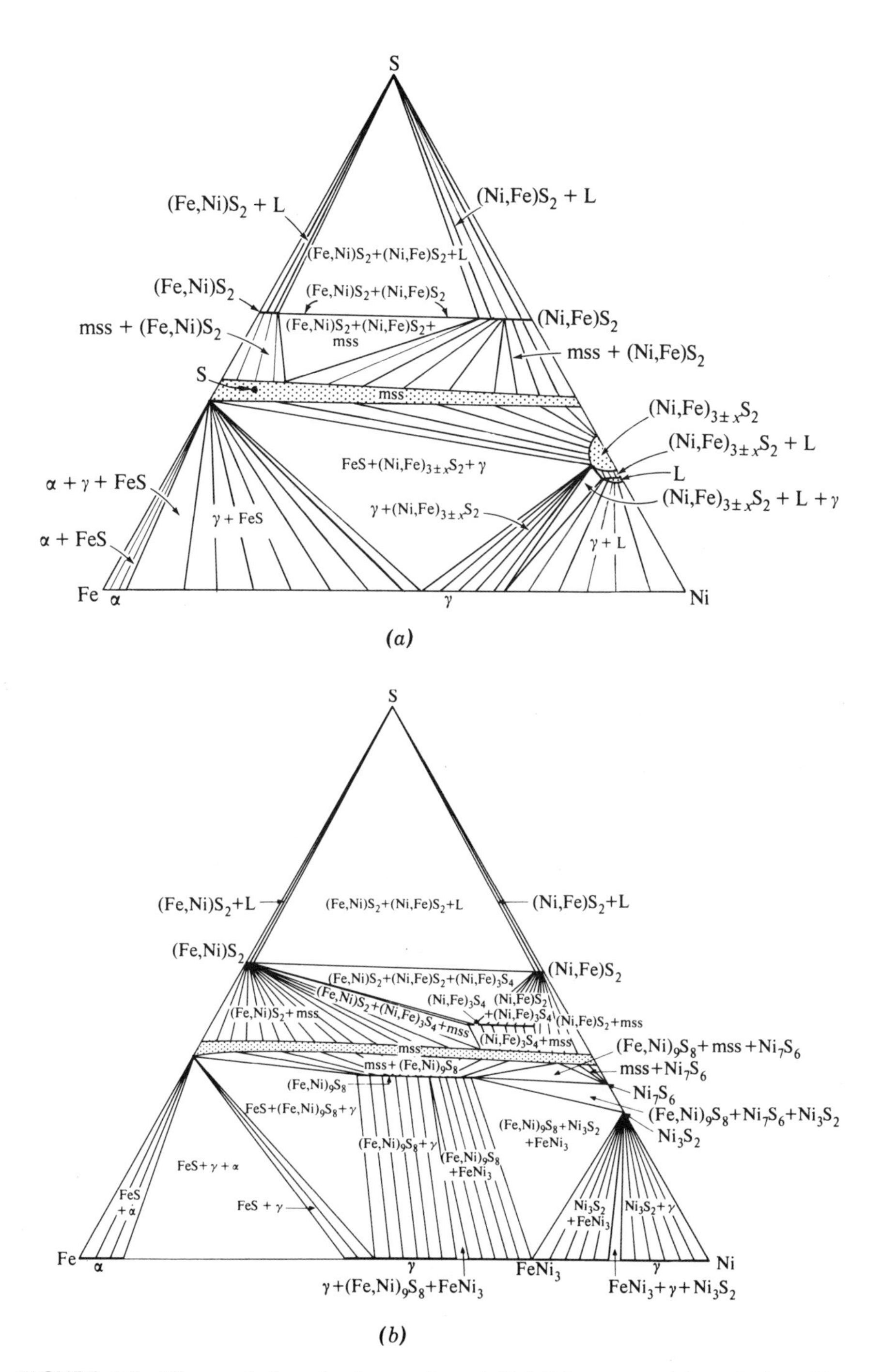

FIGURE 9.6 Phase relations in the condensed Fe-Ni-S system: *(a)* at 650°C; *(b)* at 400°C. (From D. J. Vaughan and J. R. Craig, 1978; used with permission)

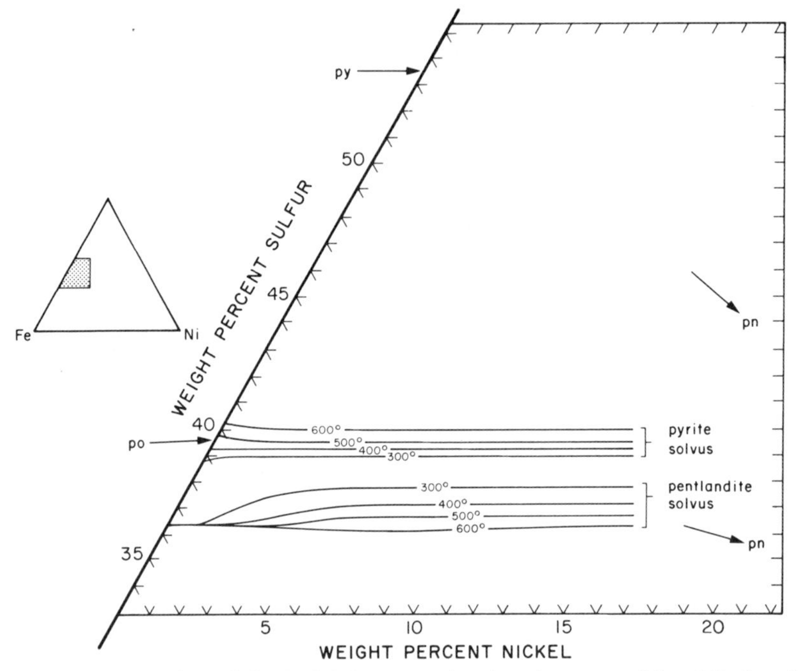

FIGURE 9.7 A portion of the Fe-Ni-S system showing the compositional limits of the monosulfide solid solution (mss) at 600, 500, 400, and 300°C. Pentlandite exsolves along the sulfur-poor boundary and pyrite along the sulfur-rich boundary.

is reduced to below 1% by 500°C; hence most chalcopyrite (or its high temperature analogue) formation occurs above this temperature. It remains as anhedral polycrystalline aggregates interstitial to the mss. Nickel, in contrast, may remain dissolved within the mss in very large amounts to temperatures below 200°C. The exsolution of the nickel in the form of pentlandite (which has a maximum thermal stability of 610°C) is controlled by the bulk composition of the mss (the compositions of most ores are such that pentlandite exsolves before pyrite). When temperature has decreased such that the composition of the ore no longer lies within the limits of the mss, pentlandite begins to exsolve. If this occurs at high enough temperature, diffusion rates are sufficiently rapid to permit segregation of the pentlandite into polycrystalline veinlets situated between the grains of mss (now becoming pyrrhotite). Diffusion of nickel from the mss to form crystallographically oriented exsolved pentlandite lamellae continues as temperature decreases, but at ever-decreasing rates. Below 100–200°C, the diffusion rates are apparently insufficient for the exsolving pentlandite to migrate to grain boundaries of the mss, and the oriented lamellae known as "flames" are held within the mss as it finally expels most of the remaining nickel (natural pyrrhotites in these ores usually contain less than 0.5 wt. % Ni). Whereas the early formed pentlandite probably grew at many points within the crystals, the last formed lamellae seem commonly to have formed along small fractures or other imperfections that provided the most favorable sites for nucleation.

Alteration Effects and Secondary Minerals

Surficial weathering of iron-nickel-copper ores characteristically leads to the formation of iron oxides and hydroxides along grain boundaries and fractures, the development of monoclinic pyrrhotite or even fine-grained porous pyrite along the boundaries of hexagonal pyrrhotite, and the replacement of pentlandite by

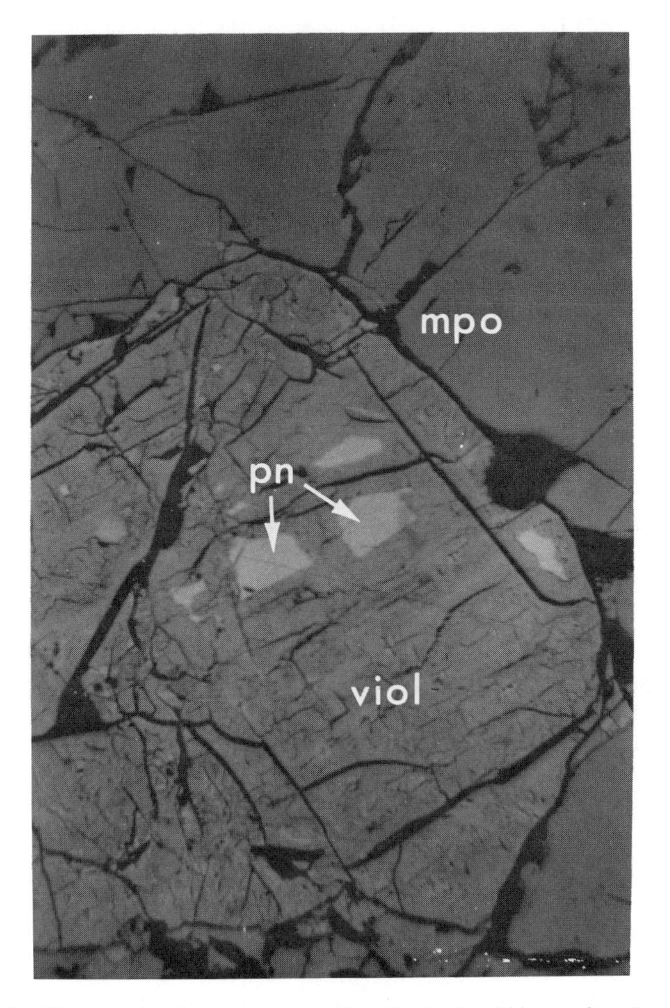

FIGURE 9.8 Violarite (viol) replacing pentlandite (pn) within unaltered monoclinic pyrrhotite (mpo). Lick Fork, Virginia. (Width of field = 40 μm.) (Reproduced from J. R. Craig and J. B. Higgins, *Am. Mineral.* **60**, 36, 1975, with permission.)

violarite, millerite, or both (see Figure 9.8). Secondary violarite is commonly porous and has a slight violet tint relative to the pyrrhotite; the millerite is distinctly yellow, accepts a fine polish, and is strongly anisotropic.

Selected References

(1979) Nickel-sulfide and platinum-group-element deposits. *Can. Mineral.* **17**, No. 2, 141–514.

(1976) An issue devoted to platinum-group elements. *Econ. Geol.* **71**, No. 7, 1129–1468.

Ewers, W. E. and Hudson, D. R. (1972) An interpretive study of a nickel-iron sulfide ore intersection, Lunnon Shoot, Kambalda, Western Australia. *Econ. Geol.* **67**, 1075–1092.

Hawley, J. E. (1962) The Sudbury Ores: Their mineralogy and origin. *Can. Mineral.* **7**, 1–207.

Naldrett, A. J., Craig, J. R. and Kullerud, G. (1967) The central portion of the Fe-Ni-S system and its bearing on pentlandite exsolution in iron-nickel sulfide ores. *Econ. Geol.* **62**, 826–847.

Vaughan, D. J., Schwarz, E. J. and Owens, D. R. (1971) Pyrrhotites from the Strathcona Mine, Sudbury, Canada: a thermomagnetic and mineralogical study. *Econ. Geol.* **66**, 1131–1144.

Wilson, H. D. B., ed. (1969) *Magmatic Ore Deposits: A Symposium*. Econ. Geol. Monograph No. 4.

9.4 IRON-TITANIUM OXIDES ASSOCIATED WITH IGNEOUS ROCKS

Mineralogy

Major: magnetite, ulvöspinel, ilmenite, rutile, hematite, apatite.

Minor: pyrite, chalcopyrite, maghemite, pyrrhotite.

Secondary: hematite, rutile, maghemite.

Mode of Occurrence

Ores as thin layers to thick massive sheets or lenses in anorthositic, gabbroic, or noritic sequences in layered complexes and plutonic intrusives. Also common as accessory minerals in acid to basic igneous rocks.

Examples

Allard Lake, Quebec; Tahawus, New York; Duluth gabbro, Minnesota; Bushveld Complex, S. Africa; Egersund, Norway.

Mineral Associations and Textures

Iron-titanium oxide ores are almost invariably composed of coarse (0.5–1.0 cm) equant grains of titaniferous-magnetite with or without ferrian ilmenite. Grains may be monomineralic, but exsolution lamellae of ulvöspinel oriented along the

(100), or of ilmenite along the (111) (Figure 9.9), planes of the host magnetite are common. Such lamellae range from 0.1 mm across to submicroscopic in a "clothlike" fabric. Hematite may be present as discrete grains but is most common as rims around, or lamellae within, magnetite or ilmenite. Magnetite and ilmenite, though similar at first glance, are readily distinguished by the faint pinkish to violet tint and anisotropism of the latter mineral. Ilmenite typically contains well-developed lamellar twinning (Figure 9.10), a feature absent in the associated minerals. Ulvöspinel is slightly darker gray-brown than magnetite and difficult to distinguish from it unless the two minerals are in contact. Hematite, on the other hand, appears nearly white in comparison with magnetite, ilmenite, or ulvöspinel. It is strongly anisotropic and, if in large enough grains, will display reddish internal reflections. Ilmenite lamellae in magnetite are usually sharp and equal in width throughout their length, but ilmenite lamellae in hematite and hematite lamellae in ilmenite tend to be much more lenslike or podlike (Figure 9.11). These intergrowths of ilmenite and hematite are extremely striking because of the marked color difference of the phases and probably will not be misidentified after having once been observed.

Phase equilibria among the iron-titanium oxides (Figure 9.12) were deter-

FIGURE 9.9 Exsolution lamellae of ilmenite along (111) planes of host magnetite. Baugsto, Norway. (Partly crossed polars, width of field = 520 μm.)

FIGURE 9.10 Lamellar twinning in ilmenite that contains lensoid exsolution bodies of hematite. The twin lamellae extend through both phases. Egersund, Norway. (Partly crossed polars, width of field = 520 μm.)

mined by Buddington and Lindsley (1964) who found that the solubility of ilmenite in magnetite is much too small, even up to magmatic temperatures, to account for most ilmenite-magnetite intergrowths by simple exsolution. They concluded that subsolidus oxidation of magnetite-ulvöspinel with subsequent formation of the ilmenite-hematite lamellae takes place during cooling of many igneous and metamorphic rocks. They further found that coexisting equilibrated pairs of titaniferous magnetite and ilmenite may permit simultaneous determination of the temperature and oxygen fugacity at the time of formation. The compositions of coexisting magnetite-ulvöspinel (solid lines) and ilmenite-hematite (dashed lines) as functions of temperature and oxygen fugacity are shown in Figure 9.13.

Titaniferous magnetites occur in a wide variety of rocks (Figure 9.14) and should always be carefully examined for the presence of exsolution lamellae. Application of the relationships illustrated in Figure 9.13 to determine temperature requires that the grains are of primary origin and that their compositions have not been altered by secondary exsolution. Occasionally, small amounts of mag-

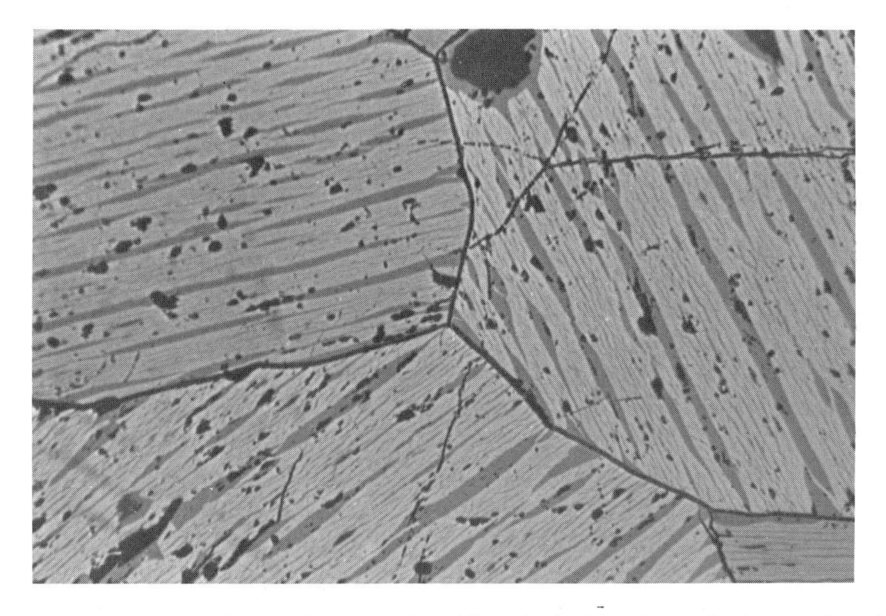

FIGURE 9.11 Lenslike exsolution bodies of ilmenite in hematite at the intersection of three grains. Wilson Lake, Labrador. (Width of field = 520 μm.)

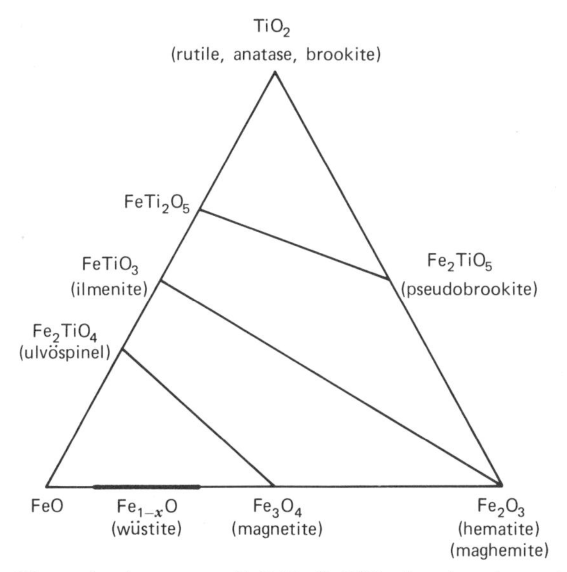

FIGURE 9.12 Phases in the system FeO-Fe_2O_3-TiO_2 showing the major solid solution series magnetite-ulvöspinel, hematite-ilmenite, and pseudobrookite-$FeTi_2O_5$. Compositions are in mole percent. (From A. F. Buddington and D. Lindsley, *J. Petrol.* **5**, 311, 1964; used with permission.)

netite occur as exsolutionlike lamellae within ilmenite-hematite; this apparently results from late-stage small-scale reduction.

Titaniferous magnetites commonly contain up to several percent vanadium, chromium, and aluminum substituting for Fe^{3+} and manganese and magnesium substituting for Fe^{2+}. Where magnetite and ilmenite have developed as a coexisting pair, the magnetite contains more V, Cr, and Al but less Mn and Mg than the ilmenite.

The oxidation of ilmenite and titaniferous magnetite leads to the development of rims and a complex variety of lamellar intergrowths of these phases with ru-

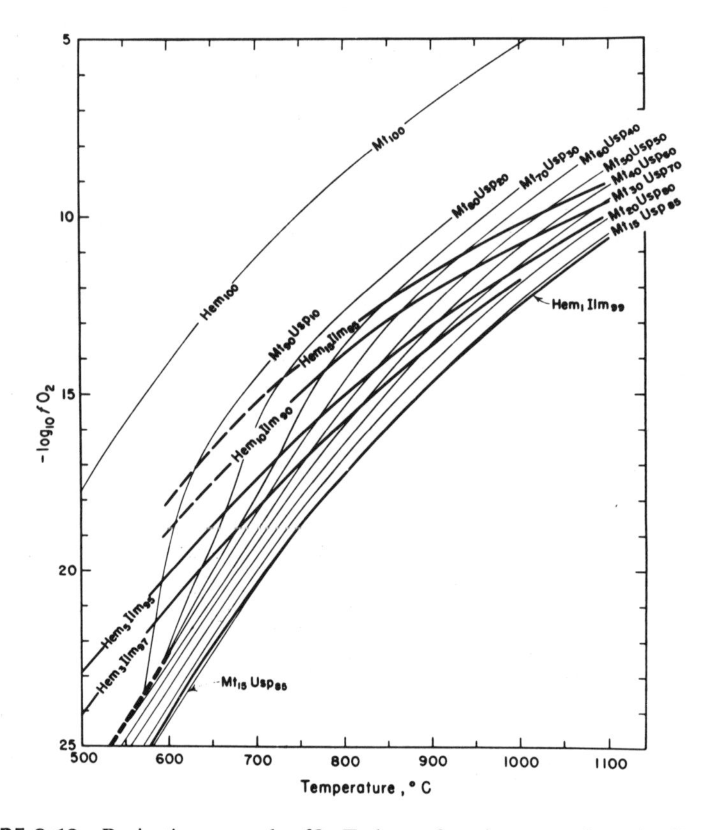

FIGURE 9.13 Projection onto the fO_2-T plane of conjugate surfaces in fO_2-T-X space. The projection is parallel to two composition axes, magnetite-ulvöspinel$_{ss}$ and the coexisting hematite-ilmenite$_{ss}$, so that intersecting contours are the projection of tie-lines connecting conjugate pairs. The temperature-composition relations of the magnetite-ulvöspinel solvus are shown in the heavy dashed line. The hematite-ilmenite solvus (not shown) lies between the curves Hem_{100} and $Hem_{15}Ilm_{85}$. Compositions are in mole percent. (From A. F. Buddington and D. Lindsley, *J. Petrol.* **5**, 316, 1964; used with permission.)

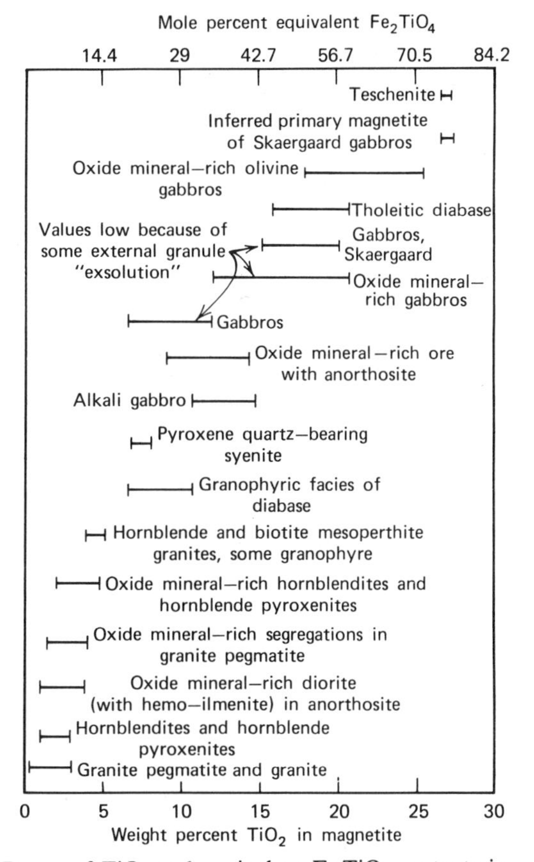

FIGURE 9.14 Range of TiO_2 and equivalent Fe_2TiO_4 contents in magnetites of some plutonic and hypabyssal rocks. (From A. F. Buddington and D. Lindsley, *J. Petrol.* **5**, 342, 1964; used with permission.)

tile, pseudobrookite, and titanohematite. The stages of development have been exhaustively described and illustrated by Haggerty (in Rumble, 1976).

Origin of the Ores

The dispersed grains of titaniferous magnetite and ilmenite in igneous rocks appear to be normal accesories which have crystallized along with the host silicates. However, the origin of the massive magnetite-ilmenite ores is not well understood but appears to be the result of fractionation from mafic complexes. The suggested modes of origin include the following.

Fractional Crystallization and Gravitative Differentiation This mechanism, which accounts for the formation of many early formed silicates in differentiating intrusions and for many chromite ores (see Section 9.2), appears satisfactorily to explain many aspects of the iron-titanium oxide layers in complexes such as the Duluth gabbro.

Residual Liquid Segregation This process has been summarized by Bateman (1951, p. 406) as follows:

> With progressive crystallization and enrichment of the residual liquid, its density would come to exceed that of the silicate crystals and its composition might reach the point of being chiefly oxides of iron and titanium. . . . At this stage three possibilities may occur: final freezing may occur to yield a basic igneous rock with interstitial oxides . . . ; the residual liquid may be filter pressed out of the crystal mush and be injected elsewhere; or the enriched residual liquid may drain downward through the crystal interstices and collect below to form a gravitative liquid accumulation.

The relationships of many large accumulations of Fe-Ti oxide layers as in the Bushveld and of discordant bodies as in the Adirondacks are consistent with this mechanism. Limited experimental investigations in the system magnetite-apatite-diorite (Philpotts, 1967) suggest that there is an eutectic at $\frac{2}{3}$ oxide to $\frac{1}{3}$ fluorapatite, a composition very similar to that found in some anorthositic oxide concentrations. These findings lend some support to a possible residual liquid segregation mechanism for the formation of iron-titanium oxide ores. A third mechanism which has been suggested is that of *metamorphic migration.* Liberation of iron and titanium from sphene, biotite, hornblende, and titanaugite and subsequent migration and concentration during granulite facies metamorphism has also been suggested as a possible mechanism to form iron-titanium oxide masses. Considerable granulation of iron-titanium oxide grains is common, but there is little evidence of massive replacement.

Selected References

Bateman, A. M. (1951) The formation of late magmatic oxide ores. *Econ. Geol.* **46**, 404–426.

Buddington, A. F. and D. H. Lindsley (1964) Iron-titanium oxide minerals and synthetic equivalents. *J. Petrol.* **5**, 310–357.

Lister, G. F. (1966) The composition and origin of selected iron-titanium deposits. *Econ. Geol.* **61**, 275–310.

Philpotts, A. R. (1967) Origin of certain iron-titanium oxide and apatite rocks. *Econ. Geol.* **62**, 303–330.

Rumble, D., ed. (1976) *Oxide Minerals.* Mineral. Soc. Am. Short Course Notes, Vol. 3 (a collection of papers, several of which deal with iron-titanium oxides).

9.5 COPPER/MOLYBDENUM SULFIDES ASSOCIATED WITH PORPHYRITIC INTRUSIVE IGNEOUS ROCKS ("PORPHYRY COPPER/MOLYBDENUM" DEPOSITS)

Mineralogy

Major: pyrite, chalcopyrite, molybdenite, bornite.

Minor: magnetite, hematite, ilmenite, rutile, enargite, cubanite, cassiterite, huebnerite, gold.

Secondary: hematite, covellite, chalcocite, digenite, native copper.

Mode of Occurrence

Sulfides are present in veinlets and as disseminated grains in, or adjacent to, porphyritic intrusions ranging in composition from quartz diorite to quartz monzonite. The host porphyry and adjacent rocks are commonly altered in concentric siliceous, potassic, phyllic, argillic and propylitic zones.

Examples

Copper: Bisbee, Ray, Ajo, Arizona; Butte, Montana; Bingham Canyon, Utah; El Salvador, Chuquicamata, Braden, Chile; Bethlehem, Endako, British Columbia; Ok Tedi, Papua–New Guinea; Cananea, Mexico.

Molybdenum: Climax, Urad, Henderson, Colorado; Questa, New Mexico.

Mineral Zoning in Porphyry Copper Deposits

Porphyry deposits of this type now constitute the world's primary sources of copper and molybdenum and also serve as significant producers of several other base metals, gold, and silver. These deposits are concisely summarized in the definition given by Lowell and Guilbert (1970, p. 374) as follows:

> A copper and/or molybdenum sulfide deposit consisting of disseminated and stockwork veinlet sulfide mineralization emplaced in various host rocks that have been altered by hydrothermal solutions into roughly concentric zonal patterns [Figure 9.15*a*]. The deposit is generally large, on the scale of thousands of feet . . . and . . . is associated with a complex, passively emplaced, stock of intermediate composition including porphyry units.

The low grades of these deposits—0.8% Cu and 0.02% Mo for a typical porphyry copper and 0.3% Mo and 0.05% Cu for a typical porphyry molybde-

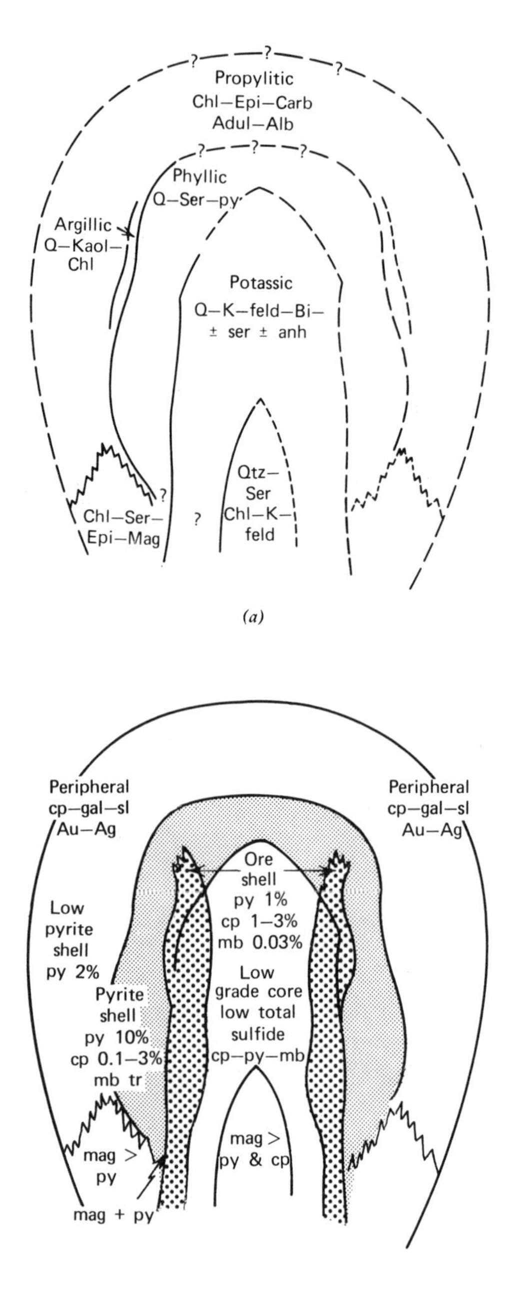

FIGURE 9.15 Idealized diagram showing concentric alteration-mineralization zones occurring in porphyry copper deposits: *(a)* alteration zones; *(b)* mineralization zones. (After J. D. Lowell and J. M. Guilbert, *Econ. Geol.* **65**, 379, 1970, with permission of the publisher.)

num—evidence the dispersed nature of the sulfides. Pyrite is generally the most abundant sulfide and may occur in association with the other sulfides or alone in barren quartz veinlets. In copper-dominant porphyry deposits, the total sulfide distribution and relative amounts of the economic minerals follow a concentric pattern coaxial with the alteration zones as shown in Figure 9.15*b* and discussed below.

Potassic Zone The inner zone of potassic alteration commonly coincides with two zones of mineralization:

1 An inner low-grade ($<0.3\%$ Cu) core characterized by pyrite and chalcopyrite in a ratio of roughly 1 : 2, minor magnetite and minor molybdenite.

2 An ore shell ($>0.5\%$ Cu), in which the pyrite to chalcopyrite ratio is roughly 1 : 1 and each mineral constitutes about 1% of the rock. Small amounts of molybdenite are present in veinlets and as dispersed grains. Bornite, as discrete grains and intimately intergrown with the chalcopyrite, is common in small amounts.

Phyllic and Argillic Zones The phyllic zone coincides with the outer portion of the "ore shell" and the surrounding low grade portion of the "pyrite shell" in which copper grade decreases to 0.1–0.5%. The pyrite to chalcopyrite ratio is roughly 10 : 1 and pyrite may constitute as much as 10% of the rock volume as coarse (0.5 mm) subhedral vein fillings and disseminated grains. The outer part of the "pyrite shell" contains up to 25% pyrite in the form of coarse anhedral pyritic-quartz veins up to 2 cm thick.

Propylitic Zone In this, the outer zone of the ore body, mineralization consists of pervasive pyrite in veinlets (2–6% of the rock) and local small veins containing typical hydrothermal base-metal assemblages, such as the pyrite-chalcopyrite-galena-sphalerite-tetrahedrite associations discussed in Section 9.6. The zonal structure of these deposits means that the level of erosion can have an important influence on the geology and mineralogy of the exposed portion of deposits as noted by Sillitoe (1973).

Molybdenum-rich deposits such as Climax and Henderson, Colorado exhibit an umbrellalike shape emplaced in or about the intrusion core (Figure 9.16). In the Climax deposit the mineralization occurs in two crudely concentric zones, an outer tungsten zone and an inner molybdenum zone. The hydrothermal alteration in and near the ore bodies consists of potassium feldspar, sericite, fluorite, and topaz with minor amounts of biotite, chlorite, and epidote.

Ore Mineral Textures

The ore minerals of the porphyry copper deposits occur in veinlets or as disseminated grains as shown in Figures 9.17 and 9.18. Pyrite is the dominant sulfide

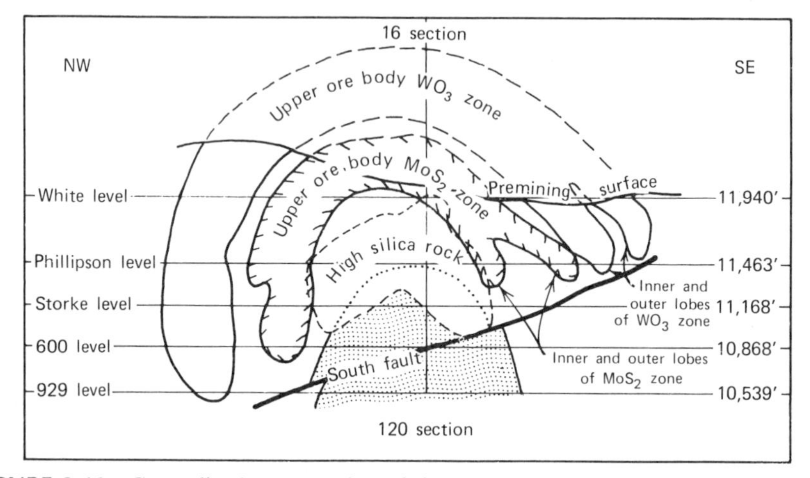

FIGURE 9.16 Generalized cross section of the Upper Ore Body at Climax, Colorado. This illustrates the concentric nature of the mineralized zones about the intrusion. (Reproduced with permission from *Ore Deposits of the United States 1933/1967*, Am. Inst. Min. Metall. Petrol. Eng., p. 628.)

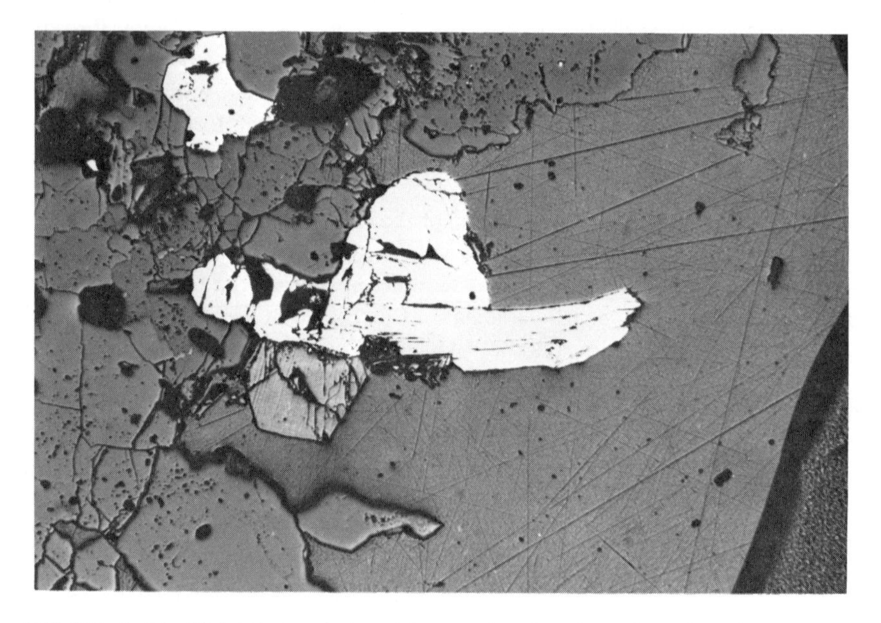

FIGURE 9.17 Molybdenite lath with associated anhedral grain of chalcopyrite. Bingham Canyon, Utah. (Width of field = 520 μm.)

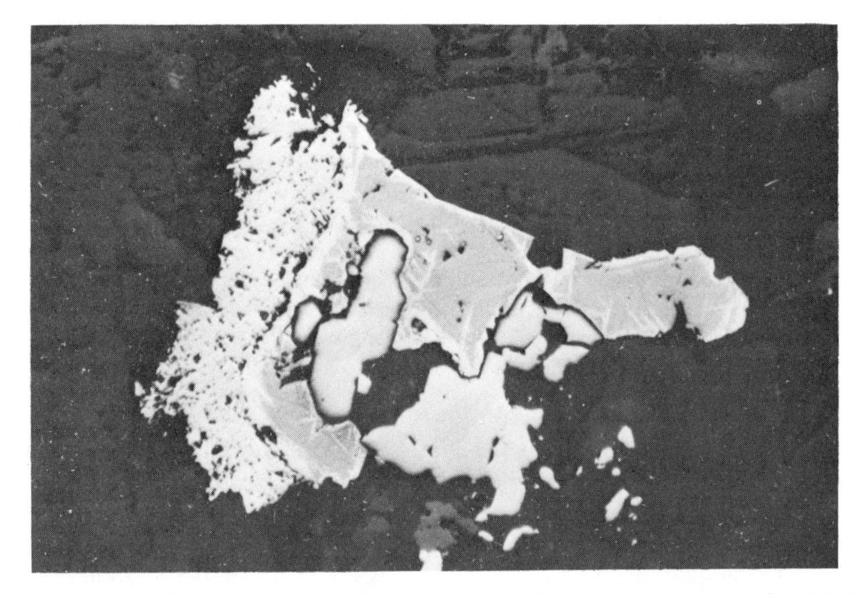

FIGURE 9.18 Central grain of rutile surrounded by bornite that shows marginal laths and rim of chalcopyrite. A polycrystalline mass of enargite is seen to the left of the field of view. Bingham Canyon, Utah. (Width of field = 520 μm.)

and is present as anhedral to euhedral grains in both types of occurrences. Chalcopyrite is the dominant copper mineral, occurring as anhedral interstitial grains and as fracture fillings in pyrite. Bornite is present as discrete anhedral grains with the pyrite and chalcopyrite and as both exsolution and oxidation lamellae within chalcopyrite. In near-surface ores, covellite, chalcocite, and digenite are commonly observed forming secondary alteration rims on chalcopyrite and bornite; native copper is sometimes present. Molybdenite is present in significant amounts in the molybdenum-rich porphyry deposits but is minor in many copper-rich examples. The molybdenite is generally observed as subparallel to crumpled blades disseminated in veinlets.

In molybdenum-rich ores such as those at Climax, Colorado, the molybdenite is present as random to subparallel tiny (<0.1 mm) hexagonal plates embedded in the vein-filling quartz. Tungsten in these ores is present as small irregularly distributed euhedral huebnerite grains in pyrite-quartz-sericite veinlets.

Enargite is occasionally present in porphyry ores as anhedral grains (Figure 9.18) or as rims on the copper-iron sulfides. Cubanite is rare but has been observed as exsolution lamellae within chalcopyrite.

Origin of the Ores

The genesis of porphyry-type deposits has been the subject of considerable study and remains, in part, conjectural. All these deposits occur within, or very close to orogenic zones; however, it appears that there is a distinction between the two major types with Cu-bearing porphyries lying along consuming plate margins whereas Mo-bearing porphyries are typical of tensional (rifting) environments. Sillitoe (1972) has suggested that the porphyry bodies represent partially melted oceanic rocks that have transported metals originally emplaced at mid-oceanic ridges, been subducted, and then been emplaced along continental margins. Mitchell and Garson (1972) have described the mechanism as follows:

> Porphyry copper deposits are emplaced in igneous belts located either on continental margins or in island arcs. These belts are related to partial melting of wet oceanic crust descending along Benioff zones at depths of 150–250 km. Deposition of the copper occurs where metal-carrying solutions rising from the descending crust meet meteoric brines.

Although this general model fits the South American and the Pacific Island arc deposits, it is difficult to reconcile with settings of many southwestern United States and Rocky Mountain occurrences (Lowell, 1979). Nevertheless, there is general agreement on the mode of emplacement once a magma has been generated. The emplacement of a porphyritic magma to within 0.5 to 2 km of surface establishes convective motions in the adjacent ground water system if the surrounding rocks have reasonable permeability. Hot, chemically potent solutions rise above the stock as new, initially cool, solutions circulate in toward the stock. As a result of the cooling, the outer shell crystallizes; the shell is subsequently highly fractured by pressure building up from H_2O released from the crystallizing magma. The central core is quenched to a porphyry, cut by a stockwork of quartz and quartz-feldspar veinlets as heat is rapidly lost to the ground water system. Precipitation of the sulfides and quartz in these veins occurs at temperatures as high as 725°C in the potassic zone and around 300–390°C in the argillized zone.

Fluid inclusion studies by several workers (e.g., Roedder 1971; Moore and Nash, 1974; Hall et al., 1979) have shown that the fluids that deposited ores in the cores of copper-molybdenum deposits were highly saline (often containing more than 60% salts), underwent extensive boiling, and were trapped at temperatures up to 725°C. In some deposits distinct zones of the hypersaline inclusions correlate not only with the ore zones but also specifically with the fracture systems believed responsible for the introduction of the metals.

Selected References

Economic Geology (1978) The Number 5 issue is devoted to "Porphyry Copper Deposits of the Southwestern Pacific Islands and Australia."

Economic Geology (1978) The Number 7 issue is devoted to "The Bingham Mining District."

Hall, W. E., Friedman, I., and Nash, J. T. (1974) Fluid inclusion and light stable isotope study of the Climax molybdenum deposits, Colorado. *Econ. Geol.* **69**, 884–911.

Lowell, J. D. (1974) Regional characteristics of porphyry copper deposits of the Southwest. *Econ. Geol.* **69**, 601–617.

Lowell, J. D. and Guilbert, J. M. (1970) Lateral and vertical alteration mineralization zoning in porphyry ore deposits. *Econ. Geol.* **65**, 373–408.

McMillan, W. J. and Panteleyer, A. (1980) Ore deposit models—1 Porphyry copper deposits. *Geosci. Can.* **7**, 52–63.

Mitchell, A. H. G. and Garson, M. S. (1972) Relationship of porphyry coppers and circum-Pacific tin deposits to palaeo-Benioff zones. *Min. Metall.* **81**, B10–B25.

Moore, W. J. and Nash, J. T. (1974) Alteration and fluid inclusion studies of the porphyry copper ore body at Bingham Utah. *Econ. Geol.* **69**, 631–645.

Roedder, E. (1971) Fluid inclusion studies on the porphyry type ore deposits at Bingham, Utah; Butte, Montana, and, Climax Colorado. *Econ. Geol.* **66**, 98–120.

Sheppard, S. M. F., Nielson, R. L., and Taylor, H. P. (1971) Hydrogen and oxygen isotope ratios in minerals from porphyry copper deposits. *Econ. Geol.* **66**, 515–542.

Sillitoe, R. H. (1972) A plate tectonic model for the origin of porphyry copper deposits. *Econ. Geol.* **67**, 184–197.

Sillitoe, R. H. (1973) The tops and bottoms of porphyry copper deposits. *Econ. Geol.* **68**, 799–815.

Wallace, S. R., MacKenzie, W. B., Blair, R. G., and Muncaster, N. K. (1978) Geology of the Urad and Henderson Molybdenite deposits, Clear Creek County, Colorado, with a section on a comparison of these deposits with those at Climax, Colorado. *Econ. Geol.* **73**, 325–368.

Wallace, S. R., Muncaster, N. K., Jonson, D. C., Mackenzie, W. B., Bookstrom, A. A., and Surface, V. E. (1968) Multiple intrusion and mineralization at Climax, Colorado. In John D. Ridge, ed., *Ore Deposits of the United States 1933/1967* (Graton-Sales Volume). A.I.M.E., New York, Vol. 1, pp. 605–640.

9.6 COPPER-LEAD-ZINC-SILVER ASSEMBLAGES IN VEIN DEPOSITS

Mineralogy

Major: pyrite, sphalerite, galena, chalcopyrite, tetrahedrite.

Minor: bornite, chalcocite, enargite, argentite, gold, hematite, pyrrhotite, proustite-pyrargyrite, Pb-Bi-Sb sulfosalts.

Secondary: cerussite, anglesite, goethite, smithsonite, azurite $[Cu_3(CO_3)_2(OH)_2]$, malachite $[Cu_2(CO_3)(OH)_2]$, argentite, covellite, chalcocite, silver, and many other phases.

Associated Gangue Minerals: quartz, calcite, dolomite, barite, fluorite, rhodochrosite, siderite, chlorite, sericite.

Mode of Occurrence

Copper-lead-zinc-silver sulfide ores occur as hydrothermal vein fillings and replacement (usually of limestone) bodies often associated with intermediate to acid intrusions.

Examples

Especially prominent in the North American Cordillera—Creede, Gilman, Leadville, Colorado; Tintic, Park City, Utah; Eureka, Nevada; Bluebell, British Columbia; Zacatecas, Mexico; Eastern Transbaikalia, Soviet Union; Casapalca, Peru.

Mineral Associations and Textures

Copper-lead-zinc-silver vein deposits are characterized by coarse grained (1 mm–2 cm), banded to massive aggregates of pyrite, galena, sphalerite, and chalcopyrite (Figure 9.19). The bulk mineralogy is thus similar to that of massive stratabound sulfide deposits except that galena tends to be much more abundant in the vein deposits. Pyrite, sphalerite, and gangue minerals such as quartz, calcite, and fluorite often tend to be euhedral with the development of well-formed faces. Galena is coarse-grained and may be present as anhedral polycrystalline aggregates interstitial to pyrite and sphalerite or as well-formed crystals. Tetrahedrite is common in small amounts and is very important as a major carrier of silver, the content of which varies directly with the Sb/As ratio. The tetrahedrite is usually present as small rounded blebs within or along the margins of galena grains (Figure 9.19*b*). Many ores of this type exhibit complex paragenesis in which multiple episodes of deposition, leaching, and replacement are evident. Incipient oxidation is evidenced by the formation of covellite or chalcocite along fractures in chalcopyrite.

The deposition of these ores in open fractures has often led to the formation of distinct bands, commonly monomineralic, parallel to walls of the fracture. Mineral growth in these bands can be either as colloform ribbons or as well-faceted subhedral crystals. Changes in the nature of the ore fluid during deposition are reflected not only in the changing mineralogy of subsequent bands and in mineralogic variations along a vein system but also in the presence of growth-zoning in individual crystals. This zoning is evident in many sulfides as rows of small inclusions or differences in polishing hardness and color, but is especially evident in coarse-grained sphalerites as distinct color bands (see Figure 7.3) which are visible if doubly polished thin sections are examined. In general, though not in every case, darker sphalerite bands are richer in iron than are lighter bands; the FeS content varies widely (0.5–20 mol %) even in individual crystals. Sphalerites in these deposits frequently contain crystallographically oriented rows of chalcopyrite blebs ($\sim$20–100 μm), which have been appropriately referred to as "chalcopyrite disease" (see Section 7.5). Some of these small blebs (as they appear in polished surfaces) are actually wormlike rods that may extend for several millimeters within a crystal. The chalcopyrite apparently forms by epitaxial growth, replacement, or by reaction of copper-bearing fluids with iron-bearing sphalerite.

Fluid inclusion studies of hydrothermal vein deposits (Roedder, 1979) have

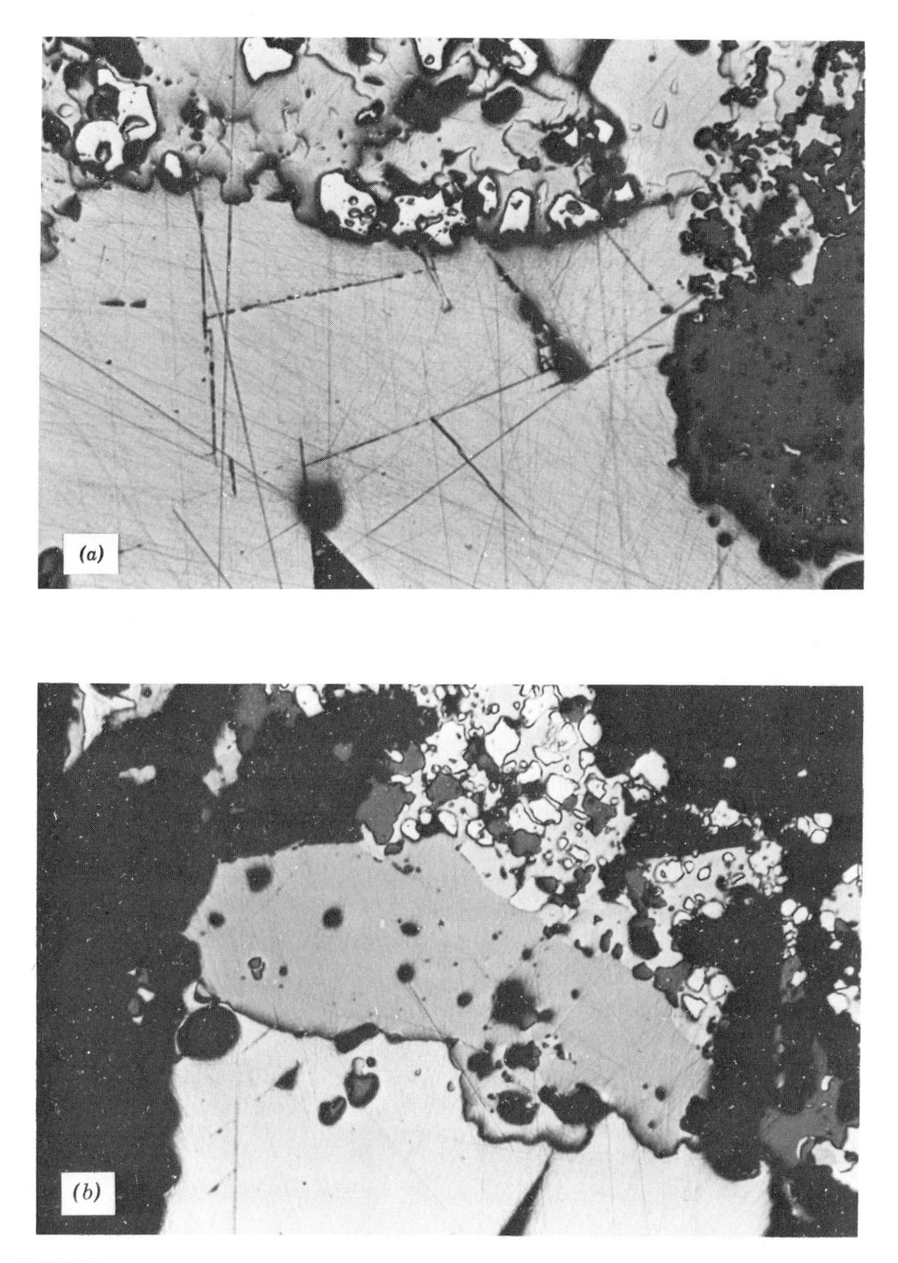

FIGURE 9.19 *(a)* Typical copper-lead-zinc hydrothermal vein assemblage of pyrite (white), chalcopyrite (medium gray at top), galena (medium gray at bottom), and sphalerite (dark gray). Wee Whistle Mine, Colorado. (Width of field = 1200 μm.) *(b)* Copper-lead-zinc-silver vein assemblage of pyrite (white), galena (light gray at bottom), silver-bearing tetrahedrite (dark gray in center), and chalcopyrite (light gray at top with pyrite). Wee Whistle Mine, Colorado. (Width of field = 1200 μm.)

revealed homogenization temperatures from 100 to 500°C and salinities which are generally less than 10 wt. % NaCl equivalent. Detailed work on the Creede, Colorado ores (Roedder 1977) yield homogenization temperatures of 200–270°C in 20 recognizable stratigraphic growth zones of sphalerite. The color zonation in these sphalerites (Figure 7.3) apparently corresponds to abrupt changes in the nature of the ore fluids as indicated by the fluid inclusions.

Supergene Alteration

Near surface portions of these ores often reveal moderate to extensive alteration by meteoric waters. The oxidation of pyrite causes the formation of sulfuric acid and ferrous sulfate, which result in the breakdown of other sulfides. The end result is that the uppermost parts of the veins consist of a boxwork gossan of iron oxides and hydroxides. At lower levels, secondary covellite, chalcocite, galena, and sometimes silver have been reprecipitated. Also much in evidence are secondary carbonates sulfates, and silicates of copper, lead, and zinc.

Origin of the Ores

Copper-lead-zinc-silver vein deposits apparently form as a result of circulating hydrothermal fluids that extract, transport, and then precipitate sulfide minerals as open space fillings and replacements. Studies of fluid inclusions, wall-rock alteration, and the sulfide ore minerals indicate that the ore-forming fluids were chloride-rich brines (see Chapter 8), often containing a very large component of recirculated meteoric water, which precipitated the sulfides in response to a decrease in temperature, a decrease in pressure (which may sometimes allow boiling), reaction with wall rocks, or mixing with other fluids. Metals originally derived from magmatic sources or from country rocks were probably transported as chloride or possibly sulfide complexes. The manner in which the sulfur of the ore minerals was transported is not known with certainity, but most workers believe that sulfate was the dominant form. Barnes (1979), however, has made out a strong case for transport of considerable metal as sulfide complexes. The flow of hot (150–300°C) reactive fluids commonly results in alteration haloes rich in quartz, feldspar, and sericite, as described by Meyer and Hemley (1967) and Rose and Burt (1979).

References

Barnes, H. L. (1979) Solubilities of Ore Minerals. In H. L. Barnes, ed., *Geochemistry of Hydrothermal Ore Deposits*, 2nd ed. Wiley-Interscience, New York, pp. 401–460.

Helgeson, H. C. (1970) A chemical and thermodynamic model of ore deposition in hydrothermal systems. Min. Soc. Am. Spec. Paper **3**, 155–186.

Helgeson, H. C. (1979) Mass transport among minerals and hydrothermal solutions. In H. L. Barnes, ed., *Geochemistry of Hydrothermal Ore Deposits*, 2nd ed. Wiley-Interscience, New York, pp. 568–610.

Meyer, C. and Hemley, J. J. (1967) Wall rock alteration. In H. L. Barnes, ed., *Geochemistry of Hydrothermal Ore Deposits*. Holt, Rinehart & Winston, New York, pp. 166–235.

Nash, T. J. (1975) Geochemical studies in the Park City District: II. Sulfide mineralogy and minor-element chemistry, Mayflower Mine. *Econ. Geol.* **70**, 1038–1049.

Ohmoto, H. (1972) Systematics of sulfur and carbon isotopes in hydrothermal ore deposits. *Econ. Geol.* **67**, 551–578.

Ohmoto, H., and Rye, R. O. (1970) The Bluebell Mine, British Columbia. I. Mineralogy, paragenesis, fluid inclusions and the isotopes of hydrogen, oxygen and carbon. *Econ. Geol.* **65**, 417–437.

Roedder, E. (1972) Composition of fluid inclusions (Data of Geochemistry Series) U.S. Geol. Survey Prof. Paper 440 JJ.

Roedder, E. (1979) Fluid inclusions as samples of ore fluids. In H. L. Barnes, ed., *Geochemistry of Hydrothermal Ore Deposits*, 2nd ed., Wiley-Interscience, New York, pp. 684–737.

Roedder, E. (1977) Fluid inclusions as tools in mineral exploration. *Econ. Geol.* **72**, 503–525.

Rose, A. W. and Burt, D. M. (1979) Hydrothermal alteration. In H. L. Barnes, Ed., *Geochemistry of Hydrothermal Ore Deposits*, 2nd ed. Wiley-Interscience, New York, pp. 173–235.

Skinner, B. J., White, D. E., Rose, H. J., and Mays, R. E. (1967) Sulfides associated with the Salton Sea geothermal brine. *Econ. Geol.* **62**, 316–330.

Taylor, H. P. (1974) The application of oxygen and hydrogen isotope studies to problems of hydrothermal alteration and ore deposition. *Econ. Geol.* **69**, 843–883.

White, D. E. (1968) Environments of generation of some base-metal ore deposits. *Econ. Geol.* **63**, 301–335.

White, D. E. (1974) Diverse origins of hydrothermal ore fluids. *Econ. Geol.* **69**, 954–973.

9.7 THE SILVER-BISMUTH-COBALT-NICKEL-ARSENIC (-URANIUM) VEIN ORES

Mineralogy

Very complex mineral assemblages, often with 30 or 40 different opaque minerals reported from a single deposit.

Major Phases: may include native silver, native bismuth, niccolite, skutterudite, rammelsbergite, safflorite, löllingite, cobaltite, gersdorffite, arsenopyrite along with such common sulfides as pyrite, marcasite, chalcopyrite and galena. Certain deposits contain substantial uranium as uraninite often associated with hematite.

Other Phases: (which may locally achieve economic importance) include a large number of other Fe, Co, Ni in combination with S, As, Sb minerals, sulfides and sulfosalts, particularly of silver (some of these are listed in Figures 9.21 and 9.22).

Gangue Phases: carbonates and quartz.

Mode of Occurrence

The ores occur in veins that are mostly fissure fillings of fault and joint planes. Commonly the veins range from a few centimeters to several meters in thickness and occur in a considerable variety of host rocks (quartzites, greywackes, conglomerates, slates, schists, diabases, and granites for example) ranging in age through most of the geological time scale.

Examples

The most famous examples occur in the Cobalt-Gowganda area, Ontario, Canada. Others include Great Bear Lake, N.W.T., Canada; the Erzgebirge, Germany; Kongsberg, Norway; Jachymov, Czechoslovakia.

Mineral Assemblages, Textures, and Parageneses

The ores of this type are characterized both by very complex mineral assemblages and the presence of delicate zonal and dendritic textures. For example, native silver commonly occurs in dendritic patterns that are surrounded by arsenides (Figure 9.20*a*) in rounded patterns termed "rosettes." Colloform textures are also commonly developed (Figure 9.20*b*) involving the native metals and arsenides. Frequently, these accretionary structures are slightly disrupted with introduction of later veining materials, and periods of direct deposition are interspersed with phases of replacement. The combination of these processes with the complex assemblages has enabled ore microscopists to unravel long, complex paragenetic sequences for these ores. The study of the Cobalt, Ontario ores (Figure 9.21) by Petruk (1971a) and of the Great Bear Lake deposits (Figure 9.22) by Badham et al. (1972) are good examples.

At Cobalt, Petruk (1971a) has subdivided the arsenide assemblages into Ni-As, Ni-Co-As, Co-As, Co-Fe-As, and Fe-As types, with Co-As being most abundant. Native silver is the main native metal and occurs as cores to arsenide rosettes, as veinlets in arsenides and carbonates and in association with sulfides. Silver at the cores of rosettes has been interpreted by Petruk (1971a) to be the earliest mineral in the ore with the arsenides being deposited around it later. In contrast, Scott (1977) interprets much of the silver in the cores of the rosettes to be a later replacement of the arsenides. The veinlet silver in arsenides and carbonates is interpreted to be a late variety and that associated with sulfides to be even later. The high grade silver ore is associated with parts of the veins containing Ni-Co-As and Co-As assemblages. The sulfides occur as disseminated grains, veinlets, and colloform masses in the veins and the wall rocks. On the basis of their textural relationships, it is suggested that most of the sulfides were in the rocks prior to ore deposition and were remobilized and redeposited in and around the veins during mineralization. Some sulfides may also represent a late stage of the main mineralization process. The paragenetic sequence determined

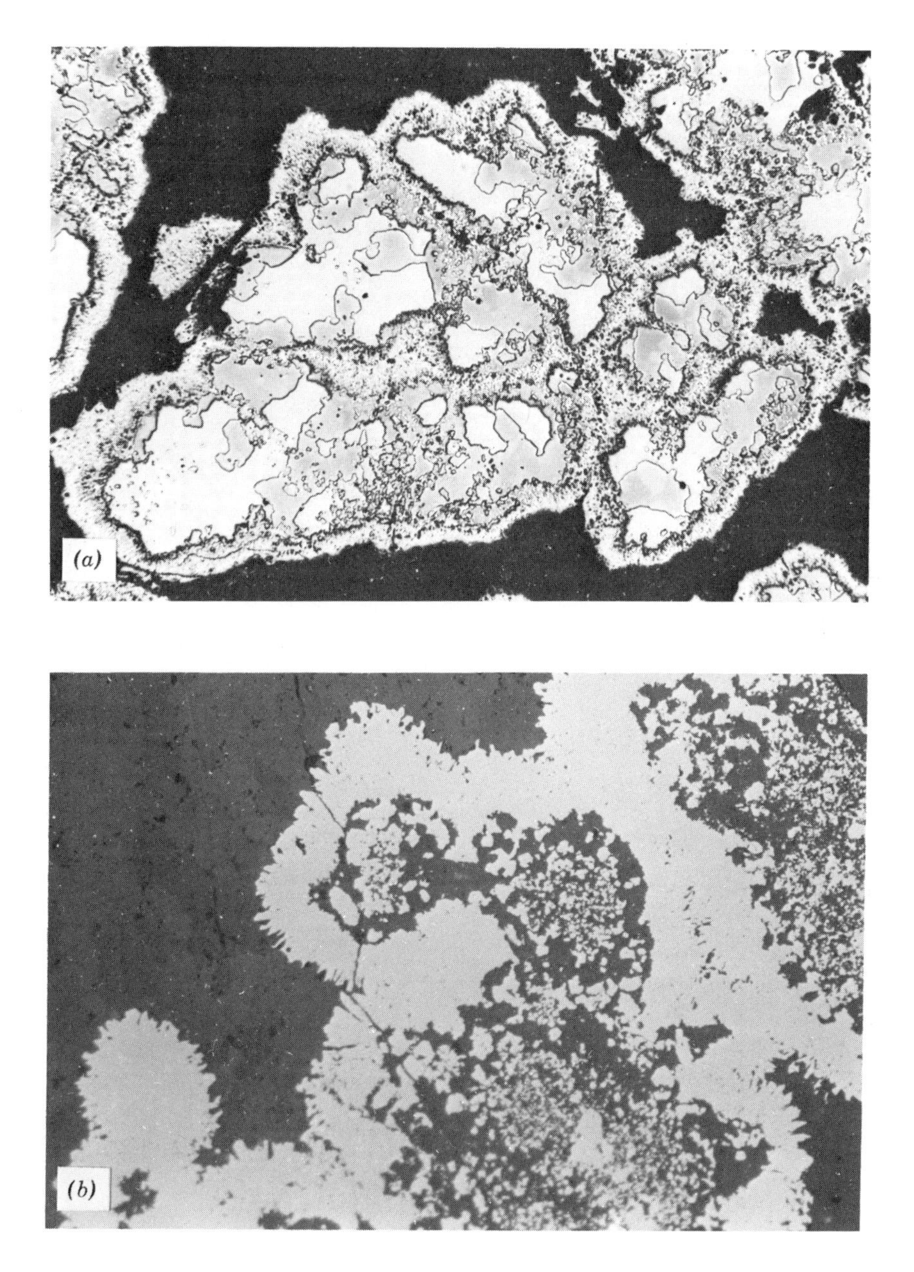

FIGURE 9.20 Vein ores from Cobalt, Ontario, Canada: *(a)* dendritic silver surrounded by breithauptite (medium gray) and cobaltite. (Width of field = 2000 μm.) *(b)* cobaltite showing colloform textures. (Width of field = 2000 μm.)

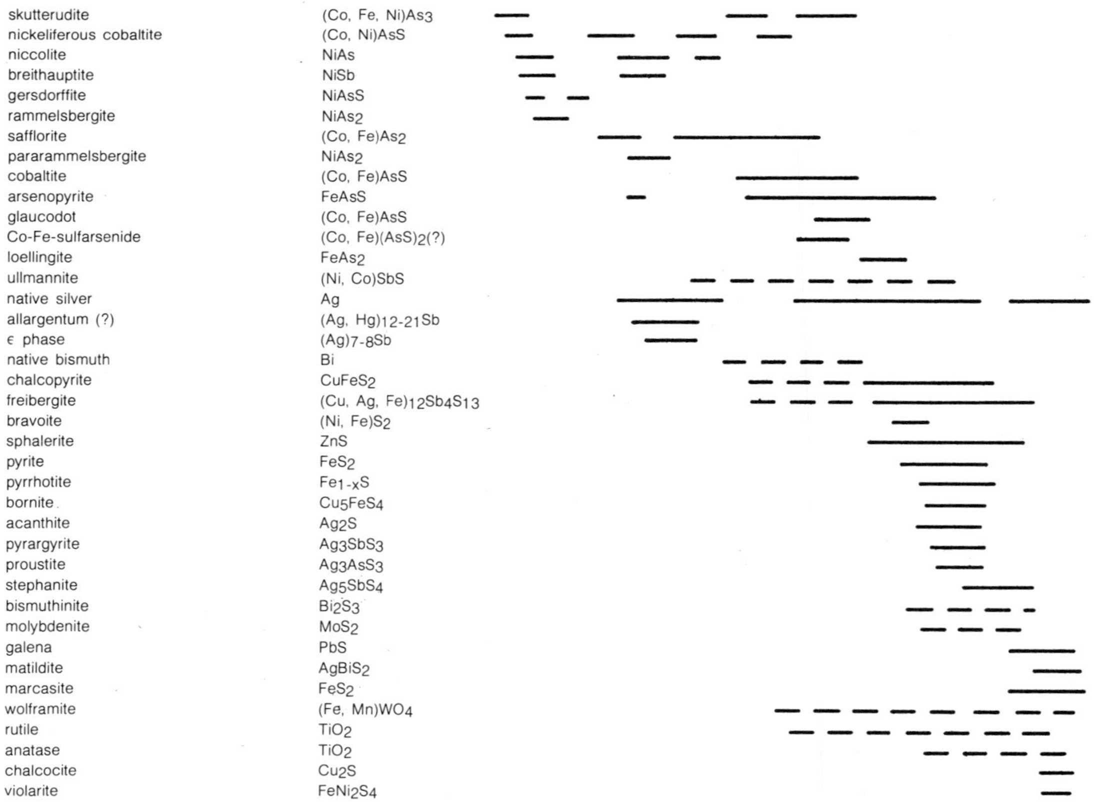

skutterudite	$(Co, Fe, Ni)As_3$
nickeliferous cobaltite	$(Co, Ni)AsS$
niccolite	$NiAs$
breithauptite	$NiSb$
gersdorffite	$NiAsS$
rammelsbergite	$NiAs_2$
safflorite	$(Co, Fe)As_2$
pararammelsbergite	$NiAs_2$
cobaltite	$(Co, Fe)AsS$
arsenopyrite	$FeAsS$
glaucodot	$(Co, Fe)AsS$
Co-Fe-sulfarsenide	$(Co, Fe)(AsS)_2$(?)
loellingite	$FeAs_2$
ullmannite	$(Ni, Co)SbS$
native silver	Ag
allargentum (?)	$(Ag, Hg)_{12-21}Sb$
ϵ phase	$(Ag)_{7-8}Sb$
native bismuth	Bi
chalcopyrite	$CuFeS_2$
freibergite	$(Cu, Ag, Fe)_{12}Sb_4S_{13}$
bravoite	$(Ni, Fe)S_2$
sphalerite	ZnS
pyrite	FeS_2
pyrrhotite	$Fe_{1-x}S$
bornite	Cu_5FeS_4
acanthite	Ag_2S
pyrargyrite	Ag_3SbS_3
proustite	Ag_3AsS_3
stephanite	Ag_5SbS_4
bismuthinite	Bi_2S_3
molybdenite	MoS_2
galena	PbS
matildite	$AgBiS_2$
marcasite	FeS_2
wolframite	$(Fe, Mn)WO_4$
rutile	TiO_2
anatase	TiO_2
chalcocite	Cu_2S
violarite	$FeNi_2S_4$

FIGURE 9.21 Mineral assemblage and approximate paragenetic sequence in the Silverfields deposit, Cobalt, Ontario. (After W. Petruk, 1968.)

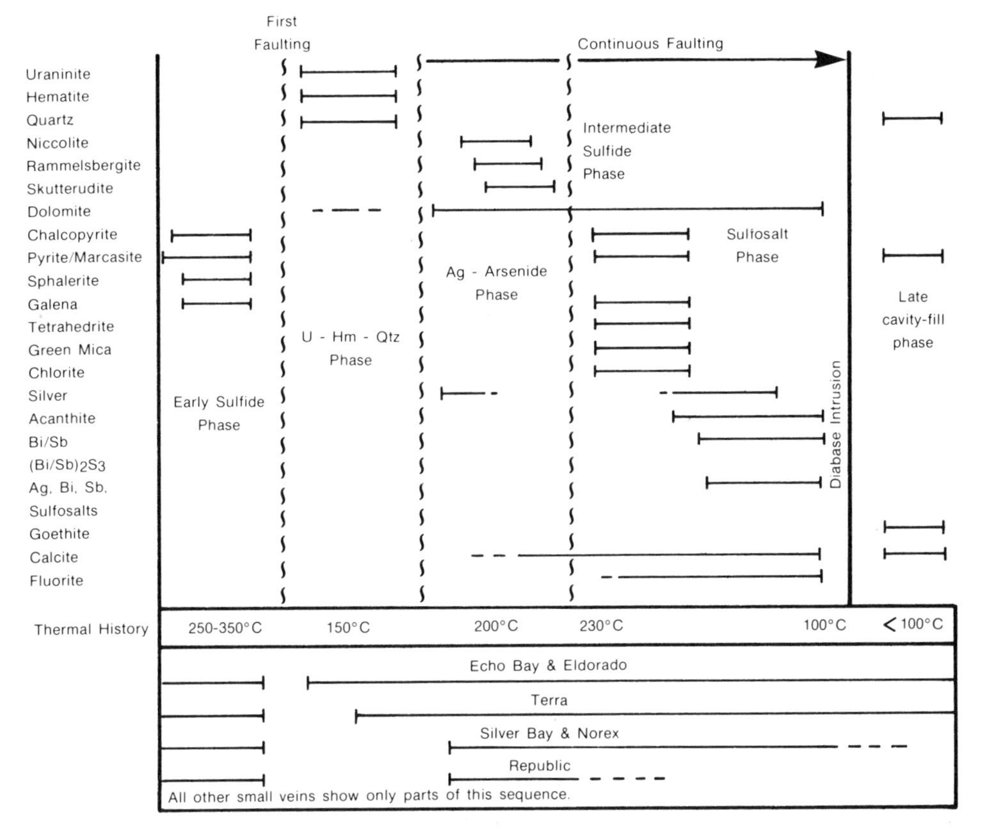

FIGURE 9.22 The paragenetic sequence, phases, and estimated temperatures of mineralization at Great Bear Lake, Northwest Territories. Also shown are the stages of mineralization occurring in particular mines and veins. (After J. P. N. Badham et al., 1972.)

by Petruk (1968) for ore formation at the Silverfields Mine, Cobalt, is illustrated in Figure 9.21.

The mineral assemblages at Great Bear Lake differ from those at Cobalt by the presence of significant uranium mineralization in the form of uraninite. Again, a complex assemblage and sequence of deposition and replacement episodes has been elucidated by ore microscopy. A general paragenetic sequence (after Badham et al., 1972) is shown in Figure 9.22. This illustrates that an early phase of sulfide mineralization was followed by uraninite-hematite-quartz mineralization, by nickel and cobalt arsenides, then by silver, by sulfides associated with dolomite, by Ag, Sb, Bi sulfosalts, and by carbonates, quartz, and fluorite. Particular mines may show only a part of this complete mineralization sequence as indicated in Figure 9.22. Studies of fluid inclusions and sulfur isotopes have made it possible to estimate temperatures of deposition as noted in the figure.

Mineral Formation

All the geological, mineralogical and textural evidence suggests that these ores have been deposited from complex hydrothermal solutions. The sites of deposition were commonly controlled by preexisting faults and fissures. For the ores of the Cobalt area, Petruk (1971b) has suggested a sequence of depositional temperatures with early quartz and chlorite deposited at ~200–360°C, arsenide deposition at temperatures around 500–590°C (above the temperature of the rammelsbergite-pararammelsbergite inversion), sulfides deposited at 400–200°C, and later gangue phases and sulfides below 200°C. Scott and O'Connor (1971) have examined two populations of fluid inclusions in ore vein quartz in the Cobalt Camp and determined depositional temperatures of 195 to 260°C and 285 to 360°C. Scott (1972) has calculated the temperatures gradients adjacent to the diabase sheet at Cobalt and concluded that the maximum temperature of ore deposition would have been 535°C. The depositional temperatures for the Great Bear Lake deposits have already been indicated in Figure 9.22. For the Echo Bay Mines, the main mineralization temperatures range between 230 and 100°C (Badham et al., 1972). Information on the chemical nature of the ore fluids has also been obtained from fluid inclusion and stable isotope studies.

The origin of the solutions producing deposits of this type is still very much a matter of debate. Origin of the Cobalt area ores is commonly linked with a diabase intrusion (Jambor, 1971) although for ores of this general type, four possible sources have been invoked (Halls and Stumpfl, 1972);

1. Direct hydrothermal evolution from granitic intrusions.
2. Direct hydrothermal evolution from basic intrusions.
3. Hydrothermal processes causing selective concentration from associated black shales or preexisting sulfide deposits.
4. Ore solutions from a deep source near the crust-mantle boundary.

Selected References

Badham, J. P. N., Robinson, B. W. and Morton, R. D. (1972) The geology and genesis of the Great Bear Lake Silver Deposits. *24th Int. Geol. Cong.* (*Montreal*), Section 4, 541–548.

Halls, C. and Stumpfl, E. F. (1972) The five elements (Ag-Bi-Co-Ni-As)—a critical appraisal of the geological environments in which it occurs and of the theories affecting its origin. *24th Int. Geol. Cong.* (*Montreal*), Section 4, 540.

Jambor, J. L. (1971) Origin of the silver veins. In *The Silver-Arsenide Deposits of the Cobalt-Gowganda Region, Ontario. Can. Mineral.* **11**, 402–412.

Petruk, W. (1968) Mineralogy and origin of the Silverfields silver deposit in the Cobalt area, Ontario. *Econ. Geol.* **63**, 512–531.

Petruk, W. (1971a) Mineralogical characteristics of the deposits and textures of the ore minerals. In *The Silver-Arsenide Deposits of the Cobalt-Gowganda Region, Ontario. Can. Mineral.* **11**, 108–139.

Petruk, W. (1971b) Depositional history of the ore minerals. Ibid., pp. 369–401.

Scott, S. D. (1972) The Ag-Co-Ni-As ores of the Siscoe Metals of Ontario Mine, Gowganda, Ontario, Canada. *24th Int. Geol. Cong.*, Section 4, 528–538.

Scott, S. D. and O'Connor, T. P. (1971) Fluid inclusions in vein quartz, Silverfields Mine, Cobalt, Ontario. In *The Silver-Arsenide Deposits of the Cobalt-Gowganda Region, Ontario. Can. Mineral.* **11**, 263–271.

9.8 TIN-TUNGSTEN-BISMUTH ASSEMBLAGES IN VEIN DEPOSITS

Mineralogy

Major: cassiterite, arsenopyrite, wolframite, bismuthinite, pyrite, marcasite, pyrrhotite.

Minor: stannite, chalcopyrite, sphalerite, tetrahedrite, pyrargyrite, bismuth, galena, rutile, gold, franckeite, molybdenite.

Gangue: quartz, tourmaline, apatite, fluorite.

Mode of Occurrence

Tin-tungsten-bismuth ores occur as open-space hydrothermal vein fillings that show several stages of vein growth. The hydrothermal veins are usually associated with granitic stocks or batholiths.

Examples

Panasqueira, Portugal; Llallagua, Tasna, Colcha, Huanuni, Sayaquira, Oruro, and Potosi Districts, Bolivia; Cornwall tin district, S. W. England.

Mineral Associations and Textures

Tin-tungsten-bismuth ores occur as open-space hydrothermal vein fillings, commonly formed from several stages of mineralization. The earliest stage is dominated by quartz and cassiterite but may also contain considerable tourmaline, bismuthinite, arsenopyrite, apatite, and wolframite. The following stage typically contains pyrrhotite, sphalerite, chalcopyrite, stannite and may also contain franckeite, cassiterite, arsenopyrite and silver-bearing minerals such as tetrahedrite, matildite, and proustite-pyrargyrite. Later stages of vein development involve the alteration of pyrrhotite to pyrite, marcasite, and siderite, minor deposition of sphalerite, and the formation of hydrous phosphates.

Early vein minerals such as quartz, cassiterite, and bismuthinite typically occur as prismatic crystals and blades that may line the vein walls or occur as pris-

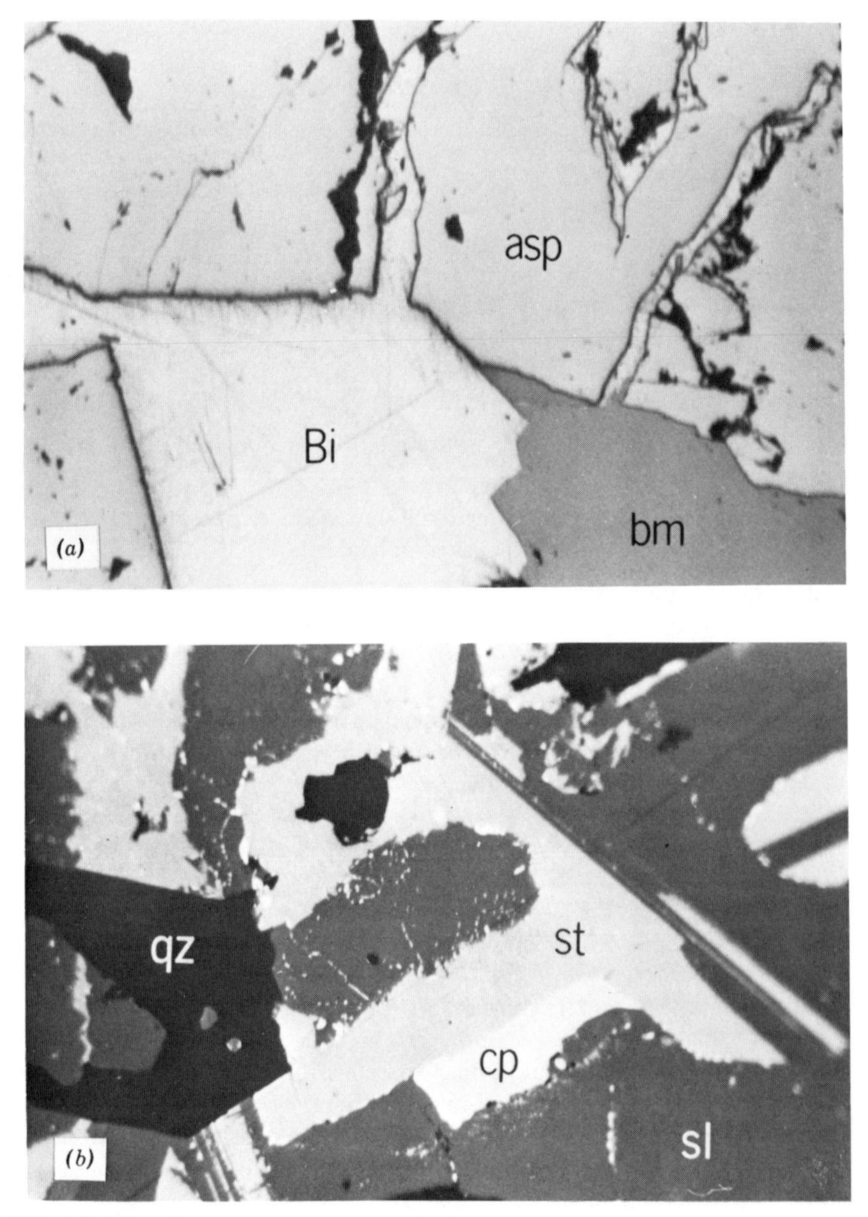

FIGURE 9.23 Ore mineral assemblages in the tin-tungsten ores of Bolivia. *(a)* Veinlets and interstitial areas of bismuthinite (bm) and native bismuth (Bi) in arsenopyrite (asp). Chacaltaya Mine. (Oil immersion, width of field = 900 μm.) *(b)* Crystallographic intergrowth of sphalerite (sl), stannite (st), and chalcopyrite (cp) probably resulting from exsolution. Chalcopyrite, the latest phase to exsolve, has collected along sphalerite-stannite contacts. Corroded quartz (qz) crystals also present. Sayaquira District. (Oil immersion, width of field = 850 μm.) (Reproduced from W. C. Kelly and F. S. Turneaure, *Econ. Geol.* **65**, 1970, with permission of authors and publisher.)

matic crystals and blades that may line the vein walls or occur as radiating bundles (see Figure 8.11*a*). Quartz usually forms a course band on the vein walls with cassiterite and bismuthinite making up the central portion of the veins. Wolframite may be present as euhedral to subhedral granular masses that overgrow quartz and cassiterite. Native bismuth may occur as alteration blebs in bismuthinite and as small patches with arsenopyrite which occurs as veinlets and scattered euhedra (Figure 9.23*a*).

Pyrrhotite and sphalerite may infill space between quartz and cassiterite crystals (see Figure 8.11*c*) and often exhibit mutual grain boundaries. Sphalerite commonly contains crystallographically oriented blebs and rods of chalcopyrite, pyrrhotite and stannite (Figure 9.23*b*). Chalcopyrite is also present as anhedral polycrystalline masses with pyrrhotite. Franckeite, if present, occurs as plates and blades. The pyrrhotite is often altered along fractures and grain boundaries to pyrite and marcasite (Figure 8.11*d*). These minerals may also be present as concentric "birds-eye" textures (see Figure 7.24) and boxwork structures. Pyrite may also be present as isolated euhedral or subhedral polycrystalline aggregates. Stannite may be present as anhedral grains (see Figure 8.11*b*) but most commonly shows an affinity for sphalerite in which it frequently occurs as crystallographically oriented intergrowths (see Figure 8.11*d*).

Origin of the Ores

Tin-tungsten-bismuth ores have generally been considered to be of magmatic-hydrothermal origin and related to intermediate to acid batholiths and stocks. The veins have formed by sequential open-space deposition of sulfides, oxides, and gangue minerals from brines that were initially NaCl-rich (up to 46 wt. %) but which were more dilute in later stages. In the Bolivian tin province, the depositional temperatures in the early stages were ~300°C and rose to 530°C; in the base-metal stage, temperatures ranged from 400 to 260°C; during pyrrhotite alteration the temperature dropped to 260–200°C. Late veinlets and crustifications formed at temperatures from 200 to 70°C (Kelly and Turneaure, 1970; Turneaure, 1971). Deposition took place at depths between 350 and 4000 m at pressures between 30 and 1000 bars. Fluid inclusion studies indicate that active boiling occurred during early stages of ore deposition and that the vapor transported some mineral matter. The Sn-rich fluids are probably generated during the late stages of fractional crystallization of a granitoid magma (see Groves and McCarthy, 1978).

Sillitoe et al. (1975) have suggested that several of the well-known Bolivian tin deposits should be designated as *porphyry tin deposits*. These authors note that such deposits are similar to porphyry copper deposits (see Section 9.5) in being associated with passively emplaced and pervasively sericite-altered stocks, in having crudely concentric zoning, in containing mineralized breccias, and in containing large volumes of rock with low grade ores (0.2 to 0.3% Sn). They propose that porphyry tin deposits represent the lower portions of stratovolcanoes.

Selected References

Groves, D. I. and McCarthy, T. S. (1978) Fractional crystallization and the origin of tin deposits in granitoids. *Mineral. Deposita* **13**, 11–26.

Kelly, W. C. and Rye, R. O. (1979) Geologic, fluid inclusion, and stable isotope studies of the tin-tungsten deposits of Panasqueira, Portugal. *Econ. Geol.* **74**, 1721–1822.

Kelly, W. C. and Turneaure, F. S. (1970) Mineralogy, paragenesis and geothermometry of the tin and tungsten deposits of the Eastern Andes, Bolivia. *Econ. Geol.* **65**, 609–680.

Kettaneh, Y. A. and Badham, J. P. N. (1978) Mineralization and paragenesis at the Mount Wellington Mine, Cornwall. *Econ. Geol.* **43**, 486–495.

Sillitoe, R. H., Halls, C., and Grant, J. N. (1975) Porphyry tin deposits in Bolivia. *Econ. Geol.* 913–927.

Turneaure, F. S. (1971) The Bolivian tin-silver province. *Econ. Geol.* **66**, 215–225.

9.9 GOLD VEIN AND RELATED MINERALIZATION

Mineralogy

Economically important ore minerals may include native gold and precious metal tellurides [e.g., sylvanite, petzite $AuAg_3Te_2$, hessite Ag_2Te, calaverite $AuTe_2$, krennerite $AuTe_2$]; associated uneconomic ore minerals may include pyrite, marcasite, arsenopyrite, pyrrhotite in major amounts and minor galena, sphalerite, chalcopyrite, stibnite, tetrahedrite-tennantite, realgar; gangue minerals are dominated by quartz with minor local carbonates (calcite, siderite, ankerite, dolomite), feldspars, tourmaline, fluorite, barite, epidote, and graphite, amorphous carbon, or carbonaceous matter.

Mode of Occurrence

Gold occurs in a wide variety of settings, ranging from volcanic sinters and breccias to skarns and hydrothermal veins that may or may not be directly associated with intrusions and from dissemination in massive sulfides to placer and palaeoplacer deposits. Veins dominated by native gold and quartz occur in ancient highly deformed and metamorphosed volcanic rocks (notably Precambrian greenstone belts). Veins dominated by gold and silver tellurides with quartz occur in this setting and in young (Tertiary) volcanic rocks of the circum-Pacific belt. In the brief account given here, it is not the intention to discuss the form, setting, and origin of the many diverse types of gold occurrence but to outline briefly the (generally rather similar) ore mineralogy. An exhaustive account of the geochemistry of gold and the nature of gold occurrences world-wide has recently been presented by Boyle (1979).

Examples

Gold-quartz vein mineralization includes deposits at Yellowknife (N.W.T., Canada), the Mother Lode (California, United States), and the Homestake Mine (S. Dakota, United States); gold and silver tellurides occur in Precambrian rocks at Kalgoorlie (W. Australia), Kirkland Lake and Porcupine (Canada); gold and silver tellurides occur in Tertiary volcanics in Fiji, the Phillipines, Japan, and California-Colorado-Nevada-New Mexico (United States). Placer gold deposits include examples in California, Alaska, United States; the Urals, Soviet Union; Otago area, New Zealand.

Mineral Associations and Textures

As is typical of fissure-filling veins, many show crustification and the development of well-formed quartz and carbonate crystals, although movement along the fractures can destroy many of these textures to leave a granular ore. The gold in the gold-quartz veins occurs within quartz or within, or marginal to, pyrite or arsenopyrite and is generally very fine-grained (Figure 9.24). In the veins which contain gold-silver tellurides, these occur as small irregular masses within quartz and are often a complex intergrowth of many minerals including, in addition to those listed already, a variety of other telluride minerals, native tellurium, and various sulfides and sulfosalts, all in very minor amounts. The mineral assemblages and the paragenetic sequences are complex but generalizations can sometimes be made. For example, the ores of Boulder County, Colorado studied in detail by Kelly and Goddard (1969) show a generalized paragenesis in which early quartz and fluorite are succeeded by sulfides, tellurides, native gold and then by carbonates and quartz. Fluid inclusion studies indicate that hydrothermal solutions with salinites of 20 to 30% precipitated the early quartz and fluorite between 250 and 375°C. Gold deposition occurred from solutions with only about 4% salinity at 205–270°C (Nash and Cunningham, 1973).

Boyle (1979) has summarized the geothermometric studies that have been carried out on epigenetic gold deposits and concluded that the bulk of quartz and sulfide mineralization took place between 500 and 150°C with much gold having been redistributed at lower temperatures. He has further given a detailed discussion of the paragenesis and zoning of epigenetic gold deposits.

The interpretation of assemblages and textures of the gold-silver tellurides has been clarified by studies of the phase relations in the Au-Ag-Te system. Markham (1960) and Cabri (1965) have presented the phase relations at 300 and 290°C, respectively, as shown in Figure 9.25; the differences between these diagrams probably reflect the rapid quench techniques of Cabri, whose results may be closer to the true relations at 290°C. The results of Markham may reflect some readjustment of compositions to lower temperatures and more closely approximate final products of cooling in the natural ores. Comparison of synthetic and natural assemblages indicate that melting phenomena or telluride melts

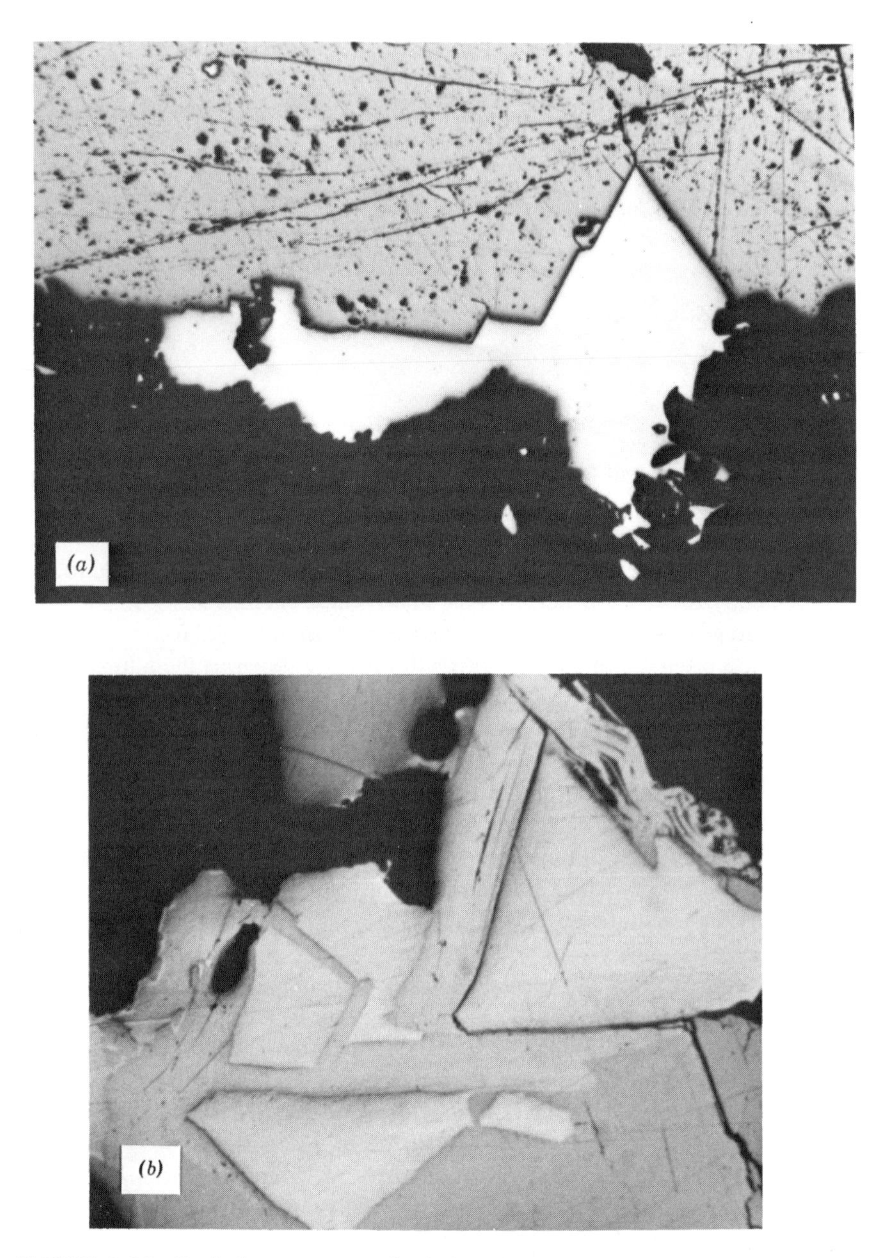

FIGURE 9.24 Typical occurrences of gold in vein and related ores. (*a*) Gold bordering arsenopyrite. Homestake Mine, South Dakota. (Width of field = 2000 μm.) (*b*) Angular gold grains with galena (darker gray) and tetradymite (light gray adjacent to gold) in a quartz vein deposit. Mineral Ridge, Fairfax Co., Virginia, U.S.N.M.N.H. 10832 Smithsonian Institution (width of field = 200 μm).

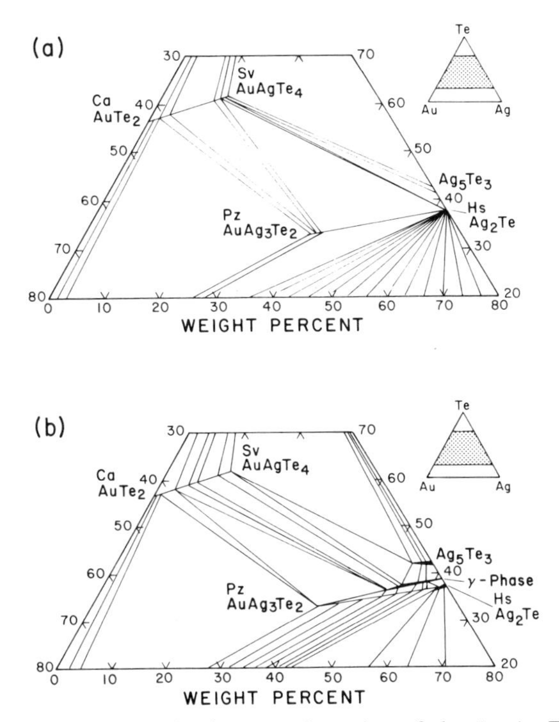

FIGURE 9.25 Phase relations in the central portion of the Au-Ag-Te system: *(a)* at 300°C (after Markham, 1960); *(b)* at 290°C (after Cabri, 1965).

(which can occur at temperatures as low as 304°C) have played no significant role in producing the observed assemblages, most of which have resulted from subsolidus processes and equilibrated at relatively low temperatures (below ~250°C).

Origin of the Ores

There is little doubt that these ores have formed by deposition from hydrothermal solutions into open fissure and fracture systems. The absence of parent intrusions and extensive hydrothermal alteration has led Worthington and Kiff (1970) to suggest that many gold deposits formed during the waning stages of volcanism. In the case of the gold-quartz veins in highly deformed and metamorphosed volcanics and sediments, it is widely thought that the gold may have been derived from the enclosing rocks and concentrated by circulating fluids during metamorphism. The feasibility of such mechanisms has been discussed by Fyfe and Henley (1973). The Tertiary deposits clearly indicate a volcanic source for the ore-bearing hydrothermal solutions, although in the case of those telluride gold deposits in older rocks, the role of volcanic or intrusive igneous rocks as sources for the hydrothermal fluids is much less certain. The compositions of these hydrothermal fluids and whether the gold is transported as gold-sulfur or gold-chloride complexes are also matters of debate (Fyfe and Henley, 1973).

Secondary Occurrences—Placer Deposits

The erosion of gold-bearing veins and sulfide ores has, in many places, resulted in the formation of placer deposits. The nature of the gold in such occurrences depends as to composition, grain size, and grade on the source rocks. The gold in placer deposits has been dispersed along a stream channel through the normal processes of erosion and transport but is concentrated because of its very high specific gravity. Sedimentological studies that have attempted to simulate gold deposition have demonstrated that the natural sites for concentration includes the confluence of streams, pools below waterfalls, and so on.

The gold grains ("nuggets") occur in a wide variety of sizes and shapes but are most commonly flattened and flakelike to crudely spherical. Immediately after liberation, the grains commonly display a faceted morphology derived from the surrounding silicate or carbonate crystals of the original lode occurrence. Transport of grains results in their being abraded, commonly flattened, and leached. The surfical leaching of gold grains results in a depletion of silver (most grains are initially an alloy of nearly equal gold and silver contents), which leaves a narrow (commonly less than 10–15 μm) rim of nearly pure gold (Desborough, 1970). This rim, and zones along fractures where leaching has also occurred, is often visible in sections because it is distinctly more yellow than the interior of the grain. Crystal faces on nuggets, which are far from the source area, have led to the speculation that the nuggets may actually have grown *in situ* by slow precipitation of gold from stream waters.

Selected References

Boyle, R. A. (1979) The geochemistry of gold and its deposits. *Can. Geol. Surv. Bull.* **280**.

Burns, V. M. (1979) Marine placer minerals. In Burns, R. G., ed., *Marine Minerals.* Min. Soc. Am. Short Course Notes, Vol. 6, 347–380.

Cabri, L. J. (1965) Phase relations in the system Au-Ag-Te and their mineralogical significance. *Econ. Geol.* **60**, 1569–1606.

Desborough, G. A. (1970) Silver depletion indicated by microanalysis of gold from placer occurrences, western United States. *Econ. Geol.* **65**, 304–311.

Fyfe, W. S., and Henley, R. W. (1973) Some thoughts on chemical transport processes with particular reference to gold. *Mineral. Sci. Eng.* **5**, 295–303.

Kelly, W. C., and Goddard, E. N. (1969) Telluride ores of Boulder County, Colorado. *Geol. Soc. Am. Mem.* 109.

Koschmann, A. H. and Bergendahl, M. H. (1968) Principal gold-producing districts of the United States. *U.S. Geol. Surv. Prof. Paper* 610.

Koshman, P. N. and Yugay, T. A. (1972) The causes of variation in fineness levels of gold placers. *Geochem. Intern.* **9**, 481–484.

Markham, N. L. (1960) Synthetic and natural phases in the system Au-Ag-Te (part 1). *Econ. Geol.* **55**, 1148–1178.

Nash, J. J. and Cunningham, C. G. (1973) Fluid inclusion studies of the fluorspar and gold deposits, Jamestown District, Colorado. *Econ. Geol.* **68**, 1247–1262.

Slaughter, A. L. (1968) The homestake Mine. In J. D. Ridge, ed., *Ore Deposits in the United States 1933–1967.* A.I.M.E., New York, Vol. 2, pp. 1436–1459.

Worthington, J. E. and Kiff, I. T. (1970) A suggested volcanogenic origin for certain gold deposits in the slate belt of the North Carolina Piedmont. *Econ. Geol.* **65**, 529–537.

9.10 ARSENIC-, ANTIMONY-, OR MERCURY-BEARING BASE METAL VEIN DEPOSITS

This broad category includes ores rich in copper-zinc but with significant amounts of arsenic (Butte, Montana), lead-zinc-silver ores rich in antimony and arsenic (Coeur d'Alene, Idaho), arsenic-antimony ores (Getchell, Nevada), and mercury sulfide ores (Almaden, Spain). The diversity of these ores requires individual mineral summaries.

Mode of Occurrence

The ores included in this section occur as massive to disseminated infillings in faults and fractures and as replacement bodies often in close proximity to acid or intermediate intrusions. Zoning of copper, zinc, arsenic, and sulfur is common both laterally and vertically in the vein systems. Wall-rock alteration in the form of feldspathization, sericitization, argillization, and bleaching is frequently developed adjacent to the sulfide ores.

Copper-Zinc-Arsenic

Mineralogy

Major: pyrite, chalcopyrite, bornite, tennantite, sphalerite, enargite.

Minor: covellite, hematite, magnetite, chalcocite, digenite, galena, molybdenite.

Gangue Minerals: quartz, siderite, calcite, barite, rhodochrosite.

Examples

Butte, Montana; Magma Mine, Arizona; Tsumeb, South Africa; Bor, Yugoslavia.

Mineral Associations and Textures

On the macroscopic scale these ores, as do other vein ores, often display a distinct zoning parallel to the axis of the vein; more or less well-defined bands rich in pyrite, chalcocite, covellite, bornite, enargite, and so on occur discontinuously with widths varying from millimeters to meters. Grain size is highly variable but is frequently 0.5 mm to 1 cm or more; hence individual minerals are readily recognized with the naked eye. Microscopically, it is apparent that many of the large grains lie in a matrix of, or are themselves composed of, intimate irregular fine-grained intergrowths of bornite, chalcocite, chalcopyrite, bornite, enargite, and copper sulfides (covellite, digenite, djurleite). Common lathlike and "basket-weave" textures in the copper sulfides and between chalcopyrite and bornite reflect exsolution from solid solutions at the higher temperatures of ore formation. As is evident from Figure 10.7, a single phase chalcocite-digenite solid solution that forms at elevated temperature can decompose to form a mixture of low-temperature copper sulfides on cooling. The resulting texture is commonly one of intersecting laths. Spectacular covellite-rich portions of these ores consist of coarse but highly fractured aggregates that are criss-crossed with fine veinlets of later covellite or other copper sulfides (Figure 9.26*a*). Pyrite is abundant in these ores, often occurring repeatedly in the paragenesis; early-formed pyrite is commonly deeply corroded and replaced by chalcopyrite, bornite, enargite, and other phases (Figure 9.26*b*).

High-temperature relationships between the copper-iron sulfides are well-characterized but this is not true of the stable low-temperature associations (Figure 8.14*b*). The frequent association of pyrite and bornite as seen in these ores was long thought to be indicative of formation above 228°C (Yund and Kullerud, 1966) but it now appears that this association reflects stable lower-temperature relationships.

Sphalerite is a major but erratically distributed mineral occurring as disseminated grains, veinlets and coarsely crystalline masses. Oriented rows of chalcopyrite blebs are common in the sphalerite. Tennantite series minerals are also irregular in distribution and are present as scattered irregular blebs and patches, as veinlets, and as pods up to meters across. Except in the pods where large (>1 cm) grains occur, the tennantite is intimately intergrown with bornite, chalcocite, and enargite. Hematite is locally abundant as randomly oriented clusters of radiating laths interspersed with pyrite, bornite, chalcopyrite, and enargite.

Lead-Zinc-Silver Ores with Antimony and Arsenic

Mineralogy

Major: galena, sphalerite, tetrahedrite, chalcopyrite.

Minor: pyrite, pyrrhotite, arsenopyrite, magnetite.

Gangue Minerals: quartz, siderite, dolomite, barite.

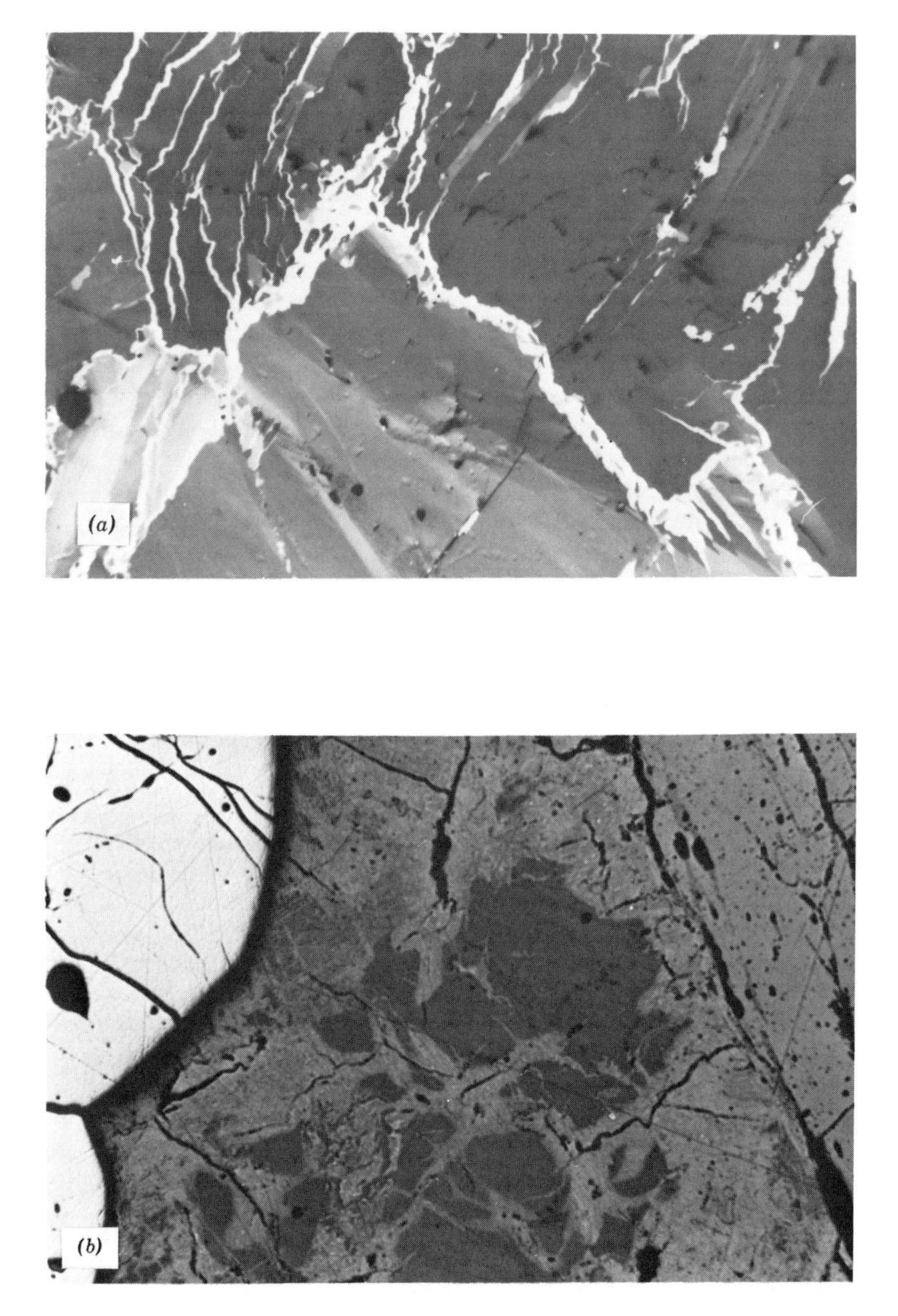

FIGURE 9.26 Typical ores of copper-zinc-arsenic affiliation. *(a)* Coarse aggregate of covellite (gray) veined by later chalcocite. Butte, Montana. (Width of field = 2000 μm.) *(b)* Islands of bornite in chalcocite flanked by enargite (on right) and pyrite (white). Butte, Montana. (Width of field = 2000 μm.)

Examples

Coeur d'Alene District, Idaho; Kapnik, Czechoslovakia; Andreasberg, Germany; Freiberg, Saxony.

Mineral Associations and Textures

These ores occur as compact, generally fine-grained vein fillings with major mineral dominance varying from one part of a district or vein to another. All ore minerals, with the exception of locally disseminated and often corroded pyrite cubes, are anhedral with grain sizes usually less than 0.5 mm (Figure 9.27*a*). Galena may contain small amounts of accessory metals, but the bulk of the silver is present in tetrahedrite, which is abundant as small rounded islands dispersed throughout (Figure 9.27*b*). In tetrahedrite-rich portions of the ore, the galena occurs as small cusplike grains. Chalcopyrite is present as small anhedral grains and irregular veinlets. Sphalerite, dispersed as grains and veinlets, is generally light in color and contains relatively little (usually less than 5%) iron and small amounts of manganese and cadmium. Occasionally, the sphalerite contains oriented rows of chalcopyrite inclusions.

The well-known Coeur d'Alene ores have been metamorphosed such that shearing, brecciation, and curved cleavages in galena are abundant.

Arsenic-Antimony Sulfide Ores

Mineralogy

Major: realgar, orpiment, pyrite, stibnite.
Minor: chalcopyrite, arsenopyrite, gold, marcasite.
Gangue Minerals: quartz, calcite, carbon, sericite, chlorite.

Examples

Getchell, Nevada; Elbrus, Caucasus; Uj-Moldava, Romania.

Mineral Association and Textures

These unique and colorful ores consist of intergrown coarse-grained (>0.1 mm) aggregates varying from realgar-rich to orpiment-rich. Overgrowth, interpenetration, and replacement of one As-sulfide by another is common and leaves an overall chaotic texture. Pyrite, where present, occurs as subhedral to euhedral grains that are either isolated or in irregular thin veinlets. Stibnite occurs locally as fine, commonly radiating, laths dispersed within the As-sulfides.

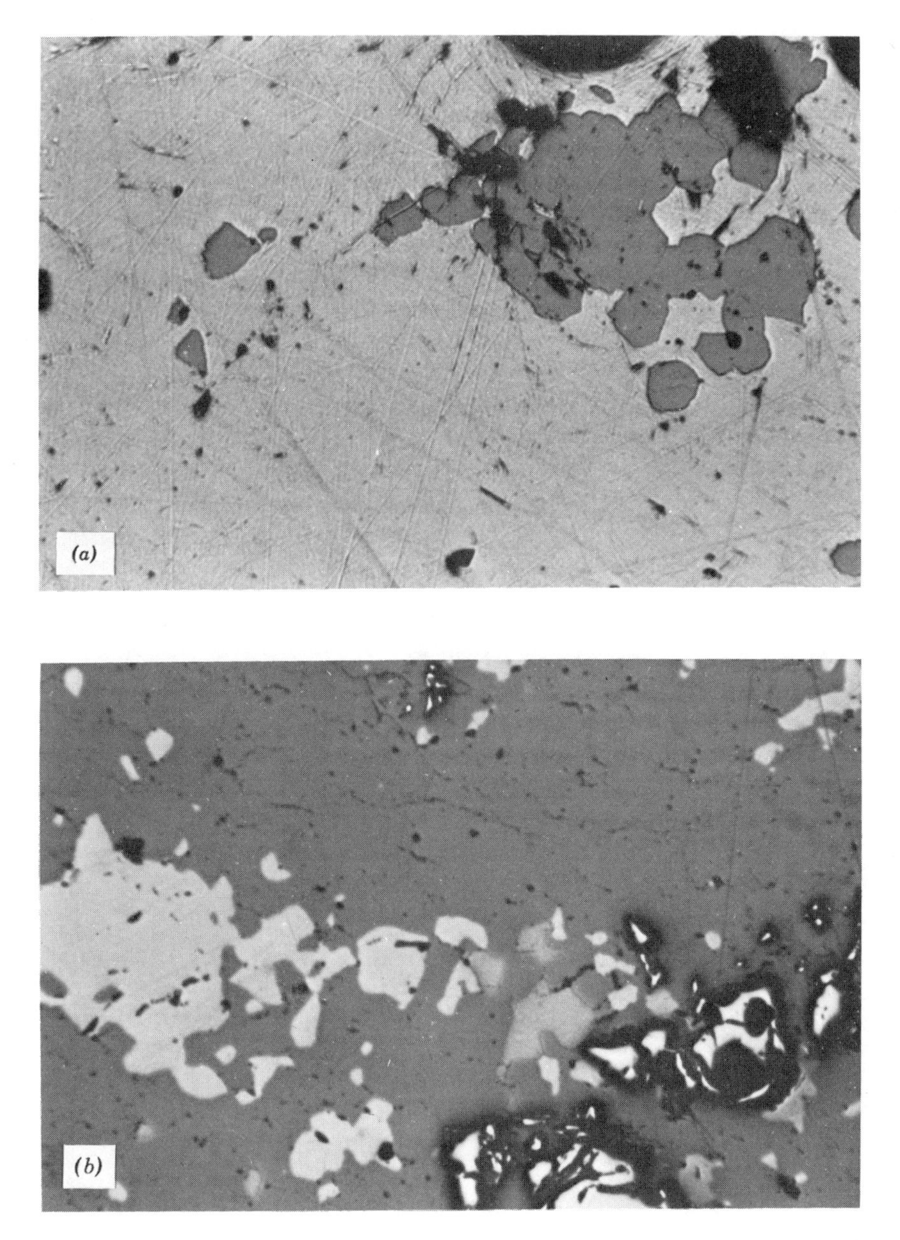

FIGURE 9.27 Lead-zinc-silver ores with antimony and arsenic from the Coeur d'Alene District, Idaho. *(a)* Galena, with inclusions of tetrahedrite. (Width of field = 2000 μm.) *(b)* Tetrahedrite with inclusions of chalcopyrite (medium gray), galena (light gray), and pyrite (white). Sunshine Mine, Idaho. (Width of field = 2000 μm.)

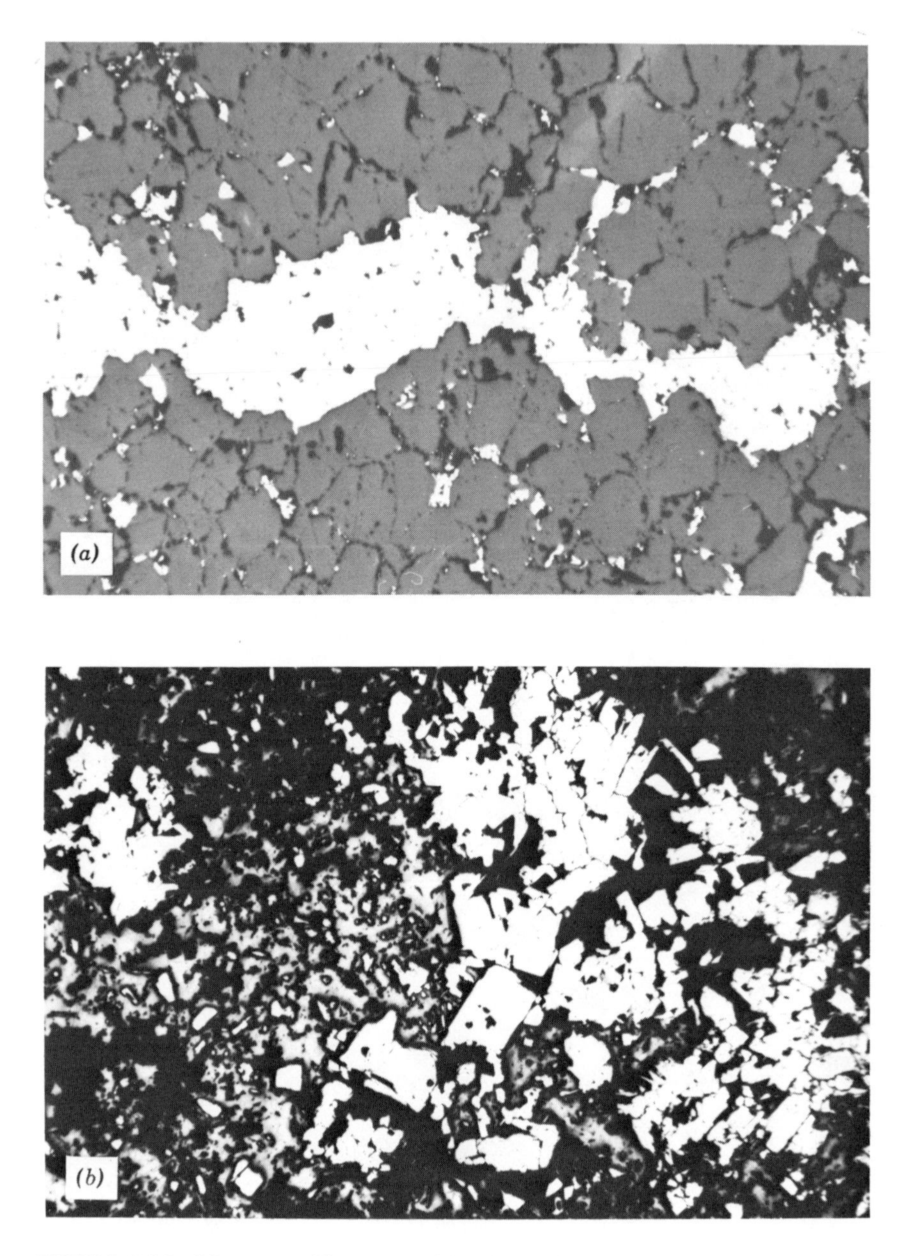

FIGURE 9.28 Mercury sulfide ores: *(a)* veinlet of cinnabar in quartzite, Almadén, Spain (width of field = 2000 μm); *(b)* subhedral crystals of pyrite and marcasite within a matrix of cinnabar, New Idria, California (width of field = 2000 μm).

Mercury Sulfide Vein Deposits

Mineralogy

Major: cinnabar, metacinnabar, pyrite, marcasite.

Minor: native mercury, stibnite, sphalerite, pyrrhotite.

Gangue Minerals: quartz, chalcedony, barite, dolomite, clay minerals.

Examples

Almaden, Spain; Idria, Trsĉe, Yugoslavia; Sulphur Bank, Amedee Hot Springs, New Idria, New Almaden, California; Steamboat Springs, Cordero, Nevada; Amiata, Italy; Huancavalica, Peru.

Mineral Associations and Textures

Cinnabar and, locally, metacinnabar occur in these ores as fine-grained (0.1–1 mm) euhedral single crystals, small polycrystalline aggregates (up to 5 mm), and as stringers, veinlets, and intergranular cements (Figure 9.28*a*). In some areas, cinnabar occurs in alternating veinlets with dolomite and quartz. Pyrite and marcasite occur with the mercury sulfides as isolated subhedral to euhedral crystals and polycrystalline veinlets (Figure 9.28*b*). Stibnite occurs locally as very fine (<0.1 mm) acicular crystals.

Cinnabar, frequently with minor amounts of metacinnabar and stibnite, is well-known as grain-coatings and vug-fillings in tufa and sinter deposits associated with hot spring activity (Dickson and Tunell, 1968; Weissberg et al., 1979). The mercury sulfides form wispy polycrystalline layers of crystals (usually less than 100 mm across) within or on which stibnite grows as individual acicular crystals and rosettelike clusters.

References

Dickson, F. W. and Tunell, G. (1968) Mercury and antimony deposits associated with active hot springs in the Western United States. In J. D. Ridge ed., *Ore Deposits in the United States 1933–1967.* A.I.M.E., New York, Vol. 2, pp. 1673–1701.

Fryklund, V. C. (1964) Ore deposits of the Coeur d'Alene District, Shoshone County, Idaho. *U.S. Geol. Surv. Prof. Paper 445.*

Hammer, D. F. and Peterson, D. W. (1968) Geology of the Magma Mine Area, Arizona. In J. D. Ridge ed., *Ore Deposits in the United States 1933–1967.* A.I.M.E., New York, Vol. 2, pp. 1282–1310.

Linn, R. K. (1968) New Idria Mining District. In J. D. Ridge ed., *Ore Deposits in the United States 1933–1967.* A.I.M.E., New York, Vol. 2, pp. 1623–1649.

Meyers, C. et al. (1968) Ore deposits at Butte, Montana. In J. D. Ridge ed., *Ore Deposits in the United States 1933–1967.* A.I.M.E., New York, Vol. 2, pp. 1374–1416.

Saupe, F. (1973) La Géologie du Gisement de Mercure d'Almaden. *Sci. Terre Mem.* **29**.

Tunell, G. (1964) Chemical processes in the formation of mercury ores and ores of mercury and antimony. *Geochim. Cosmochim. Acta* **28**, 1019–1037.

Weissburg, B. G., Browne, P. R. L., and Seward, T. M. (1979) Ore metals in active geothermal systems. In H. L. Barnes, ed., *Geochemistry of Hydrothermal Ore Deposits,* 2nd ed. Wiley-Interscience, New York, pp. 738–780.

White, D. E. (1967) Mercury and base-metal deposits with associated thermal and mineral waters. In. H. L. Barnes, ed., *Geochemistry of Hydrothermal Ore Deposits.* Holt, Rinehart & Winston, New York, pp. 575–631.

Yund, R. A. and G. Kullerud (1966) Thermal stability of assemblages in the Cu-Fe-S system. *J. Petrol.* **7**, 454–488.

10

Ore Mineral Assemblages Occurring in Sedimentary, Volcanic, Metamorphic, and Extraterrestrial Environments

10.1 INTRODUCTION

In this chapter, the discussion of characteristic associations is continued and extended to ores occupying a variety of sedimentary, volcanogenic, and metamorphic environments, concluding with a brief account of the occurrence of ore minerals in meteorites and lunar rocks. As in Chapter 9, the common associations are described and their modes of origin are briefly discussed with, again, no attempt at a comprehensive account of their geology and petrology.

The first associations described represent part of the continuum of sedimentary processes and include syngenetic placer-type gold deposits, Eh-pH controlled chemical precipitates, coal, and base metal accumulations spatially related to submarine hydrothermal vents and volcanism. Lead-zinc deposits in carbonates (and arenites), although normally regarded as epigenetic, are included here because of their stratabound nature and the increasingly prevalent belief that they are related to diagenetic processes and to the migration of connate brines. The effects of regional metamorphism on ores, especially massive sulfides of sedimentary-volcanic affiliation, are treated in Section 10.9 and the contact metamorphic skarn deposits are discussed in Section 10.10. The unique mineralogical characteristics of extraterrestrial materials, which are becoming increasingly available for study, are described in Section 10.11. References are given to relevant literature at the end of each section; many additional articles appear regularly in the major periodicals such as *Economic Geology*, *Mineralium Deposita*, and *Transactions of the Institution of Mining and Metallurgy*.

10.2 IRON AND MANGANESE ORES IN SEDIMENTARY ENVIRONMENTS

Iron

Most sedimentary rocks contain significant quantities of iron and there is a complete range up to those of ore grade. Sedimentary iron ores can broadly be considered as occurring in three major classes: bog iron ores, ironstones, and (banded) iron formations; this also being the increasing order of their economic importance.

10.2.1 Bog Iron Ores

Mineralogy

Goethite, limonite, siderite; minor carbonates, vivianite [$Fe_3(PO_4)_2 \cdot 8H_2O$].

Mode of Occurrence

As lake or swamp sediments often in temperate or recently glaciated areas or in volcanic streams and lakes; also in association with coal measures in older sedimentary sequences ("blackband ironstones").

Examples

In tundra areas of Canada and Scandinavia; temperate coastal areas of the eastern United States and Canada; in volcanic provinces such as Japan and the Kurile Islands; in Carboniferous and Permian sedimentary sequences in the eastern United States, Northern England, and so on.

Mineral Associations and Textures

Goethite, the major phase of many bog iron deposits, occurs as oolitic or pisolitic grains (1–10 mm) cemented to form disks ($\sim$3–30 cm diameter), which in turn form bands or lenses of ore. Other examples are comprised of more earthy limonite material with substantial carbonate and phosphate. The blackband ironstones are largely siderite.

10.2.2 Ironstones

Mineralogy

Chamosite $(Fe,Al)_6(Si,Al)_4O_{10}(OH)_8$, hematite, limonite or goethite, siderite; minor magnetite, pyrite, collophane $Ca_5(PO_4)_3$ greenalite [$(Fe^{2+},Fe^{3+})_6Si_4O_{10}$-$(OH)_8$].

Mode of Occurrence

As the "minette-type" (or Clinton-type) ores deposited in shallow-water marine sequences associated with a variety of sedimentary rocks including limestones, siltstones, shales, sandstones. Also to a minor extent in volcanic sediments.

Examples

Minette- or Clinton-type ores occur in Phanerozoic sediments of Jurassic age in Alsace-Lorraine and other areas of France, Germany, Belgium, Luxembourg; in the Jurassic of England and in the Silurian (Clinton Beds) of the United States; in volcanic sediments, Lahn-Dill, Germany.

Mineral Associations and Textures

Megascopically, ironstones appear as dull-reddish earthy sandstones or as oolitic or pelletal accumulations that may be distinctly green or red-brown in color. Microscopically, they appear as oolitic, pisolitic, or pelletal grains that may have been entirely or partially replaced by interlayered limonite and hematite (Figure 10.1). As in modern sediments, most ooids and pellets have formed around sand grains or fragmental fossil or mineral material. Micro- and macrofossils are commonly present and may also be replaced by iron minerals. Siderite may occur, chiefly as cementing material, and the silicate, oxide, and carbonate may occur in a wide range of ratios and detailed textural relationships. In sandy facies, the iron minerals occur dominantly as concentric quartz grain coatings and interstitial filling.

10.2.3 Banded Iron Formations

Mineralogy

Major hematite and magnetite, in some cases iron carbonates and silicates may be important, pyrite usually minor. Major gangue mineral is chert.

Mode of Occurrence

As well-bedded, elongate bodies, often with alternating chert-iron mineral stratification. The ores occur as laterally and vertically extensive sequences in Precambrian strata throughout the world. In some cases, the sedimentary setting appears to have been marine, in others estuarine or fresh water, and sometimes an association with volcanic rocks is evident.

Examples

Widespread in Precambrian sequences but major examples include the Lake Superior district of Minnesota-Wisconsin-Michigan-Ontario-Quebec and the Lab-

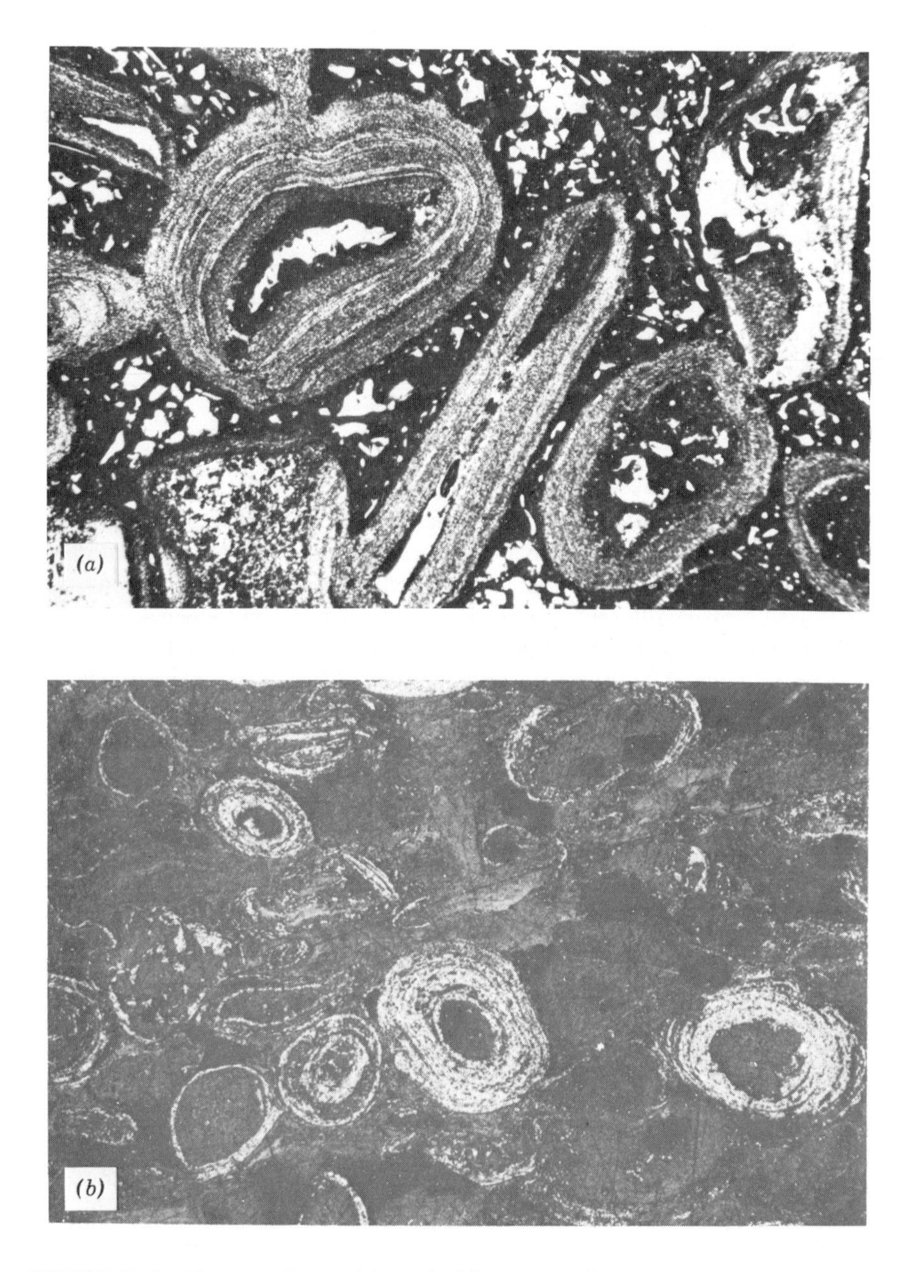

FIGURE 10.1 Textures observed in typical ironstones: *(a)* distorted ooliths comprised of fine-grained chamosite, hematite, and goethite, Red Mountain Formation, Alabama (width of field = 2000 μm); *(b)* ooliths of chamosite and goethite, Rosedale, Northern England (width of field = 2000 μm).

rador-Quebec belts of North America, the Hamersley Basin of Western Australia, and deposits in Brazil, India, and South Africa.

Mineral Associations and Textures

Characterized by banding resulting from the interlayering of oxides and silica both on a coarse and a fine scale (see Figure 10.2); the units may be lenses rather than layers giving a "wavy" appearance to the stratification. Hematite and magnetite constitute the major ore minerals but iron in carbonate, silicates (greenalite, minnesotaite, stilpnomelane), and sulfides (pyrite and very minor pyrrhotite) is also found. In the classic work of James (1954, 1966) the occurrence of the ores dominantly as oxide, carbonate, silicate, or sulfide has been related to a chemical "facies" of precipitation dependent on Eh and pH conditions as indicated in Figure 10.3.

In the oxide facies, two major variants occur—a banded hematite/chert ore and a banded magnetite/chert ore. In the former category, the hematite may occur in the form of ooliths or pisoliths, as finely crystalline laminae, or as laths oriented subparallel to the bedding (Figure 10.2*b*). The latter subdivision shows more variation, with carbonate and silicate phases often present. Magnetite occurs as disseminated to massive bands of subhedral to euhedral grains (Figure 10.2*a*). The carbonate facies has interlayered chert and carbonate that may grade into interlayered silicate/magnetite/chert or carbonate/pyrite. The sulfide facies is characteristically associated not with chert but with black carbonaceous shales in which the sulfide is normally very fine pyrite having clear crystal outlines. Any pyrrhotite is normally found as fine plates with the long axis parallel to bedding. The silicate facies can occur in association with any of the other facies types.

10.2.4 Origins of Iron-Rich Sediments

The bog iron ores can be observed in process of formation so their origin is clearly understood. In the closed drainage systems in tundra areas, these ores are derived by subsurface weathering and leaching with transport of the iron as bicarbonates and humates in ground water of low pH and Eh and subsequent concentration and deposition in lakes and marshes by loss of CO_2 and, commonly, oxidation. The origin of blackband ores is less certain; some may be primary and others of diagenetic origin.

The origins of *ironstone* deposits pose major problems regarding both the source of the iron and the methods by which it is concentrated (see Turner, 1977, for further discussion). In particular cases, the ultimate source of iron may have been continental erosion, submarine volcanic springs, or upwelling ocean currents. The site of final deposition was commonly a shallow water marine environment, but since this was a relatively high energy, oxygenated environment, formation of chamosite is unlikely to have occurred here. Possibly the initial for-

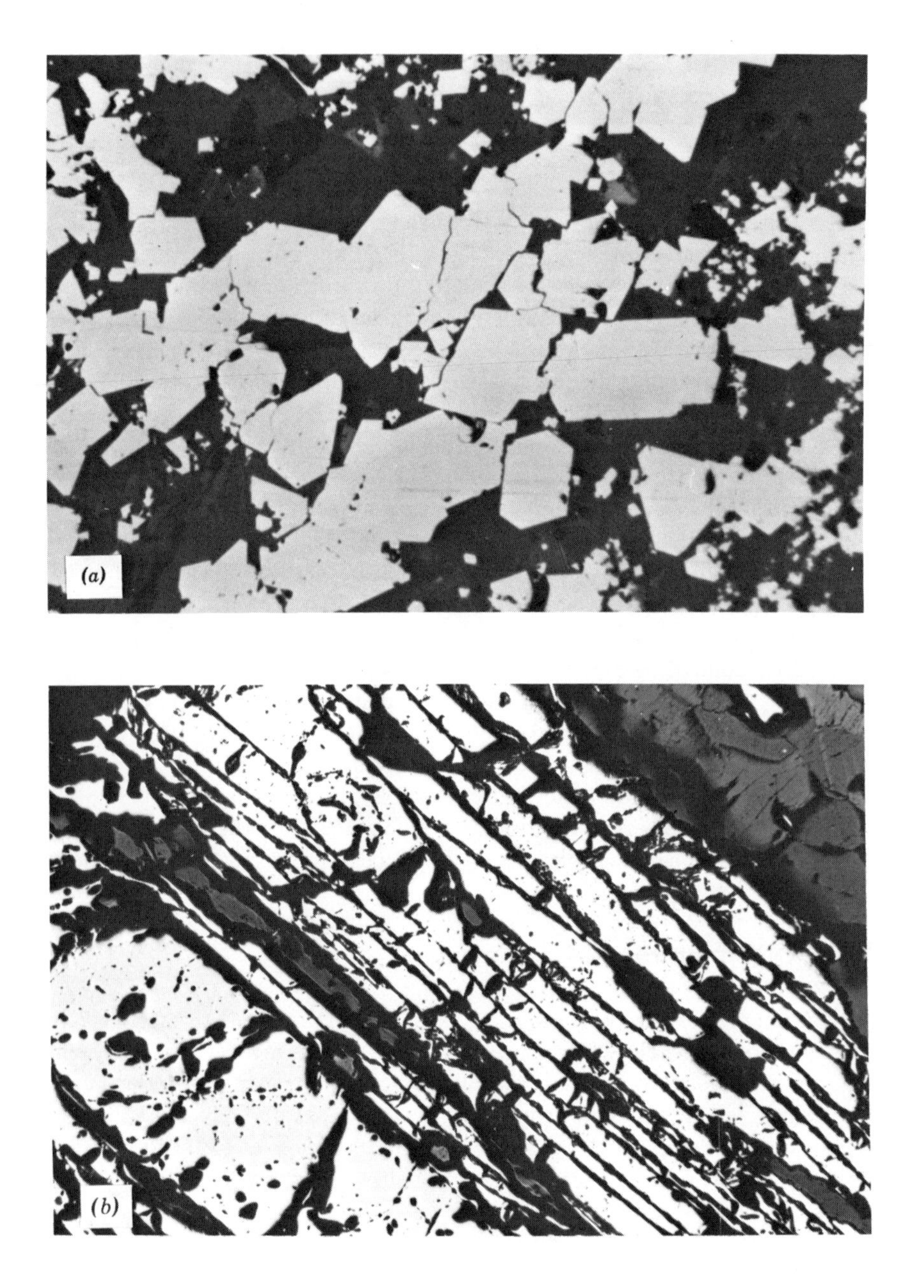

FIGURE 10.2 Typical banded iron formation assemblages and textures: *(a)* layered subhedral magnetite, Eastern Gogebic Range, Michigan (width of field = 300 μm); *(b)* bladed specular hematite with blocky magnetite, Rana Gruber Mines, Norway (width of field = 2000 μm).

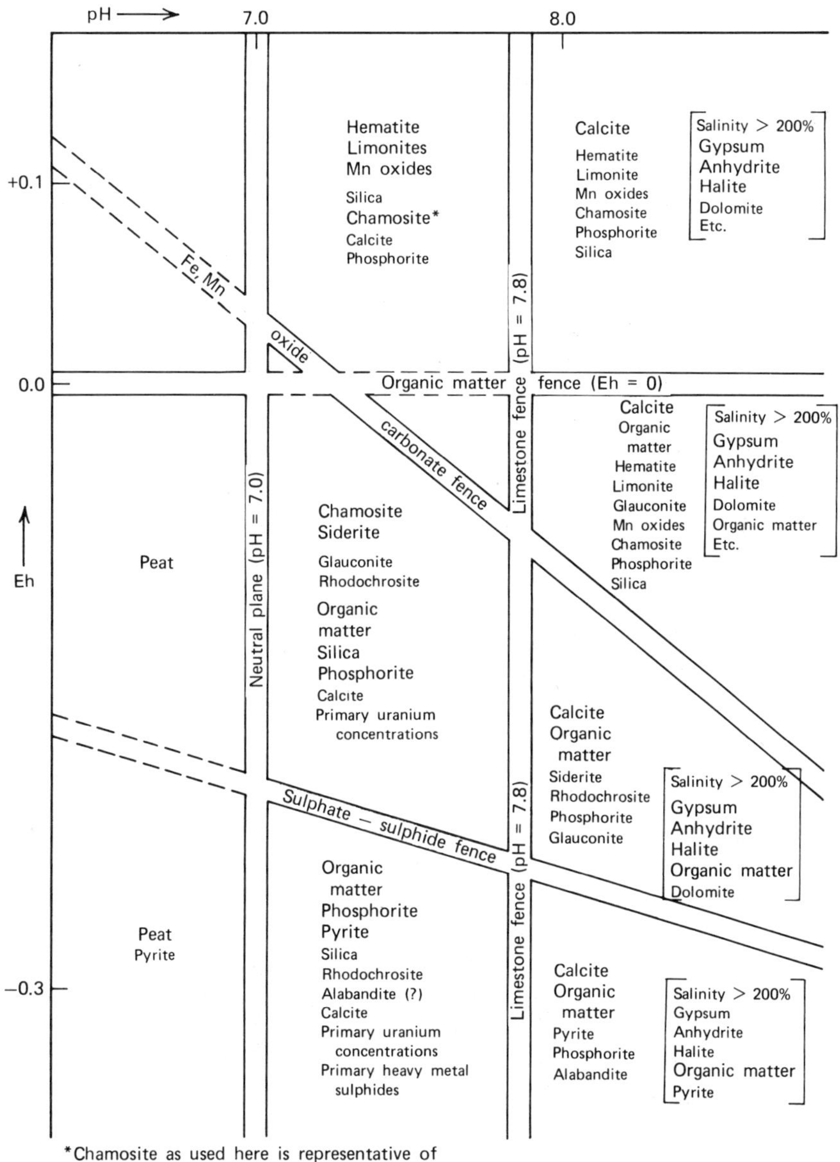

*Chamosite as used here is representative of the sedimentary iron silicates.

FIGURE 10.3 Fence diagram showing Eh-pH fields in which end-member minerals important in sedimentary iron ores and other chemical sediments are formed under normal seawater conditions. Associations in brackets are for hypersaline conditions (salinity > 200‰). (Reprinted from W. C. Krumbein and R. M. Garrels, "Origin and classification of chemical sediments in terms of pH and oxidation-reduction potential," *Jour. Geol.* **60**, 26, 1952, by permission of the University of Chicago Press.)

mation of chamosite occurred in a reducing environment beneath the sediment surface with subsequent transport to the site of accumulation. Ooid formation may have occurred diagenetically during initial formation or during the processes of transport and final deposition. Any understanding of the origin of these ores is further hindered by their complex diagenetic history, which often involves total or partial replacement of ooliths and fossil fragments. These replacement features have led to suggestions that the ores result from wholesale replacement of oolitic limestones by iron minerals, but detailed study of the textures and geological settings shows many flaws in this model.

The banded iron formations are very different from the ironstones in that the former have very low contents of Al_2O_3 and P_2O_5 and a high silica (chert) content. Their restriction to the Precambrian (and Cambrian) is also considered significant in some theories, and requires explanation, as does the regular interbedding of iron ore and chert. Various theories have proposed cyclical continental erosion, seaboard and submarine volcanism, or sea-floor leaching as the origin of the iron, with deposition in either a marine or lacustrine environment. Many workers believe that the absence of free oxygen in the early to mid-Precambrian atmosphere was an important factor in the transport and precipitation of the iron. There is, however, no concensus of opinion regarding the mechanisms of formation. For the arguments in favor and against various conflicting theories the reader is referred to the standard texts of ore geology and to *Economic Geology,* Vol. 68, No. 7, 1973, "Precambrian Iron Formations of the World" or the other original papers.

Manganese

Most sedimentary rocks also contain detectable concentrations of manganese although generally this is an order of magnitude lower in concentration than iron; as with iron-rich sediments, there is a complete range from minor amounts to ore grades. The major classes of sedimentary manganese ores are: marsh and lake deposits, deposits of the orthoquartzite-glauconite-clay association, deposits of volcanic affiliation, and modern marine deposits (including *manganese nodules*). The marsh and lake deposits are commercially insignificant and are undoubtedly the analogues of bog iron ores, being derived by the same processes. The manganese occurs as poorly crystalline hydrous oxides. The other classes of sedimentary manganese deposits will be briefly discussed in turn.

10.2.5 Manganese Deposits of Orthoquartzite-Glauconite-Clay Association

Mineralogy

Pyrolusite, psilomelane, manganite, manganocalcite, and rhodochrosite ($MnCO_3$).

Mode of Occurrence

Thin lens-shaped beds conformable with enclosing sedimentary strata, the sediments having generally been deposited on a stable platform area and being estuarine through to shallow marine deposits (sandstones, silts, clays, glauconitic beds, limestones, and even coaly beds).

Examples

Southern Soviet Union (Nikopol, Chiatura, etc.); Timna region of Southern Israel; in Turkey and Bulgaria and other localities of the Northern Mediterranean (all in Tertiary strata); Northern Australia (L. Cretaceous beds of Groote Eylandt).

Mineral Associations and Textures

The mineral associations and their spatial relationships appear to reflect different facies of chemical sedimentation as suggested for certain iron deposits; in this case a sequence of oxide, oxide-carbonate, carbonate ore appears to represent successively greater distance from a palaeo-shoreline. The ores generally consist of irregular concretions and nodules and earthy masses of the oxides or carbonates in a silt or clay matrix (see Figure 10.4*a*).

10.2.6 Manganese Deposits of the Limestone-Dolomite Association

Mineralogy

Dominantly either oxides (pyrolusite, hausmannite, bixbyite, psilomelane, etc.) in type (1) below, or carbonate (rhodochrosite, etc.) in type (2).

Mode of Occurrence

Subdivided by Varentsov (1964) into:

1 Manganiferous limestone—dolomite formations developed on stable platforms ("Moroccan type").
2 Manganiferous limestone—dolomite formations of geosynclinal zones ("Appalachian-type" and "Usinsk type").

The former type consists of a sequence of manganese oxide ores interlayered with dolomites, limestones, and sometimes gypsum, with both underlying and overlying red terrigenous sediments, all deposited on the eroded surface of a stable platform. The latter type consists either merely of manganiferous limestones or more complex manganese carbonate ores in limestone-dolomite sequences associated with volcanic deposit.

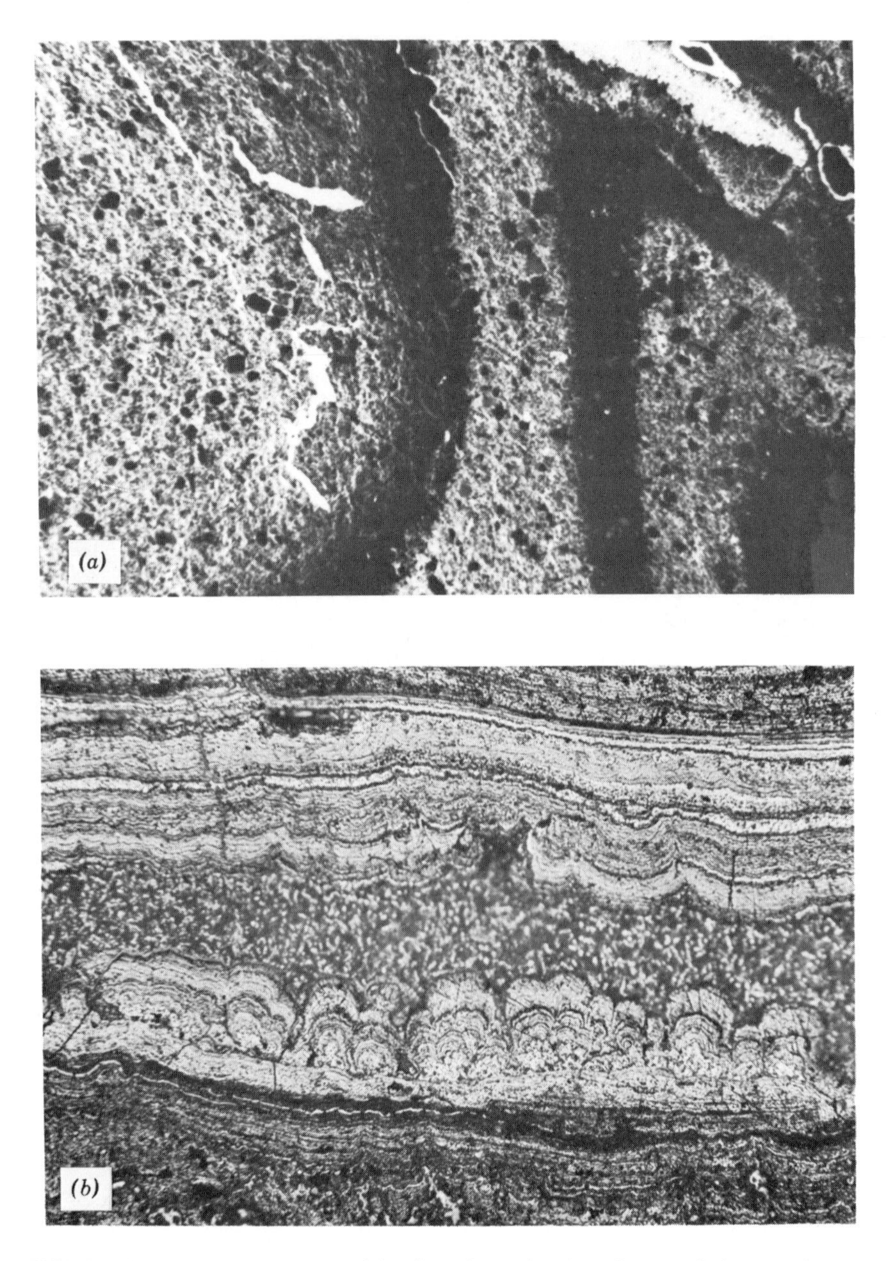

FIGURE 10.4 Manganese ores: *(a)* psilomelane showing characteristic growth texture (width of field = 2000 μm); *(b)* textures of a characteristic manganese nodule, Blake Plateau, Atlantic Ocean (width of field = 520 μm).

Examples

North Africa (Morocco), Appalachian area (United States), Usinsk deposit (S.W. Siberia, Soviet Union).

Mineral Associations and Textures

In both types of deposits the ore zones vary from small lenses to continuous beds of Mn-rich sediment. The associations of type (1) are composed almost entirely of oxides and are low in iron, aluminum, and phosphorus, although heavy metal impurities are characteristic (BaO $< 7\%$, PbO $< 6.5\%$). The ores of type (2) are dominated by calcian and ferroan rhodocrosites and also make up $> 8\%$ of the rock. The carbonates occur as ooliths and very fine laminae intercalated with manganoan stilpnomelane. Algal and sponge remains commonly occur and may be replaced by Mn carbonate. The type (2) ores, although low in Ba and Pb, may contain $> 15\%$ iron oxide and minor phosphorus.

10.2.7 Manganese Deposits of Volcanic Affiliation

Mineralogy

Dominantly the manganese oxides (hausmannite, jacobsite, etc.). Also braunite and associated iron oxides (hematite, magnetite) and minor sulfides (pyrite, arsenopyrite, chalcopyrite, galena, sphalerite, tetrahedrite); also quartz and chalcedony.

Mode of Occurrence

Very widespread in volcanic-sedimentary sequences as concordant lenses of ore (generally rather small).

Examples

In Palaeozoic pyroclastic sequences in Western North America, the Urals (Soviet Union), East Australia; in Tertiary volcanic sequences in Japan, Indonesia, the West Indies.

Mineral Associations and Textures

The manganese occurs as layered and colloform to botryoidal masses of intermixed pyrolusite, psilomelane, todorokite, and jacobsite. Grain-size varies from the micron scale to radiating fibrous masses in which the fibers exceed one centimeter in length.

10.2.8 Modern Marine Deposits ("Manganese Nodules")

Mineralogy

In manganese nodules and marine ferromanganese crusts the dominant minerals are Mn^{4+} oxides related to the terrestrial minerals todorokite, δ-MnO_2 (sometimes termed vernadite) or, more rarely, birnessite. Although most nodules contain significant concentrations of iron, the iron-bearing phases in most nodules are poorly crystalline; recently, the mineral feroxyhyte (δ'-FeOOH) has been claimed to occur as a precursor to goethite (α-FeOOH). Common minor phases are goethite, quartz, feldspar, clays, and zeolites. Manganese-rich sediments and oozes have been less well studied but are also dominated by fine-grained oxide and hydroxide phases.

Mode of Occurrence

Manganese nodules are widely distributed on the floors of the major oceans and are generally spherical and from 1 to 30 cm in diameter. Their importance as a resource is related not only to the manganese and iron they contain, but also to the presence of significant cobalt, nickel, and copper taken up within the structures of the manganese minerals (see Section 11.4.5). Manganese-rich sediments and oozes also occur in ocean regions with a clear link to submarine volcanism in some cases and in confined seas as those of the arctic regions.

Mineral Associations and Textures

The internal structure of manganese nodules is of porous concentric, often colloform, zoning with layering of different widths clearly visible under the microscope (see Figure 10.4*b*). The zones have been classified as laminated, massive, columnar, compact, and mottled and contain simple, arcuate, or chaotic layering (Sorem and Foster, 1972). The zones of high reflectance are those rich in Mn and containing Ni and Co; the darker bands are Fe-rich. The porosity, fractures, irregular layering, and included organic matter require impregnation of materials prior to preparation of polished sections.

10.2.9 Origins of Manganese-Rich Sediments

The manganese deposits of volcanic affiliation almost certainly owe their origin to direct discharge of manganese from submarine volcanoes or hot springs and subsequent accumulation as chemical sediments along with other sedimentary and volcanic detritus. A similar origin can be invoked for certain manganese concentrations in modern marine environments.

In contrast, the manganese ores of limestone-dolomite and of ortho-quartziteglauconite-clay association show no clear indication of a volcanic origin for the manganese although this has been invoked by some authors. Both the source of manganese and mechanism of its precipitation are more problematic.

The origin of the manganese and other metals concentrated in nodules and oozes may derive from volcanic and partly from terrigenous sources. The growth of nodules, apparently takes place by the release of ions in the reducing environment beneath the sediment surface, their upward migration when thus mobilized, and their subsequent fixation following oxidation at the sediment surface and attachment as layers on a detrital particle (see Glasby 1977 and Burns 1979).

10.2.10 Gossans

General

Gossans, or "iron-caps," develop on many types of iron sulfide-bearing deposits as a result of surface or near-surface weathering and oxidation.

Mineralogy

Goethite, limonite, lepidocrocite, hematite in varying proportions, sometimes with minor amounts of manganese oxides and residual base metal sulfides.

Examples

Developed world-wide at surficial exposures of sulfide containing deposits.

Mineral Associations and Textures

Gossans generally consist of porous irregular laminae and colloform bands of mixed iron oxides with minor but variable amounts of residual sulfide minerals. Remnant textures or primary grain shapes, cleavages, and fractures are commonly preserved in the iron oxides. The grain size is usually very fine ($< 10\ \mu m$) but radiating bundles of fibrous crystals (like those in Figure 7.13) up to a millimeter or more in length may be locally developed.

Origin of Gossans

Gossans develop as iron sulfides decompose during surface and near-surface oxidation to yield sulfuric acid and soluble ferrous sulfate in reactions such as:

$$FeS_2 + 3\tfrac{1}{2}O_2 + H_2O = FeSO_4 + H_2SO_4$$

The acid dissolves base metal sulfides and the ferrous sulfate oxidizes to leave mixed iron oxides. The end result is a porous network of filimentous, layered to concentrically developed iron oxides. A detailed discussion of the nature and origin of gossans has been recently presented by Blain and Andrew (1978).

Selected References

Blain, C. F. and Andrew, R. L. (1977) Sulphide weathering and the evaluation of gossans in mineral exploration. *Mineral. Sci. Eng.* **9**, 119–150

Burns, R. G., ed. (1979) *Marine Minerals.* Mineral. Soc. Am. Short Course Notes, Vol. 6.

Economic Geology, Vol. 68, No. 7 (1973) "Precambrian Iron-Formations of the World."

Glasby, G. P., ed. (1977) *Marine Manganese Deposits.* Elsevier Oceanography Series (15). Elsevier, Amsterdam. (note particularly articles by Cronan, Sorem and Fewkes, Burns and Burns)

James, H. L. (1954) Sedimentary facies of iron formation. *Econ. Geol.* **49**, 235–293.

James, H. L. (1966) Chemistry of the iron-rich sedimentary rocks. *U.S. Geol. Surv. Prof. Paper 440.*

Krumbein, W. C. and Garrels, R. M. (1952) Origin and classification of chemical sediments in terms of pH and oxidation-reduction potentials. *J. Geol.* **60**, 1–33.

Roy, S. (1968) Mineralogy of the different genetic types of manganese deposits, *Econ. Geol.* **63**, 760–786.

Sorem, R. K. and Foster, A. R. (1972) Internal structure of manganese nodules and implications in beneficiation. In D. R. Horn, ed., *Ferromanganese Deposits on the Ocean Floor.* NSF, Washington, D.C., pp. 167–179.

Turner, P. (1977) Ironstones. Brit. Assoc. Adv. Sci. Ann. Mtg. Aston. Univ. Sept. 1977.

Varentsov, I. M. (1964) *Sedimentary Manganese Ores.* Elsevier, Amsterdam.

10.3 OPAQUE MINERALS IN COAL

Mineralogy

Major: pyrite, marcasite.

Rare: arsenopyrite, chalcopyrite, bornite, sphalerite, galena, millerite, linnaeite, rutile, pyrrhotite.

Associated Minerals: quartz, calcite, dolomite, siderite, kaolinite, illite, gypsum, and a variety of secondary iron sulfates and iron oxides.

Mode of Occurrence

The sulfides are generally present in coal as: (1) veins that are thin or filmlike on the vertical joints (cleats); (2) lenses that range from millimeters to tens of centimeters across; (3) nodules or balls in which sulfides are intergrown with variable amounts of carbonates and clays; (4) disseminated crystals and globules replacing organic matter.

Examples

Virtually all known coal deposits contain sulfides although the amount present is highly variable. Work in several coal fields has demonstrated a correlation of sulfide occurrence with the proximity of overlying marine strata suggesting that sulfur may have been derived by bacterial reduction of seawater sulfate.

Mineral Associations and Textures

Sulfide minerals are the most well-known contaminants in coals because they are major contributors to the total sulfur content of the coal and because they are often macroscopically visible. There is a strong tendency in the coal industry to refer to all sulfides generally as pyrite, and, although it is the dominant sulfide mineral, it is often not the only one present. Marcasite, the dimorph of pyrite, is often present and intergrown with pyrite; minor amounts of sphalerite, galena, arsenopyrite, chalcopyrite, bornite, millerite, and linnaeite have been reported locally. Pyrite in the typical veins, lenses, and nodules is generally fine-grained and anhedral as shown in Figure 10.5*a*. Individual layers within the lenses and nodules are usually composed of roughly equant, generally fine, anhedral grains. In some occurrences on cleats, especially where the sulfide is thin and disseminated, pyrite is intergrown with marcasite and exhibits a poorly-developed radial growth structure. Marcasite appears to have been the primary phase in many occurrences and to have been subsequently converted in part to pyrite. Disseminated pyrite may be present either along cleats or within the residual structure of the organic matter. In the latter case, the pyrite frequently occurs as framboids, tiny spherical aggregates of pyrite euhedra as shown in Figure 10.5*b*. Sphalerite has been extensively studied in the coals of Illinois (United States) where it occurs as cleat fillings (Hatch et al., 1976). The sphalerite contains up to 2.5 wt. % Fe and as much as 1.3 wt. % Cd. Its color is anomalous and includes gray-white and purple varieties; the cause of the coloration is not known.

Origin of the Opaque Minerals in Coals

The sulfides in coal constitute approximately one-half of the total sulfur content of the coal and are believed to have formed, for the most part, through the activity of sulfate-reducing bacteria during diagenesis. Studies of sulfur isotopes support this mode of origin of sulfur (Price and Shieh, 1979). The original sulfur content of the plant matter remains trapped in the organic substances now constituting the coal and was clearly insufficent to account for the bulk of the sulfide present in many coals. Although some disseminated pyrite, especially that

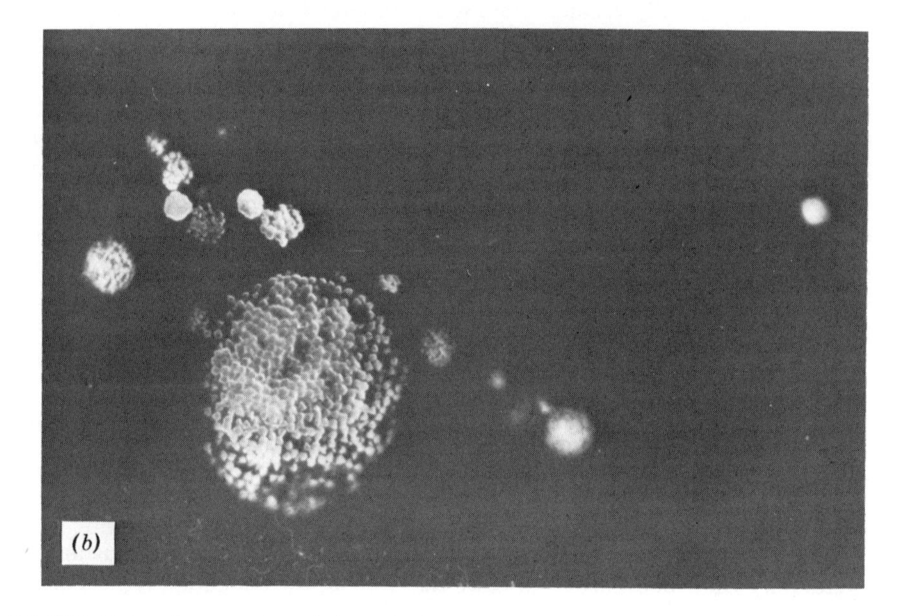

FIGURE 10.5 Opaque minerals in coal: *(a)* anhedral grains of pyrite and pyrite infilling cleats, Minnehaha Mine, Illinois (reproduced from F. T. Price and Y. N. Shieh, *Econ. Geol.* **74**, 1448, 1979 with permission of authors and publisher); *(b)* typical framboidal pyrite in an Appalachian coal (width of field = 25 μm) (photograph courtesy of Dr. F. Caruccio).

occurring as framboids, may have formed at the time of burial, most of the sulfide apparently formed during later diagenesis or metamorphism.

Coal Petrography

Coal itself is readily studied by means of reflected light microscopy. It is a heterogeneous, noncrystalline material, the components of which are called "macerals." There are three major groups—vitrinite, exinite, and inertinite—which are in turn subdivided as a function of the nature and shape of the material (e.g., collinite, cutinite, fusinite). The International Committee for Coal Petrology has established standards for coal microscopy and defined the morphological and reflectance characteristics of each maceral. For a comprehensive treatment of coal macerals and the techniques employed in their examination, the student is referred to Strach's (1975) *Coal Petrology.*

Selected References

Boctor, N. Z., Kullerud, G., and Sweany, J. L. (1976) Sulfide minerals in Seelyville Coal III, Chinook Mine, Indiana. *Mineral. Deposita* **11**, 249–266.

Gluskoter, H. J. (1975) Mineral matter and trace elements in coal. In *Trace Elements in Fuel, Advances in Chemistry Series No. 141,* pp. 1–22.

Gluskoter, H. J. (1977) Inorganic sulfur in coal. *Energy Sources* **3**, 125–131.

Hatch, J. R., Gluskoter, H. J. and Lindahl, P. C. (1976) Sphalerite in Coals from the Illinois Basin. *Econ. Geol.* **71**, 613–624.

Price, F. T. and Shieh, Y. N. (1979) The distribution and isotopic composition of sulfur in coals from the Illinois Basin. *Econ. Geol.* **74**, 1445–1461.

Strach, E. et al. (1975) *Coal Petrology,* 2nd ed. Gebrüder Borntraeger, Berlin.

10.4 URANIUM-VANADIUM-COPPER ORES ASSOCIATED WITH SANDSTONES

Mineralogy

Different deposits show varying relative concentrations of minerals of the three metals, often with only one or two of the metals present or greatly dominant. *Major minerals* from each metal are:

Uranium: uraninite (or its cryptocrystalline-equivalent, pitchblende), coffinite [$U(SiO_4)_{1-x}(OH)_{4x}$], various types of asphaltic organic matter bearing uranium.

Vanadium: roscoelite (vanadium mica), montroseite [$VO(OH)$] and vanadium bearing mixed-layer clay minerals.

Copper: chalcocite (or related minerals of $\sim Cu_2S$ composition), bornite, chalcopyrite, covellite, native copper.

Other (usually minor) minerals include a very large number of sulfides, notably pyrite, galena, sphalerite, gersdorffite, molybdenite, native gold and silver; silver sulfides. *Secondary minerals* include a very wide range of oxides, hydrated oxides, sulfates, and carbonates produced from the primary assemblages.

Mode of Occurrence

Within conglomerates, sandstones and siltstones (particularly the reduced zones within red beds) as irregular masses of ore occurring as fillings of pore spaces, veinlets, and replacing organic materials, particularly fossil plants. Also as veins and veinlets closely associated with a major unconformity recording a period of continental weathering.

Examples

The Colorado Plateau area of Colorado, Arizona, Utah, and New Mexico; Wyoming; Texas; the Athabasca Sandstone, Northern Saskatchewan; Darwin Area, Australia (the most important uranium-bearing examples). Copper-rich examples are very widespread—Corocoro, Bolivia; Udokan, Siberia. Many examples of limited or no economic importance are also known (e.g., Pennsylvania, United States; Alderley Edge, Cheshire, England).

Mineral Associations and Textures

The deposits of this group commonly occur in continental and marginal-marine clastic sediments that are considered as having been deposited under fluviatile-deltaic conditions. The bulk of these sediments have the distinctive red color due to fine-grained hematite and ferric oxyhydroxides which lead to the term *red beds.* However, in contrast to the surrounding red sediments, the mineralized zones are often grey-greenish, containing a predominance of ferrous iron with relatively high sulfur and carbon concentrations. The mineralization occurs as lenses, pods, layers, or concave "rolls" that are grossly conformable with enclosing sediments although cross-cutting in detail. Most concentrations of mineralization are associated with organic debris and some significant uranium deposits have resulted from the replacement of fossil logs.

The amounts of uranium, vanadium, and copper mineralization vary enormously both within and between deposits. The copper ores may also contain significant concentrations of silver. The ore minerals occur as veinlets and pore-space fillings in the sediments and as replacements of fossil plant matter. Often the cell structure of wood may be preserved, although totally replaced by uranium minerals and copper and iron sulfides (Figure 10.6). Although the major copper sulfide described from these deposits is "chalcocite," more recent descriptions of the complex phase relations in the Cu-S system (and the system Cu-Fe-S) suggest that much may remain to be learned regarding the detailed mineralogy of these ores and the roles played by such phases as digenite, djurleite, anilite (see Figure 10.7). A sulfur-rich bornite is common in many red bed deposits where it may occur interstitially between sand grains or where it may selectively replace organic structures. The stability and mode of formation of this type of bornite remains an unresolved problem in phase relations of the Cu-Fe-S system. Pyrite is common in the reduced portions of the ores but is largely replaced by secondary hematite and hydrated iron-oxides in more oxidized areas.

At the Rabbit Lake Deposit, Northern Saskatchewan, mineralization is spatially related to the unconformity that underlies the Athabasca Formation, a fluviatile sedimentary sequence that includes red sediments with reduced zones and also carbonaceous materials. The mineralization has been studied in detail (Hoeve and Sibbald in Kimberley, 1978), and several stages, associated with oscillating episodes of oxidation and reduction, have been recognized. The earliest mineralization occurs as fracture or breccia infillings with pitchblende as colloform encrustations and later massive mineralization associated with coffinite and sulfides. A second stage is of veins of complex paragenesis including euhe-

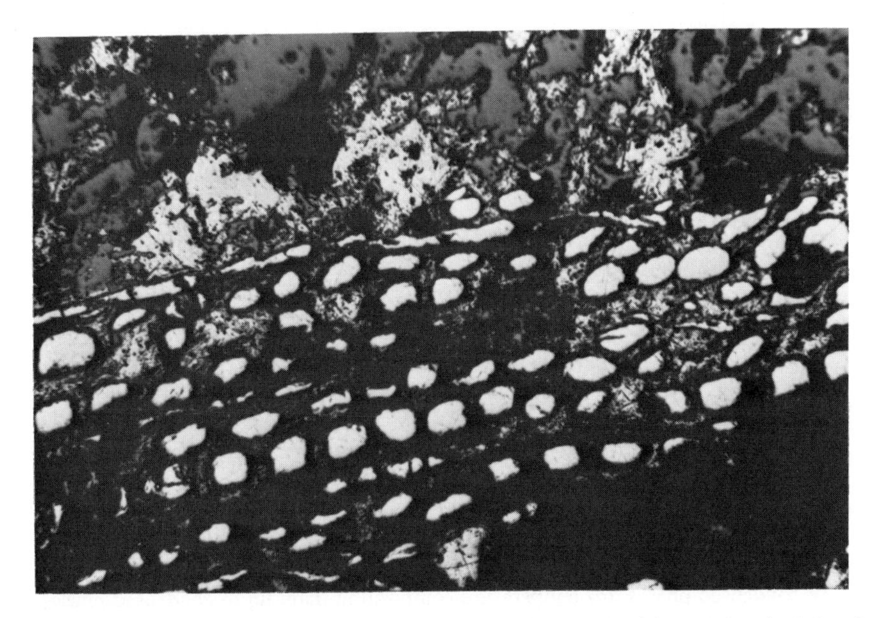

FIGURE 10.6 Pyrite replacing cellular structure in wood with associated chalcocite. Marysvale, Utah. (Width of field = 520 μm.)

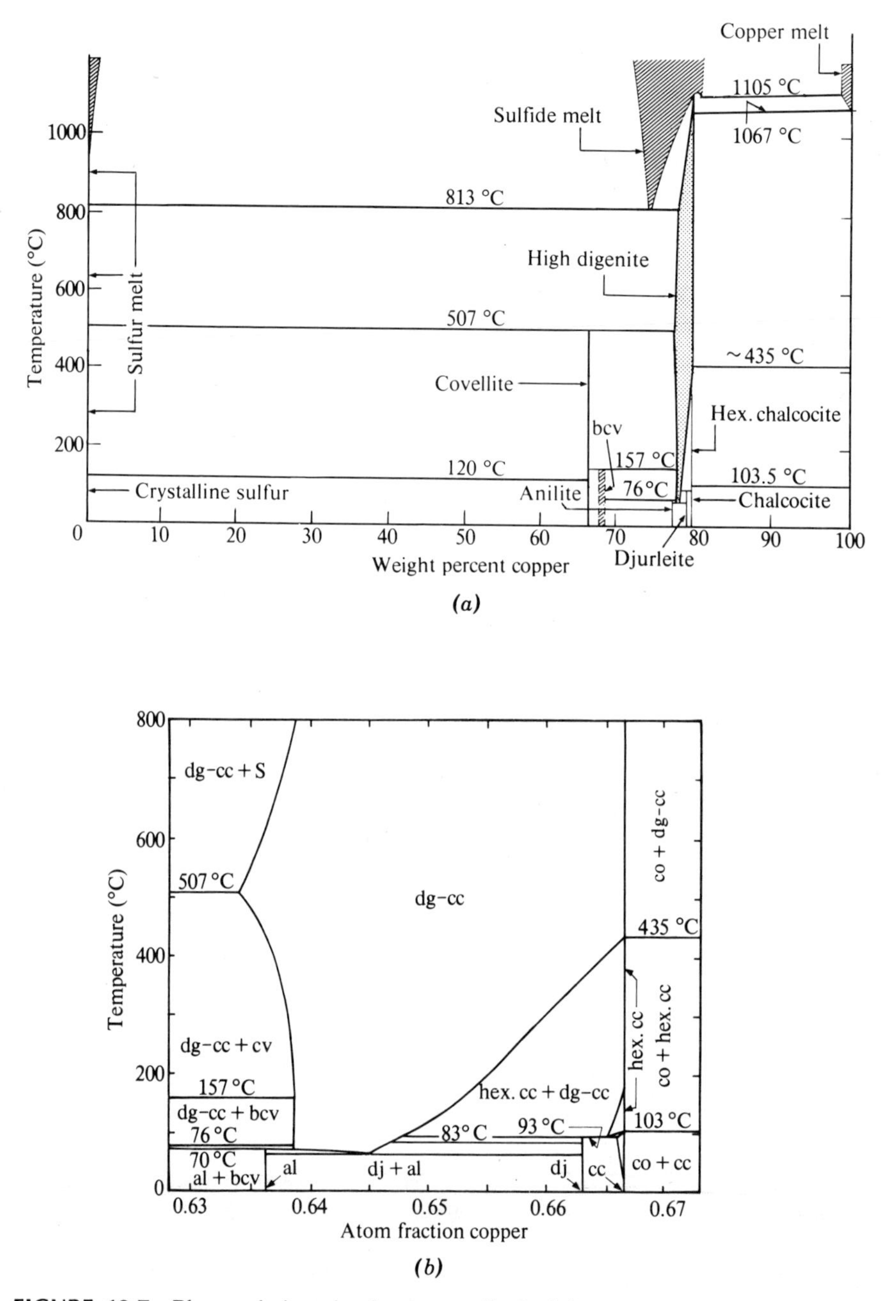

FIGURE 10.7 Phase relations in the system Cu-S: *(a)* temperature-composition diagram of condensed phases; *(b)* temperature-composition diagram of condensed phases in central portion of the system. Abbreviations: dg-cc, digenite-chalcocite; cv, covellite; al, anilite; dj, djurleite; co, copper. (From D. J. Vaughan and J. R. Craig, 1978.)

dral quartz, carbonates, pitchblende, coffinite, sulfides, arsenides, and native copper. The final stage involves impregnation by sooty pitchblende and coffinite along fractures and joints and appears to be a reworking of preexisting of mineralization.

Origin of the Ores and Textures

It is widely accepted from overwhelming evidence that the ore minerals of this group were introduced later than the deposition of the host sediments. These ores are the result of precipitation from solutions passing through the sediments. The problems of origin concern the nature and source of the solutions and their mode of transport and of precipitation. Three main suggestions have been made for the origin of the solutions.

It is widely accepted the solutions were groundwater that leached metal ions and SO_4^{2-} ions from associated strata at low temperature, and that the ore minerals were precipitated on encountering local reducing environments. The field relations of the major uranium-bearing deposits allow for large-scale leaching of underlying granites and other arkosic rocks and interspersed volcanic ash, which could supply the necessary metals. In oxidizing solutions, uranium can be transported as the relatively soluble uranyl $(U^{6+}O_2)^{2+}$ion or as a carbonate complex such as $UO_2(CO_3)_3^{4-}$ or $UO_2(CO_3)_2^{-2}$. The relatively insoluble uranous (U^{4+}) ion is stabilized under reducing conditions where it may precipitate out to form uraninite (UO_2) as shown in Figure 10.8. Since the mineralized zones are commonly grey-greenish (Fe^{2+}-rich) reduced areas, the concept of precipitation on encountering such areas appears sound. The formation of such localized reducing areas is commonly linked with the presence of organic matter. Bacterial activity in these areas could cause reduction of sulfate in the pore waters, which could react with introduced copper in solution to precipitate the highly insoluble copper sulfides. Vanadium could be transported as the V^{4+} ion and precipitated

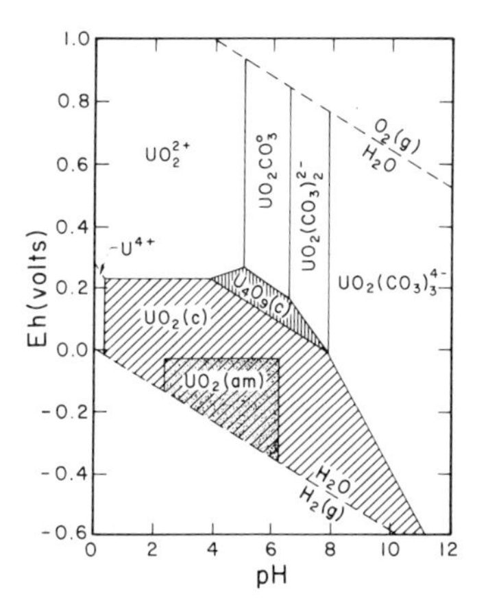

FIGURE 10.8 Eh-pH diagram in the U-O_2-CO_2-H_2O system at 25°C for $P_{CO_2} = 10^{-2}$ atm, showing the stability fields of amorphous $UO_2[UO_2(am)]$, ideal uraninite $[UO_2(c)]$, and $U_4O_9(c)$. Solid solution boundaries are drawn at 10^{-6} M (0.24 ppm) dissolved uranium species. (After D. Langmuir in M. M. Kimberley, 1978; used with permission of the author.)

by reduction in the mineralized zone. The theory that organic reduction is important in the formation of the sulfide ores of this association is supported by data from sulfur isotopes. Fluid movement through porous sandstones often results in development of a "roll-type" deposit in which solution of U-V minerals occurs along an oxidizing surface that follows behind a reducing front where precipitation is occurring (Figure 10.9).

An alternative suggestion for the origin of the mineralizing solutions is that they were derived directly from igneous rocks at depth (i.e., of igneous/hydrothermal origin). Such solutions would have to pass up through fractures into the sedimentary hosts in which precipitation could occur through the same processes already outlined. The general absence of large feeder veins, the vast lateral extent of such examples as the great Colorado Plateau deposits, and the absence of a universal association with igneous sources are major arguments against this theory. A third view combines the first two theories by deriving the solutions from mixing of hydrothermal and meteoric solutions.

Work by Rose (1976) on the copper deposits of this association has drawn attention to their geological association with evaporites, which could have furnished chloride-rich groundwaters. In most normal oxidizing groundwaters, the solubility of copper is less than 1 ppm at reasonable pH values. However, in chloride-containing solutions, the cuprous ion forms the complexes $CuCl_2^-$ and $CuCl_3^{2-}$ which allow solubilities of ~100 ppm copper in 0.5 m Cl^- at pH 7.0 and intermediate Eh. As illustrated in Figure 10.10, the $CuCl_3^{2-}$ complex is stable under pH/Eh conditons compatible with the presence of hematite. The solutions transporting copper are believed to be in equilibrium with hematite,

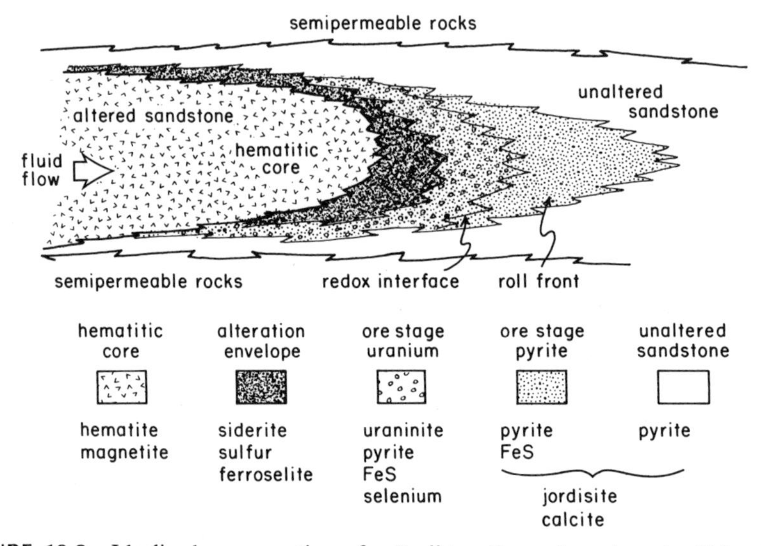

FIGURE 10.9 Idealized cross section of a "roll-type" uranium deposit. (After H. C. Granger and C. G. Warren, 1974.)

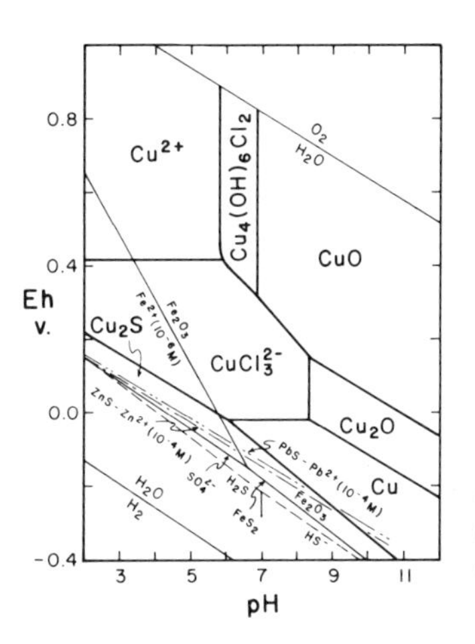

FIGURE 10.10 Eh-pH diagram for the system Cu-O-H-S-Cl at 25°C ($\Sigma S = 10^{-4}$ M, $Cl^- = 0.5$ M as NaCl, boundaries of Cu species at 10 M). Boundaries are also shown for some Fe and S-bearing species; the stability of copper sulfides other than chalcocite is not shown. (After A. W. Rose, *Econ. Geol.* **71,** 1041, 1976; used with permission.)

quartz, feldspar, and mica and at temperatures below ~75°C. The general relationships illustrated in Figure 10.10 (at 25°C) therefore apply, so that reduction will bring about precipitation of copper as copper sulfides.

Selected References

Fischer, R. P. (1968) The uranium and vanadium deposits of the Colorado Plateau region. In J. D. Ridge, ed., *Ore Deposits of the United States 1933/67. A.I.M.E.,* New York, pp. 735–746.

Granger, H. C. and Warren, C. G. (1974) Zoning in the altered tongue associated with roll-type uranium deposits. In *Formation of Uranium Ore Deposits,* Proceedings of the Symposium on the Formation of Uranium Ore Deposits, International Atomic Energy Agency, Vienna. International Atomic Energy Agency, pp. 185–200.

Kimberley, M. M., ed. (1978) *Uranium Deposits, Their Mineralogy and Origin.* Mineral. Assoc. Canada Short Course Hdbook, Vol. 3.

Rose, A. W. (1976) The effect of cuprous chloride complexes in the origin of Red-bed copper and related deposits. *Econ. Geol.* **71**, 1036–1044.

Woodward, L. A., Kaufman, W. H., Schumacher, O.L., and Talbott, L. W. (1974) Stratabound copper deposits in Triassic sandstone of Sierra Nacimiento, New Mexico. *Econ. Geol.* **69**, 108–120.

10.5 GOLD-URANIUM ORES IN CONGLOMERATES

Mineralogy

Major pyrite; gold (as native metal with some silver) may be a major ore mineral; uranium in uraninite, "thucholite" (a uraniferous hydrocarbon) or brannerite (a complex uranium-bearing silicate) may also be major in importance. Osmiridium (OsIr) may be of minor economic importance.

Associated sulfides of no economic significance may include marcasite, pyrrhotite, sphalerite, galena, molybdenite; also arsenides and sulfarsenides containing Co and Ni may occur.

Detrital heavy minerals may include chromite and zircon, ilmenite, magnetite, rutile and various mafic silicates. The remainder of the ore material is largely made up of quartz with significant feldspar, sericite, chlorite, and chloritoid.

Mode of Occurrence

The gold and uranium ores occur disseminated in beds and lenses of coarse conglomerate which are parts of arenaceous sequences. Generally, the conglomerates are of quartz pebbles and appear to be of fluiviatile or shallow-water deltaic origin. The major examples are Pre-Cambrian in age.

Examples

Two very important examples are the Witwatersrand, South Africa (major gold and uranium deposit) and Elliot Lake (Blind River), Ontario, Canada (major uranium deposit); Jacobina, Brazil is a less important gold deposit.

Mineral Associations and Textures

The ore minerals of gold and uranium are generally very fine-grained (commonly not visible except under the microscope). They commonly occur interstitially to the conglomeratic fragments with gold as grains and pore-space fillings, although they occur occasionally as fine veinlets. Uraninite, thucholite and brannerite occur as detrital grains and in some cases as colloform sheets or as veinlets, and gold and uranium show a strong tendency to occur together. Osmiridium in the Witwatersrand ores is partly intergrown with gold. Pyrite, found in concentrations of 2–12% by volume in the conglomerates has been described by Ramdohr (1958) as being of three textural types:

1. Allogenic. Having rounded outlines and a smooth homogeneous interior.
2. Concretionary authigenic. Having a structure composed of loosely aggregated fragments.
3. Reconstituted authigenic.

A detailed account of the mineralogy of Witwatersrand ores is given by Feather and Koen (1975). Examples of both the textural varieties of pyrite and of the occurrence of gold in the Witwatersrand ores are shown in Figures 10.11*a*, and 10.11*b*. Ore concentrations are usually greatest where the conglomerates are thickest, and, in the Witwatersrand ores, the silver content of the gold (which ranges from ~5 to 16 wt. %) shows systematic variation, decreasing with depth in any single ore lens ("reef"). High gold concentrations also occur in very thin carbonaceous seams and much of the gold in the conglomerates may represent the reworking of such material.

Origin of the Ores

Historically three main theories have developed regarding the origin of these ores:

1. Placer theory. The ore minerals have been derived by erosion of adjacent areas, transported, and deposited by streams along with the conglomerates.
2. Hydrothermal theory. The gold, uranium, and some of the other metals have been introduced in hot aqueous solutions derived from an external source such as an igneous intrusion.
3. Modified placer theory. The ore minerals, having been deposited as placers, have been locally redistributed within a particular orebody.

The occurrence of gold and of uranium mineral veinlets suggest at least some textural modification so the major disagreement on origin has been between a modified placer and a hydrothermal theory. The bulk of the evidence available at present favors a modified placer origin for these deposits.

Specific arguments concerning the origin of the Elliot Lake and the Witwatersrand deposits are outlined by Derry (1960) and by Pretorious (1975, 1976).

Selected References

Derry, D. R. (1960) Evidence of the origin of the Blind River uranium deposits. *Econ. Geol.* **55**, 906–927.

Feather, C. E. and Koen, G. M. (1975) The mineralogy of the Witwatersrand reefs. *Mineral. Sci. Eng.* **7**, 189–224.

Hallbauer, D. K. and Utter, T. (1977) Geochemical and morphological characteristics of gold particles from recent river deposits and the fossil placers of the Witwatersrand. *Mineral. Deposita* **12**, 293–306.

Minter, W. E. L. (1978) A sedimentological synthesis of placer gold, uranium and pyrite concentrations in Proterozoic Witwatersrand sediments. In A. D. Maill, ed., *Fluvial Sedimentology,* Can. Soc. Petrol. Geol. Mem. **5**, 801–829.

Pretorius, D. A. (1975) The depositional environment of the Witwatersrand goldfields: a chronological review of speculations and observations *Mineral. Sci. Eng.* **7**, 18–47.

Pretorius, D. A. (1976) The nature of the Witwatersrand gold-uranium deposits. In K. H. Wolf, ed., *Handbook of Stratabound and Stratiform Ore Deposits.* Elsevier, Amsterdam-Oxford-New York, Vol. 7, pp. 29–88.

Ramdohr, P. (1958) New observations on the ores of the Witwatersrand in South Africa and their genetic significance. *Trans. Geol. Soc. S. Afr.* **61**, 1–50.

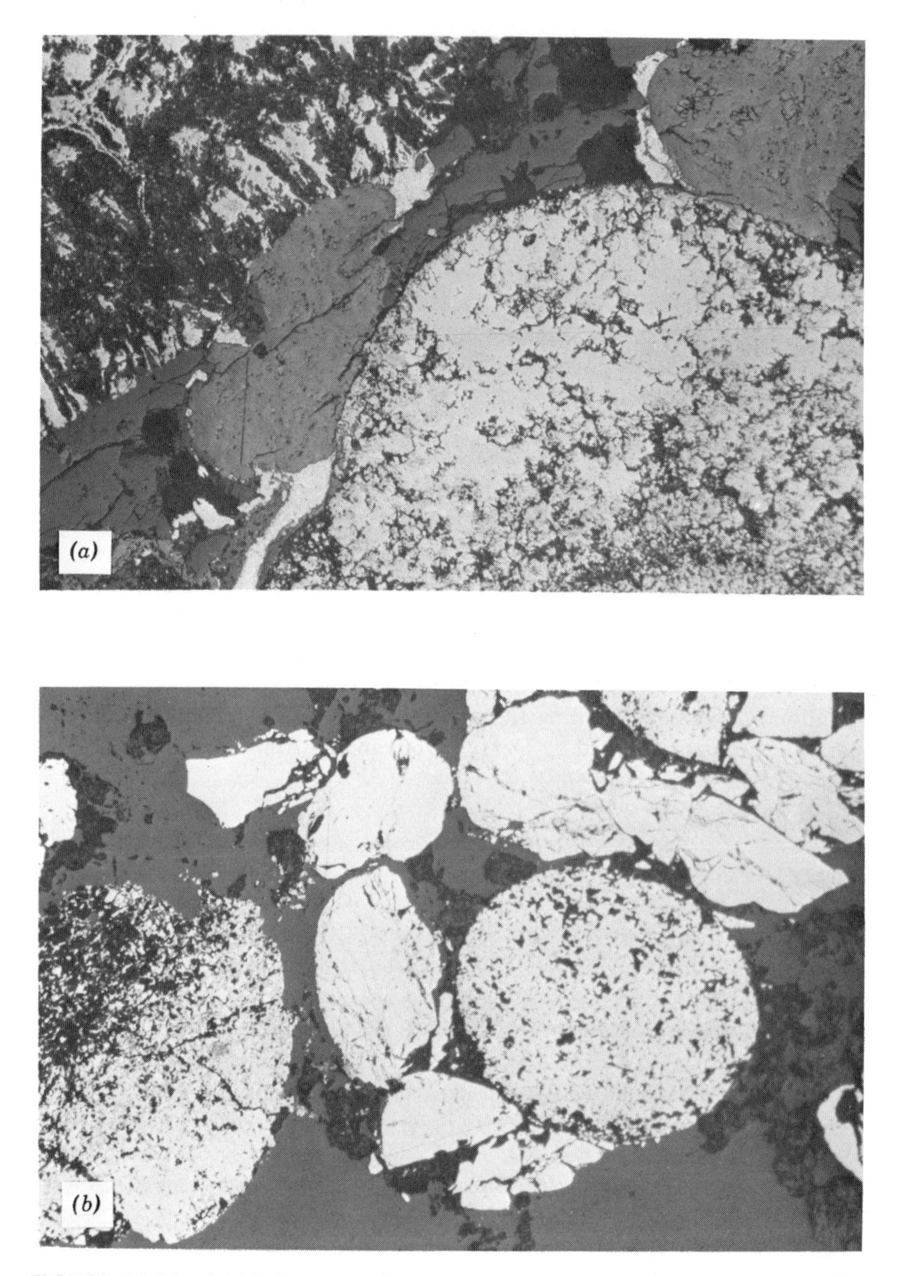

FIGURE 10.11 *(a)* Native gold (light gray) occurring around the margins of quartz grains and pyrite aggregates. Witwatersrand, South Africa. (Width of field = 2000 μm.) *(b)* Pyrite of allogenic and concretionary authigenic textural types. Witwatersrand, South Africa. (Width of field = 2000 μm.) *(c)* Colloform uraninite with associated fine pyrite. (Width of field = 2000 μm.)

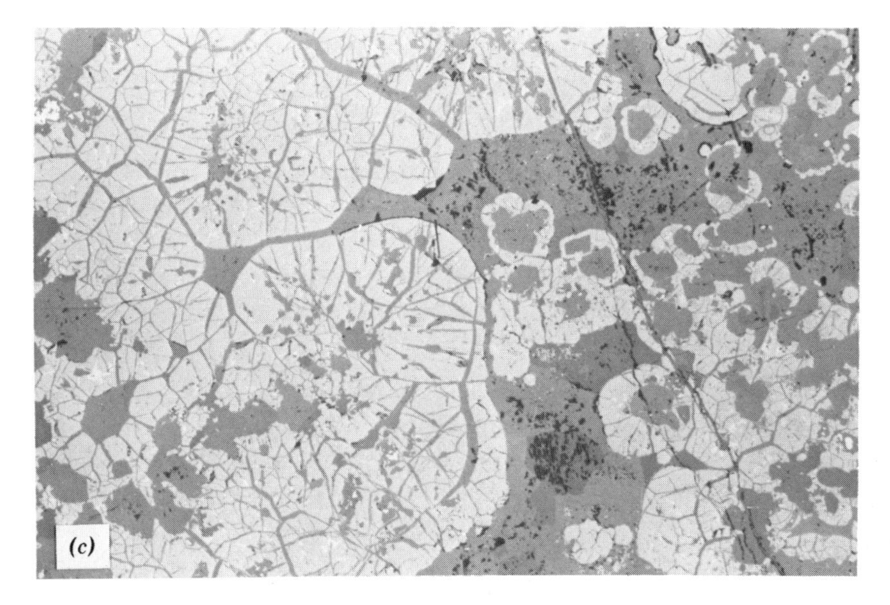

FIGURE 10.11 (*Continued*)

10.6 LEAD-ZINC DEPOSITS IN CARBONATE ROCKS AND OTHER SEDIMENTS

Mineralogy

Major: Galena and sphalerite are the metalliferous ore minerals. Barite and fluorite are locally economically important. Pyrite and marcasite are widespread and so is chalcopyrite although it is rarely significant as a source of copper.

Minor: Sulfides that may occur in minor amounts include wurtzite, greenockite, millerite, bravoite, siegenite, bornite, covellite, enargite, luzonite. A variety of oxides (hematite, cuprite, limonite), carbonates (e.g., smithsonite, cerussite, malachite) and sulfates (e.g., anglesite, jarosite [$KFe_3(SO_4)_2(OH)_6$] result largely from the alteration of the major sulfides.

Gangue Minerals: dominantly calcite, dolomite, aragonite, and quartz. Siderite, ankerite, and colloform silica may also occur.

Mode of Occurrence

The ores nearly always occur in sedimentary carbonate host rocks, particularly dolomites but also limestones and magnesian limestones. They may also be found in associated sandstones, shales, or conglomerates. Many of the larger ore bodies occur as stratiform masses (i.e., parallel, or nearly so, with the bedding) but others occur as vein infillings (or sometimes replacements) along fissures (commonly faults or joints) which may cross the bedding. Often the carbonate host rock constitutes part of a reef (i.e., is biohermal) and the ores are localized relative to a particular reef facies. In other cases, the ores may occur in association with a solution collapse breccia related to karst and paleoaquifer development.

Examples

The most important examples occur in Paleozoic and Mesozoic carbonate sediments, particularly in North America, Europe, Russia, and North Africa. Principal areas of early exploitation were in the Triassic sediments of the Eastern Alps (hence the label "alpine-type") and in the upper and middle regions of the Mississippi Valley of the United States (hence "Mississippi-Valley-type") within Paleozoic sediments. Within the latter area, the "Tri-State" field (around the Missouri-Kansas-Oklahoma border), an important early mining district, has been superceded by the Old and New (=Virburnum Trend) Lead belts of S.E. Missouri. Other major North American occurrences are in the Southern Appalachians, especially Tennessee, and the Pine Point area, North West Territories, Canada. The deposits of the English Pennines, although now little exploited, have been the subject of much scientific study. Other notable European examples occur in Central Ireland and in Silesia. Similar lead-zinc ores (e.g., Laisvall) occur in the Lower Cambrian and uppermost Precambrian sandstones along the western border of the Baltic Shield in Norway and Sweden.

Mineral Compositions, Textures, and Parageneses

The simple mineral assemblages characteristic of these deposits are made up of a few well crystallized phases of simple composition. The galena is characteristically lower in silver content than is the galena of certain other ores (see Section 9.6) and the sphalerite is commonly pale in color with highly variable but generally small amounts of iron and manganese substituting for zinc. The sphalerite does commonly carry relatively high cadmium contents (in some cases greenockite, CdS, being an accessory mineral).

The textures exhibited by the ores in polished section and on a larger scale, although varying with the particular occurrence, are also relatively simple. In the larger orebodies, the phases occur as large irregular polycrystalline aggregates within the host rock. When the sulfides occur in veins or as breccia fillings, they may be massive or may provide beautiful examples of crustification with delicately developed symmetric or asymmetric bands occurring parallel to the margins of the veins (see Figure 8.4*a*). Alternating bands of very fine-grained sphalerite and wurtzite found in these ores are termed *schalenblende.* Colloform textures are also common in both vein ores and in the *breccia ores* associated with solution collapse brecciation. These textures may involve the layered intergrowth of sphalerite and wurtzite, galena, pyrite and marcasite, or other sulfides and nonsulfides. The free growth of crystals on the walls of solution cavities can also produce large, well-formed crystals. Much of the ore in these deposits consists, therefore, of crystals deposited one upon another in open space. Examples of characteristic textures are shown in Figure 10.12.

The paragenetic sequences reported for these deposits can be illustrated through a number of examples. In Figure 10.13, a paragenetic diagram for the Tri-State deposits based on the work of Hagni and Grawe (in Hagni, 1976) is shown. Although the sequence is generalized and a number of aspects are controversial, it illustrates the characteristically repetitive nature of this style of mineralization (several periods of sphalerite deposition in this case). Another example is the Magmont Mine (S.E. Missouri) where Hagni and Trancynger (1977) recognized three phases of mineralization—early disseminated mineralization, followed by colloform sulfides, followed by crystalline sulfides, quartz, and calcite in fractures and vugs. Overlap and repeated deposition of sulfides occurs throughout the episodes of mineralization with galena and chalcopyrite deposited during six intervals; deposition of sphalerite, marcasite, and pyrite occurs during four; deposition of dolomite and quartz occurs during three. In the South Pennine Ore-Field (England), the primary sulfide mineralogy is remarkably uniform and displays a consistent paragenetic sequence of bravoite, nickel-rich and nickel-poor pyrite and marcasite, chalcopyrite, galena, and sphalerite (Ixer and Townley, 1979) similar to that discussed and illustrated in detail in Section 8.3.3, whereas greater diversity is exhibited in the Northern Pennine Orefield (Vaughan and Ixer, 1980). The importance of nickel as a minor element in "Mississippi-Valley-type" mineralization is also highlighted by these authors. Although the major mineral assemblages of these ore deposits are both simple and uniform, there is commonly evidence for a zonal distribution in terms of minor elements.

Along the western border of the Baltic Shield in Norway and Sweden, most notably at Laisvall in Sweden, lead-zinc mineralization is confined to sandstones. The assemblage includes pyrite, calcite, barite, fluorite, galena, and sphalerite filling the interstices of the sand grains (Figure 10.14). Although the host rock is different, these ores appear to be closely related to the lead-zinc ores of carbonate rocks.

Ore Formation

The literature describing lead-zinc ore deposits in carbonate rocks and discussing their genesis is very considerable. Important examples of the more recent literature include the monograph edited by Brown (1967) and studies by Beales and Jackson (1966), Heyl (1969), Brown (1970), Anderson (1975), and Sangster (1976). Despite the considerable amount of work undertaken on these ores, their origin remains controversial.

The characteristically simple mineralogy and the localization of ore in carbon-

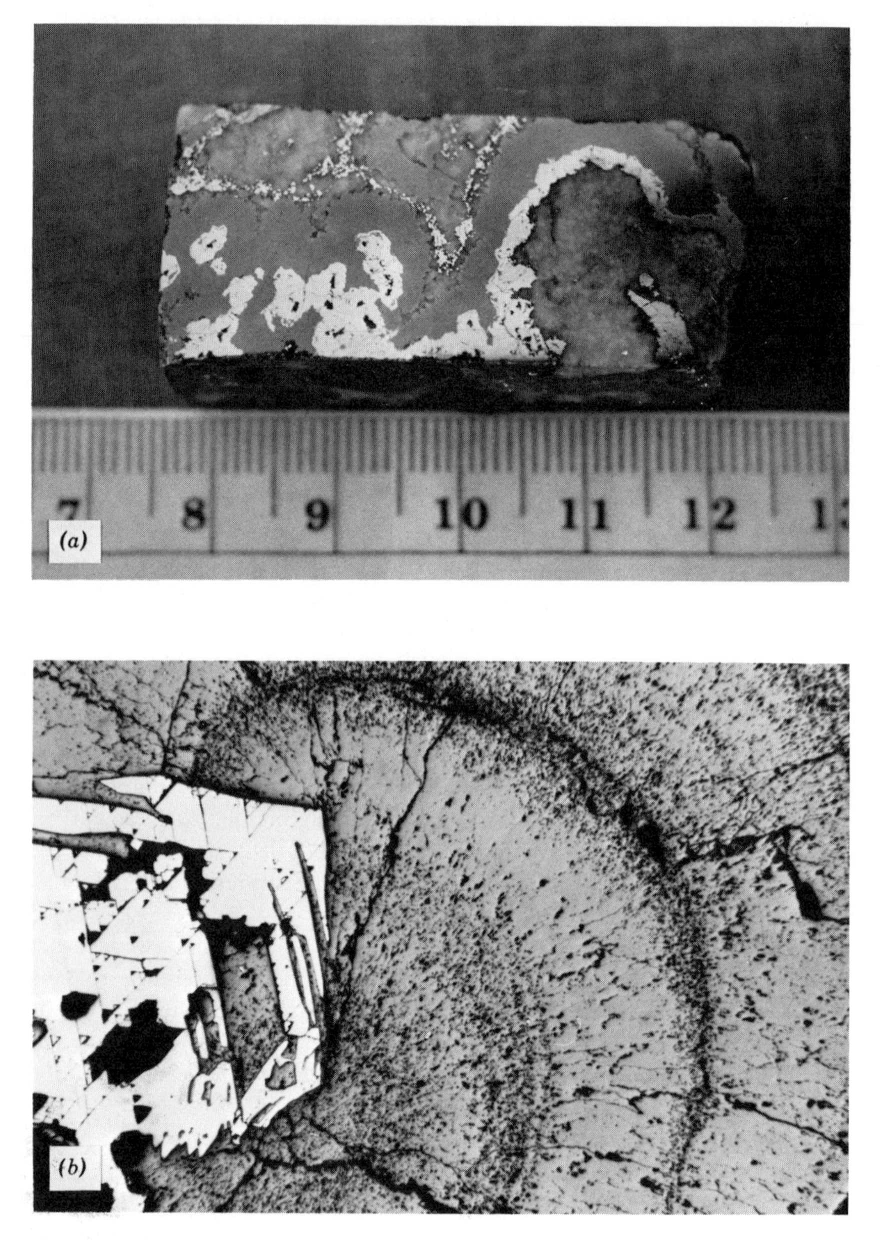

FIGURE 10.12 Characteristic textures of lead-zinc ores in carbonate rocks: *(a)* crustiform pyrite (white) overgrown by sphalerite and a later generation of pyrite flanked by pre- and post-ore dolomite, Austinville, Virginia (scale is in cm); *(b)* skeletal galena crystal (white) overgrown by concentric bands of sphalerite, Pine Point, Northwest Territories (width of field = 2000 μm); *(c)* early pyrite subhedra associated with a later veinlet of galena in a dolomite host, Austinville, Virginia (width of field = 2000 μm).

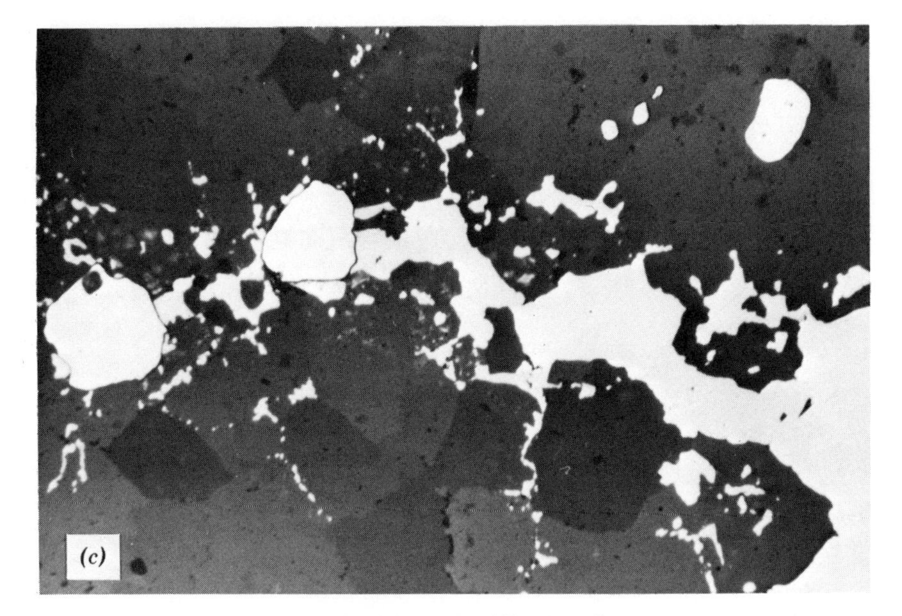

FIGURE 10.12 (*Continued*)

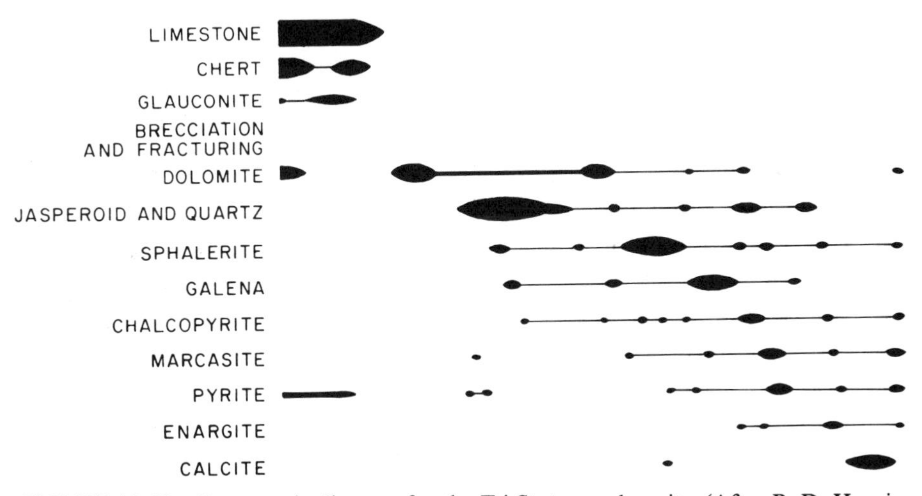

FIGURE 10.13 Paragenetic diagram for the Tri-State ore deposits. (After R. D. Hagni and O. R. Grawe, *Econ. Geol.* **59**, 455, 1964; with the publisher's permission.)

ate host rocks (sometimes in particular facies, sometimes in specific tectonic or karstic structures) have already been emphasized. The vein-type ores clearly have resulted, in many cases, from successive deposition from introduced (predominantly aqueous) solutions. Fluid inclusion studies have indicated that these were saline solutions which deposited the minerals at temperatures below ~200°C, frequently below ~100°C, but the origin of these fluids remains con-

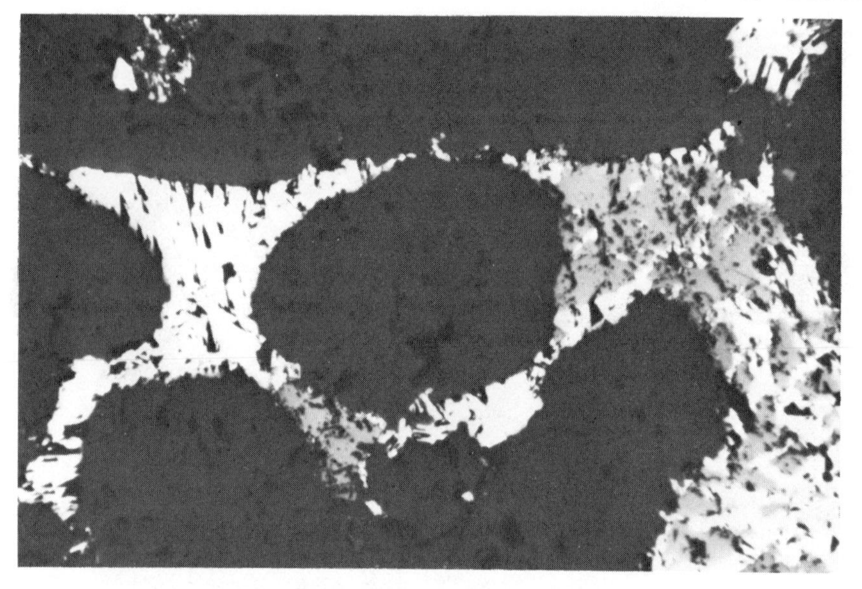

FIGURE 10.14 Galena (light gray) and sphalerite (medium gray) infilling around quartz grains. Laisvall, Sweden. (Width of field = 2000 μm.)

troversial. As to the origin of the "stratiform" orebodies, theories ranging from syngenetic to wholly epigenetic have been propounded. In 1970, Brown stated that North American opinion favored a "dominantly connate marine but epigenetic ore fluid with probably minor additions from deeper sources" and that European opinion was "divided almost equally between proponents of syngenesis-diagenesis and of magmatic epigenetic origin."

Since 1970, particular emphasis has been placed on the problems of transport and deposition of the metal sulfides. Many have supported the view that connate brines have acted as a means of transporting metals (as chloride complexes) and that mixing with a separate H_2S-rich fluid in the limestone environment has caused sulfide precipitation (e.g., Beales and Jackson, 1966; Anderson, 1975). Hydrocarbons in the limestones and sulfate-reducing bacteria have been cited as important in the conversion of dissolved sulfate to sulfide and the spacial link of Mississippi Valley deposits to oilfieldlike brine sources emphasized. The conflict between the European proponents of a syngenetic-diagenetic origin (largely those workers studying the Eastern Alps) and the proponents of an epigenetic origin has been somewhat clarified by Sangster (1976). He has proposed a division into two major classes of lead-zinc deposits in carbonates:

1 Mississippi-Valley-type, which are "stratabound" and were emplaced after lithification of the host rocks (i.e., epigenetically) into "open space" provided by a variety of structures (e.g., Mississippi Valley, Pine Point, English Pennines).

2 Alpine-type, which are "stratiform" and also synsedimentary in large part. Here the original source of the ores is regarded as contemporaneous with the host rocks and linked to submarine volcanism. Remobilization may have resulted in concentration of these ores and the formation of epigenetic features (e.g., the Eastern Alps, Central Ireland).

Selected References

Anderson, G. M. (1975) Precipitation of Mississippi Valley-type ores. *Econ. Geol.* **70**, 937–942.

Beales, F. W. and Jackson, S. A. (1966) Precipitation of lead-zinc ores in carbonate reservoirs as illustrated by Pine Point orefield. *Trans. Inst. Min. Metall.* **B75**, 278–285.

Brown, J. S., Ed. (1967) *Genesis of Stratiform Lead-Zinc-Barite-Fluorite Deposits: A Symposium.* Econ. Geol. Monograph 3.

Brown, J. S. (1970) Mississippi Valley-type lead-zinc ores. *Mineral. Deposita* **5**, 103–119.

Hagni, R. D. (1976) Tri-state ore deposits: the character of their host rocks and their genesis. In K. H. Wolf, ed., *Handbook of Stratabound and Stratiform Ore Deposits,* Vol. 6. Elsevier, Amsterdam-Oxford-New York, pp. 457–494.

Hagni, R. D. and Trancynger, T. C. (1977) Sequence of deposition of the ore minerals at the Magmont Mine, Viburnum Trend, Southeast Missouri. *Econ. Geol.* **72**, 451–463.

Heyl, A. V. (1969) Some aspects of genesis of zinc-lead-barite-fluorite deposits in the Mississippi Valley, U.S.A. *Trans. Am. Inst. Min. Metall. Eng.* **78**, B148–B160.

Ixer, R. A. and Townley, R. (1979) The sulphide mineralogy and paragenesis of the South Pennine Orefield, England. *Mercian. Geol.* **7**, 51–63.

Sangster, D. F. (1976) Carbonate hosted lead-zinc deposits. In K. H. Wolf, ed., *Handbook of Stratabound and Stratiform Ore Deposits.* Elsevier, Amsterdam-Oxford-New York, Vol. 6, pp. 447–456.

Vaughan, D. J. and Ixer, R. A. (1980) Studies of the sulphide mineralogy of North Pennine ores and its contribution to genetic models. *Trans. Inst. Min. Metall.* **89**, B99–B109.

10.7 STRATIFORM BASE METAL SULFIDE ORES IN SEDIMENTARY ROCKS

Mineralogy

Major: pyrite; chalcopyrite or galena and sphalerite; in some ores pyrrhotite, bornite, chalcocite (also digenite, djurleite) or even native copper may be major and cobalt sulfides (carrollite) or sulfarsenides may be economically important. Cobalt also substitutes in pyrite.

Minor: arsenopyrite, tetrahedrite, native bismuth, bismuthinite, argentite, niccolite; molybdenite, covellite, and other sulfides may occur.

Gangue Minerals: carbonates, barite, fluorite.

Mode of Occurence

These are disseminated to massive stratiform sulfide ores that are generally conformable within sedimentary sequences in which they occur and grade into the ores discussed in Section 10.8. The host rock may be a black shale, dolomite or, more rarely, an arenaceous unit (e.g., quartzite) and may be of considerable lateral extent. The host sediments may be undisturbed or may have undergone mild folding and metamorphism.

Examples

Kupferschiefer-Marl Slate of Northern Europe; Copperbelt of Zambia and Zaire; White Pine Michigan, United States; Selwyn Basin, Yukon, Canada, Mt. Isa, Broken Hill, Australia.

Mineral Associations and Textures

The ore minerals of this association are characteristically fine-grained and disseminated in the host rock often as lenses comformable with the bedding (Figure 10.15). A characteristic texture found in the sulfides, particularly pyrite, is the framboid (Figure 10.16). Also common are colloform textures in pyrite, galena,

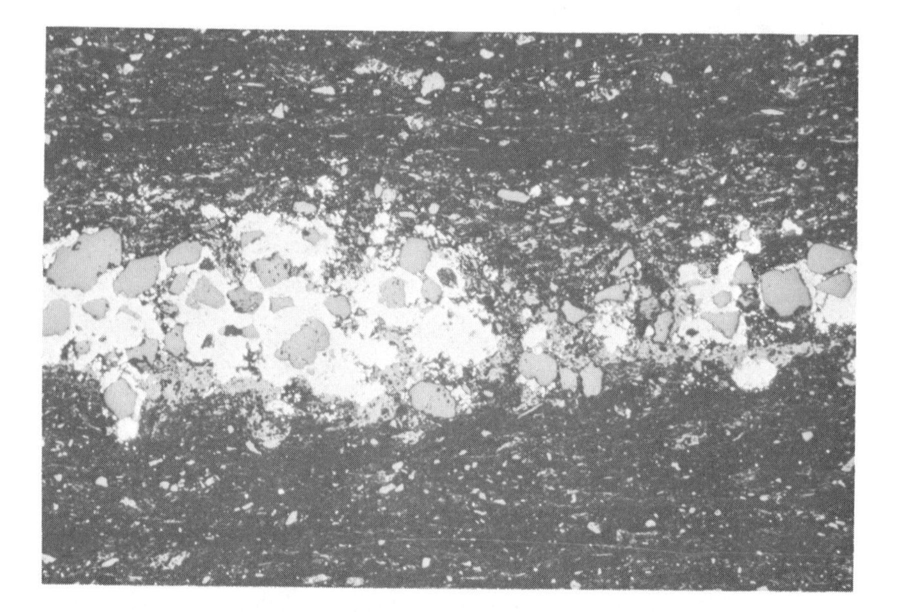

FIGURE 10.15 Fine-grained lens of pyrite with minor chalcopyrite; the lighter gray subhedra are of quartz. Marl Slate, Northern England. (Width of field = 300 μm.)

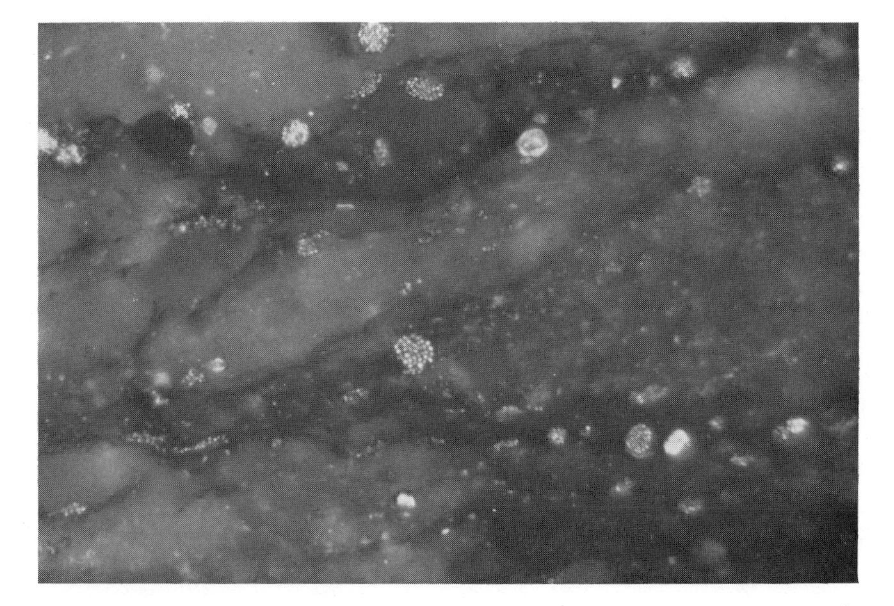

FIGURE 10.16 Finely dispersed pyrite framboids in dolomite. Marl Slate, Northern England. (Width of field = 300 μm.)

and sphalerite. Most of the sulfides occur as anhedral grains but pyrite is one of the few sulfides that may be euhedral. The ore minerals generally occur as random aggregates, although intimate intergrowth textures involving laths, intersecting spindles, or myrmekitic fabrics occur between the copper and copper-iron sulfides. The intergrowths of bornite and chalcopyrite are clearly a result of exsolution and other intergrowths involve bornite with chalcocite or other copper sulfides (Figure 10.17) (the precise identity of many of the copper sulfides in these ores has never been checked). Some of the more deformed ores of this type contain very minor development of mineralized veins.

An important characteristic of these ores is the presence of a zonal distribution of ore metals on a regional or a more local scale. In the White Pine deposit, for example, in passing stratigraphically upwards through the host shales, a sequence of copper, chalcocite, bornite, chalcopyrite, pyrite is observed (Brown, 1971). In certain Zambian deposits, a sequence bornite → chalcopyrite → pyrite has been related to syngenetic sulfide concentrations in shales deposited in progressively deeper water (Fleischer, Garlick, and Haldane, 1976).

Metamorphism of stratified synsedimentary ores often results in their recrystallization while preserving their intimate stratified nature. This is evidenced by the development of coarser equigranular annealed textures in the sulfides and the growth of micas (Figure 10.18). Intense metamorphism results in disruption of the finely laminated structure and the development of the textures described in Section 10.9.

FIGURE 10.17 Intergrowth of bornite (dark gray) and chalcocite (light gray), which has formed as a result of exsolution of an originally homogeneous solid solution. Kolwezi, Katanga, Zaire. (Width of field = 300 μm.)

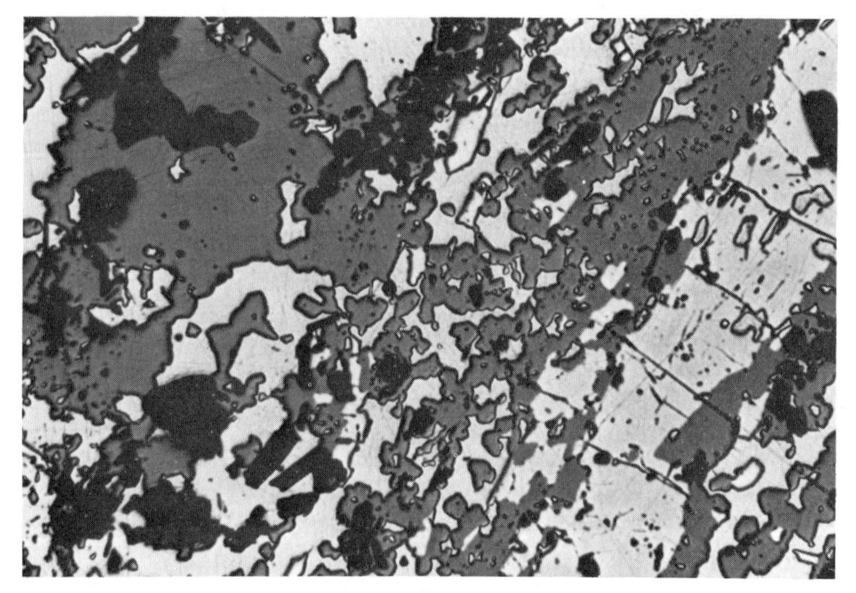

FIGURE 10.18 Primary depositional banding retained through metamorphism. Recrystallization of the ores, primarily galena but with minor pyrrhotite and sphalerite in the band shown, has been accompanied by the growth of micas (dark laths). Sullivan Mine, British Columbia. (Width of field = 2000 μm.)

Origin of the Ores

Ore deposits of this type are still among the most controversial as regards their origin. The pronounced bedded character of the ores has led authors to propose that they are directly deposited sulfide-rich sediments (i.e., are syngenetic) and that euxinic conditions in the depositional basin were combined with an influx of metals from an erosional source to produce the ores and their zonal distribution. Problems of introducing sufficient metals have commonly led to suggestions of submarine volcanic springs as a source.

The opposing view regarding origins is that the ores were introduced by mineralizing solutions after formation of the sediments (i.e., the ores are epigenetic) and were selectively precipitated to replace and pseudomorph characteristically sedimentary structures. Commonly the metals in this case have been regarded as derived through leaching of associated rocks by saline solutions derived from associated evaporite sequences. Thus Brown (1971) has suggested that the copper mineralization in the Nonesuch shale of the White Pine deposit results from replacement of preexisting iron sulfides by copper-rich solutions that migrated upward from the underlying Copper Harbor Conglomerate. Detailed arguments for and against the various theories, which incorporate almost every possibility between the two extremes, can be obtained from the relevant literature (e.g., Fleischer et al., 1976; Bartholomé, 1974; Jung and Knitzschke, 1976; Brown, 1971). There is little reason to suspect that all deposits of this group have the same origin, although it is worth noting that in even the most apparently undisturbed ores of this type, an appreciation of the role of chemical transformation and replacement during diagenesis is growing (e.g., Turner, Vaughan, and Whitehouse, 1978).

Selected References

Bartholomé, P., ed. (1974) *Gisements Stratiforms et Provinces Cupriferes.* Société Geologique de Belgique, Liege.

Brown, A. (1971) Zoning in the White Pine copper deposit, Ontonagon County, Michigan. *Econ. Geol.* **66**, 543–573.

Fleischer, V. D., Garlick, W. G., and Haldane, R. (1976) Geology of the Zambian Copperbelt. In K. H. Wolf, ed., *Handbook of Stratabound and Stratiform Ore Deposits.* Elsevier, Amsterdam, Vol. 6, pp. 223–352.

Jung, W. and Knitzschke, G. (1976) Kupferschiefer in the German Democratic Republic (GDR) with special reference to the Kupferschiefer Deposit in the Southeast Harz Foreland. In K. H. Wolf, ed. *Handbook of Stratabound and Stratiform Ore Deposits.* Elsevier, Amsterdam, Vol. 6, pp. 353–406.

Turner, P., Vaughan, D. J., and Whitehouse, K. I. (1978) Dolomitization and the mineralization of the Marl Slate (N. E. England) *Mineral. Deposita* **13**, 245–258.

10.8 COPPER-IRON-ZINC ASSEMBLAGES IN VOLCANIC ENVIRONMENTS

Mineralogy

Major: pyrite, sphalerite, chalcopyrite; in some examples pyrrhotite or galena.

Minor: bornite, tetrahedrite, electrum, arsenopyrite, marcasite, cubanite, bismuth, copper-lead-bismuth-silver-sulfosalts, cassiterite, plus many others in trace amounts.

Mode of Occurence

Massive to disseminated stratiform sulfide ores in volcano-sedimentary sequences ranging from ophiolite complexes (Cyprus-type deposits) felsic tuffs, lavas and sub sea floor intrusions (Kuroko-type deposits) to mudstones and shales with little immediately associated recognizable volcanic material (Besshi-type deposits).

Examples

Kuroko- and Besshi-type deposits of Japan; Timmins, Ontario; Bathurst, New Brunswick; Sullivan, British Columbia; Flin-Flon, Manitoba-Saskatchewan; Noranda, Quebec; Mt. Lyell, Australia; Rio Tinto, Spain; Scandinavian Calidonides; Avoca, Ireland; Parys Mountain, Wales; Troodos Complex deposits, Cyprus; Bett's Cove, Newfoundland; Modern Red Sea and East Pacific Rise deposits.

Mineral Associations and Textures

The deposits range from ores in thick volcanic sequences such as the Kuroko ores of Japan and ores directly associated with a volcanic vent (Vanna Levu, Fiji) to ores associated with ophiolite sequences (Cyprus; Bett's Cove, Newfoundland) to distal ores that are emplaced in dominantly sedimentary sequences (Besshi deposits of Japan) and sequences containing no recognizable volcanics (Sullivan, British Columbia). They thus grade into ores of the type described in Section 10.7. In spite of the different settings in which these ores are found, there are similarities among the ore types observed. Zoning within many of these deposits is recognizable and three major ore types occur; the distribution of the primary minerals in the Kuroko ores is shown in Figures 10.19 and 10.20. Although the major ore types described in the following are those commonly observed in the Kuroko deposits, they appear in most or all of the ores of this class with only minor variation. These ores, which appear to grade into the ores described in Section 10.7, have frequently been considered in terms of

Cu-Pb-Zn ratios as shown in Figure 10.21. Plimer (1978) has suggested that a trend in ore-type from Cu-dominant to Zn-dominant to Zn-Pb-dominant corresponds to a progression in time and distance from the volcanic source (i.e., proximal to distal in nature). Jambor (1979) has enlarged on this theme and proposed a classification of the Bathurst-area (Canada) deposits based on their established or assumed displacement from feeder conduits (proximal versus distal) and position of sulfide crystallization (autochthonous versus allochthonous).

Although the ores of the volcanic deposits are members of a continuum, several specific ore types are observed most commonly; the following is a brief discussion of these ore types.

Pyritic (= Cyprus type) These ores, associated with ophiolite complexes are composed of massive banded to fragmental pyrite with small amounts of interstitial chalcopyrite and other base metal sulfides. The pyrite is present as friable masses of subhedral to euhedral, commonly zoned, grains, as colloform banded masses, and as framboids. Marcasite is admixed with the pyrite and often appears to have replaced the pyrite. Chalcopyrite occurs as anhedral interstitial grains and as inclusions in the pyrite; sphalerite occurs similarly but is less abun-

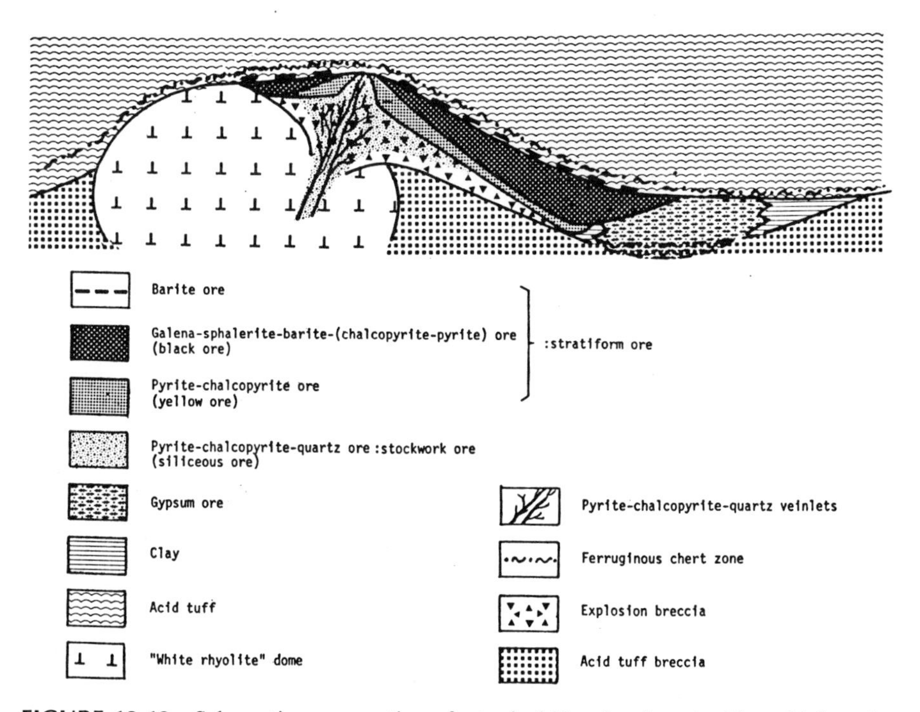

FIGURE 10.19 Schematic cross section of a typical Kuroko deposit. (From T. Sato, in *Geology of Kuroko Deposits*, Soc. Mining Geol. Japan, 1974, p. 2; used with permission.)

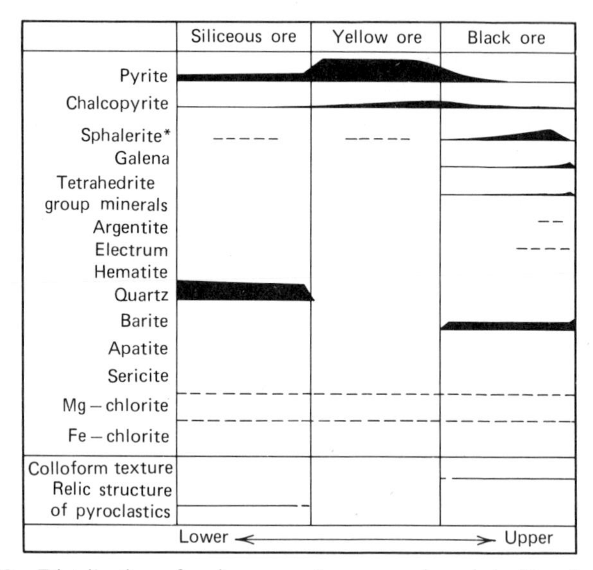

FIGURE 10.20 Distribution of major ore and gangue minerals in Kuroko-type deposits.

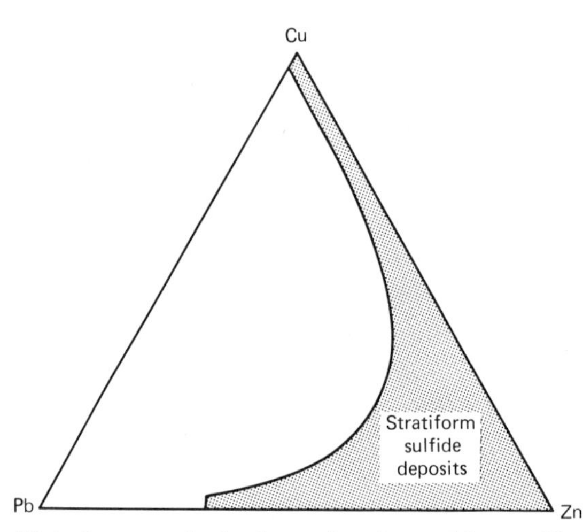

FIGURE 10.21 Plot of copper : lead : zinc ratios observed in stratiform sulfide ore deposits. A trend has been observed from copper to zinc to lead-zinc dominated ores with distance from the volcanic source.

dant. Secondary covellite, digenite, chalcocite, and bornite occur as rims on, and along fractures in, pyrite and chalcopyrite.

Siliceous Ore (=Keiko-type of Kuroko Deposits) These ores apparently represent feeder veins and stockworks and consist primarily of pyrite, chalcopyrite, and quartz with only minor amounts of sphalerite, galena, and tetrahedrite. The pyrite occurs as euhedral grains, subhedral granular stringers, and colloform masses. The other minerals are minor and occur as anhedral interstitial grains in pyritic masses and gangue. Scott (pers. commun., 1980) has noted that a black siliceous ore composed of sphalerite and galena is not uncommon in Kuroko deposits.

Yellow Ore (=Oko-type of Kuroko Deposits) This ore type is characterized in both hand sample and polished section by the conspicuous yellow color resulting from the presence of chalcopyrite interstitial to the dominant euhedral to anhedral pyrite (Figure 10.22*a*). Minor amounts of sphalerite, galena, tetrahedrite, and lead sulfosalts and trace amounts of electrum are dispersed among the major sulfides. In unmetamorphosed bodies, the pyrite is often quite fine (<0.1 mm), but in metamorphosed ores pyrite commonly recrystallizes to form euhedral grains which are several millimeters across. These ores and the black ores described later commonly exhibit extensively developed clastic textures that apparently formed at the time of ore deposition or immediately therefter as a result of slumping.

Black Ore (Kuroko-type) The black ores (Figures 10.22*b* and 7.4*c*), the most complex of the common volcanogenic ore types, were so named because of the abundant dark sphalerite within them. Galena, barite, chalcopyrite, pyrite, and tetrahedrite are common but subsidiary to the sphalerite. Bornite, electrum, lead sulfosalts, argentite, and a variety of silver sulfosalts are customary accessory minerals. The black ores are usually compact and massive but primary sedimentary banding is often visible and brecciated and colloform textures are not uncommon. In ores unmodified by metamorphism, pyrite occurs as framboids, rosettes, colloform bands, and dispersed euhedral to subhedral grains. Pyrite grain size increases during metamorphism but growth zoning is often visible either after conventional polishing or after etching. In polished sections, sphalerite appears as anhedral grains that frequently contain dispersed micron-sized inclusions of chalcopyrite. Barton (1978) has shown, by using doubly polished thin sections in transmitted light (see Figure 2.7) that this "chalcopyrite disease" consists of rods and thin vermicular, myrmekiticlike growths, probably formed through epitaxial growth or replacement. He has also shown the presence of growth-banding and overgrowth textures in sphalerite and tetrahedrite. During metamorphism, the sphalerite is commonly recrystallized and homogenized, and the dispersed chalcopyrite is concentrated as grains or rims along sphalerite grain-boundaries.

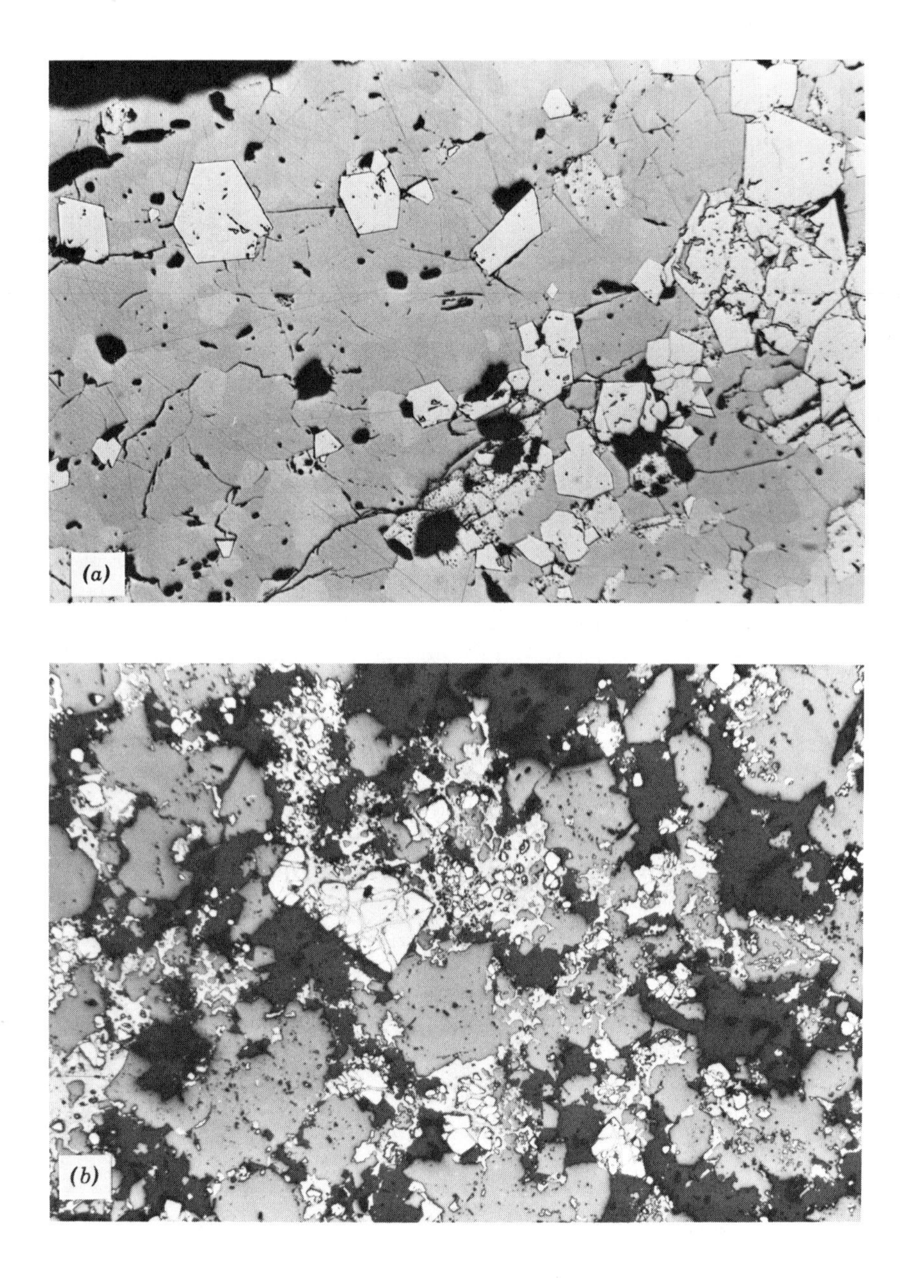

FIGURE 10.22 Typical Kuroko-type ores: *(a)* yellow ore composed of euhedral pyrite crystals within a matrix of chalcopyrite, Ainai Mine, Japan (width of field = 2000 μm); *(b)* black ore composed of a matrix of irregularly intergrown sphalerite and galena containing euhedral and anhedral grains of pyrite, Furutobe Mine, Japan (width of field = 2000 μm).

Barite Ore, Gypsum Ore (Sekkoko) and Ferruginous Chert (Tetsusekiei) These three zones are often present in the Japanese volcanogenic ores but are difficult to recognize in many older volcanogenic ores; frequently they contain few ore minerals. The barite zone usually overlies the black ore and consists of stratified barite. The gypsum zone consists of gypsum and anhydrite with minor amounts of pyrite, chalcopyrite, sphalerite, and galena. The uppermost part of many volcanogenic deposits is a complex mixture of tuff and cryptocrystalline quartz, containing chlorite, sericite, and pyrite and colored red by small amounts of flaky hematite.

The terminology used above (Siliceous ore, Black ore etc) was developed to describe the little altered Japanese Kuroko-type ores and thus does not apply without some modification to their metamorphosed equivalents of other parts of the world. Probably the principal changes during metamorphism are the development of significant amounts of pyrrhotite and the modification of textures (see Section 10.9). Nevertheless the same general ore types (e.g., pyrite with chalcopyrite; sphalerite, pyrite, galena, chalcopyrite) are encountered in many deposits.

Fluid inclusion studies (Roedder 1976) indicate that the ore-forming fluids were generally of low salinity (less than 5wt. % NaCl equivalent) and ranged in temperatures up to about 300°C.

Origin of the Ores

The ores considered in this section have been variously described as: massive pyrite deposits related to volcanism, stratabound massive pyritic sulfide deposits, and stratiform sulfides of marine and marine-volcanic association (Stanton, 1972). Although the degree of volcanogenic affinity varies from ores within a volcanic vent (Vanua Levu, Fiji), to intercalation of ores with volcanic clastics and flows (Kuroko ores, Japan), to the occurrence of ores within dominantly terrigenous sediments (Besshi-type deposits, Japan; Sullivan, British Columbia), the origin of the ores appears to be related to submarine exhalative or hydrothermal activity associated with volcanism or seafloor fracture zones. Early views held that all or most of the ores accumulated as a result of a "snowfall" of very fine-grained sulfides that formed as hot solutions issued onto the seafloor, as observed in modern sulfide formation at the island of Volcano and along the crest of the East Pacific Rise (Francheteau et al., 1979). Ore breccia textures have been interpreted as resulting from steam explosions and soft sediment slumping. Barton (1978) has pointed out that it is difficult to envisage the maintenance of seafloor temperatures of 200–300°C (as indicated by fluid inclusion studies) over wide areas for periods of time long enough to allow the growth of coarse-grained, zoned sphalerites. He has suggested that at least some sulfide formation must have occurred beneath a crust, either by recrystallization of earlier primary syngenetic sulfide or by introduction of a hot, saline, hydrothermal fluid into a mass of fine-grained sulfide. Fracturing, healing of cracks, overgrowth, and breccia textures suggest that crystal growth continued episodically and was interspersed with periods of slumping, boiling, or explosive activity. The deeper-seated fracture-filling siliceous pyrite-chalcopyrite ores appear to have formed by precipitation from hydrothermal solutions in feeder zones.

Selected References

Barton, P. B. (1978) Some ore textures involving sphalerite from the Furutobe Mine, Akita Prefecture, Japan. *Min. Geol. (Jap.)* **28**, 293–300.

Constantinou, G. and Govett, G. J. S. (1973) Geology, geochemistry, and genesis of Cyprus sulfide deposits. *Econ. Geol.* **68**, 843–858.

Francheteau, J. et al. (1979) Massive deep-sea sulphide ore deposits discovered on the East Pacific Rise. *Nature* **277**, 523–528.

Ishihara, S., ed. (1974) *Geology of Kuroko Deposits*, Special Issue No. 6, *Soc. Min. Geol. Japan*.

Jambor, J. L. (1979) Mineralogical evaluation of proximal-distal features in New Brunswick massive-sulfide deposits. *Can. Mineral.* **17**, 649–664.

Lambert, I. B. and Sato, T. (1974) The Kuroko and associated ore deposits of Japan: A review of their features and metallogenesis. *Econ. Geol.* **69**, 1215–1236.

Mining Geology (Japan) (1978) Vol. 28, No. 4 and 5 are devoted to studies of Kuroko-type ores.

Ohmoto, H. and Rye, R. O. (1974) Hydrogen and oxygen isotopic compositions of fluid inclusions in the Kuroko deposits, Japan. *Econ. Geol.* **69**, 947–953.

Plimer, I. R. (1978) Proximal and distal stratabound ore deposits. *Mineral. Deposita* **13**, 345–353.

Roedder, E. (1976) Fluid inclusion evidence on the genesis of ores in sedimentary and volcanic rocks. In K. H. Wolf, ed., *Handbook of Stratabound and Stratiform Ore Deposits*. Elsevier, Amsterdam-Oxford-New York, Vol. 2, pp. 67–110.

Shimazaki, Y. (1974) Ore minerals in the Kuroko-type deposits, in S. Ishihara, ed. *Geology of Kuroko Deposits*, Special Issue No. 6, *Soc. Min. Geol. Japan*, 311–322.

Stanton, R. L. (1972) *Ore Petrology*. McGraw-Hill, New York.

Tatsumi, T., ed. (1970) *Volcanism and Ore Genesis*. University of Tokyo Press.

10.9 OPAQUE MINERALS IN METAMORPHOSED MASSIVE SULFIDES

Mineralogy

Major: pyrite, pyrrhotite (hexagonal and monoclinic forms), sphalerite, chalcopyrite, galena, tetrahedrite.

Minor: cubanite, marcasite, arsenopyrite, magnetite, ilmenite, mackinawite.

Mode of Occurrence

In regionally metamorphosed rocks, especially volcanic sequences, at moderate to high metamorphic grades.

Examples

Ducktown, Tennessee; Ore Knob, North Carolina; Great Gossan Lead, Virginia; Flin Flon, Manitoba; Sullivan, British Columbia; Mt. Isa, Broken Hill, Australia; Skellefte District, Sweden; Sulitjelma and Røros, Norway.

Mineral Associations and Textures

Few metalliferous ores owe their existence to regional metamorphism but countless massive sulfide ores have been significantly altered by metamorphic effects. The mineral associations in these ores are largely dependent on the original (premetamorphic) mineralogy and the textures are dependent on the original structure and the extent of thermal and dynamic metamorphism. The macroscopic effects of regional metamorphism include a general coarsening of grain size, development of schistosity, drag folds, isoclinal folding with attenuation of fold limbs and thickening in hinges, rupturing of folds, brecciation, and boudinage. The same deformation features are seen on a smaller scale under the ore microscope (Figure 10.23*a*), but additional structural details and certain mineralogic changes may also be evident. Frequently during dynamic metamorphism, lathlike silicates are locally fractured and ductile sulfides, especially galena, chalcopyrite, pyrrhotite, and sphalerite, are forced into the resulting relatively low pressure areas (Figure 10.23*b*). In micaceous ores, the sulfides may be forced along the basal cleavage planes of mica crystals (Figure 10.23*c*). In contrast to the ductile sulfides, the more brittle sulfides such as pyrite and arsenopyrite deform by fracturing and thus may be observed as shattered crystals, infilled by more ductile sulfides, or even as drawn out lenslike polycrystalline aggregates. The effects of stress may be evident in the development of twinning (especially if twins are deformed), curved cleavage-traces (especially visible in galena), kinkbanding, undulose extinction (see Figures 7.17, 7.18, and 7.19) and the presence of curved rows of crystallographically oriented inclusions (e.g., chalcopyrite in sphalerite). In pyrite, mild strain effects such as the development of micromosaic structures, invisible after normal polishing, may be brought out by etching (conc. HNO_3 followed by brief exposure to 6*M* HCl).

Thermal metamorphism, even of ores that have previously or synchronously undergone intense deformation, commonly results in an increase in grain-size and the development of 120° triple junctions in monomineralic masses. If small amounts of other phases are present, recrystallization may result in the entrapment of small lenslike grains that outline original grain boundaries (Figure 10.23*d*). In heterogeneous iron sulfide-bearing ores, pyrite tends to recrystallize as euhedral cubic porphyroblasts, whereas chalcopyrite, pyrrhotite, and sphalerite tend to develop equant anhedral forms. Although unmetamorphosed pyrite grains commonly exhibit well-developed growth zoning, metamorphosed pyrite commonly exhibits irregular overgrowths and only incomplete zoning (Shadlun, 1971). In dominantly iron sulfide ores, pyrite porphyroblasts are commonly 1

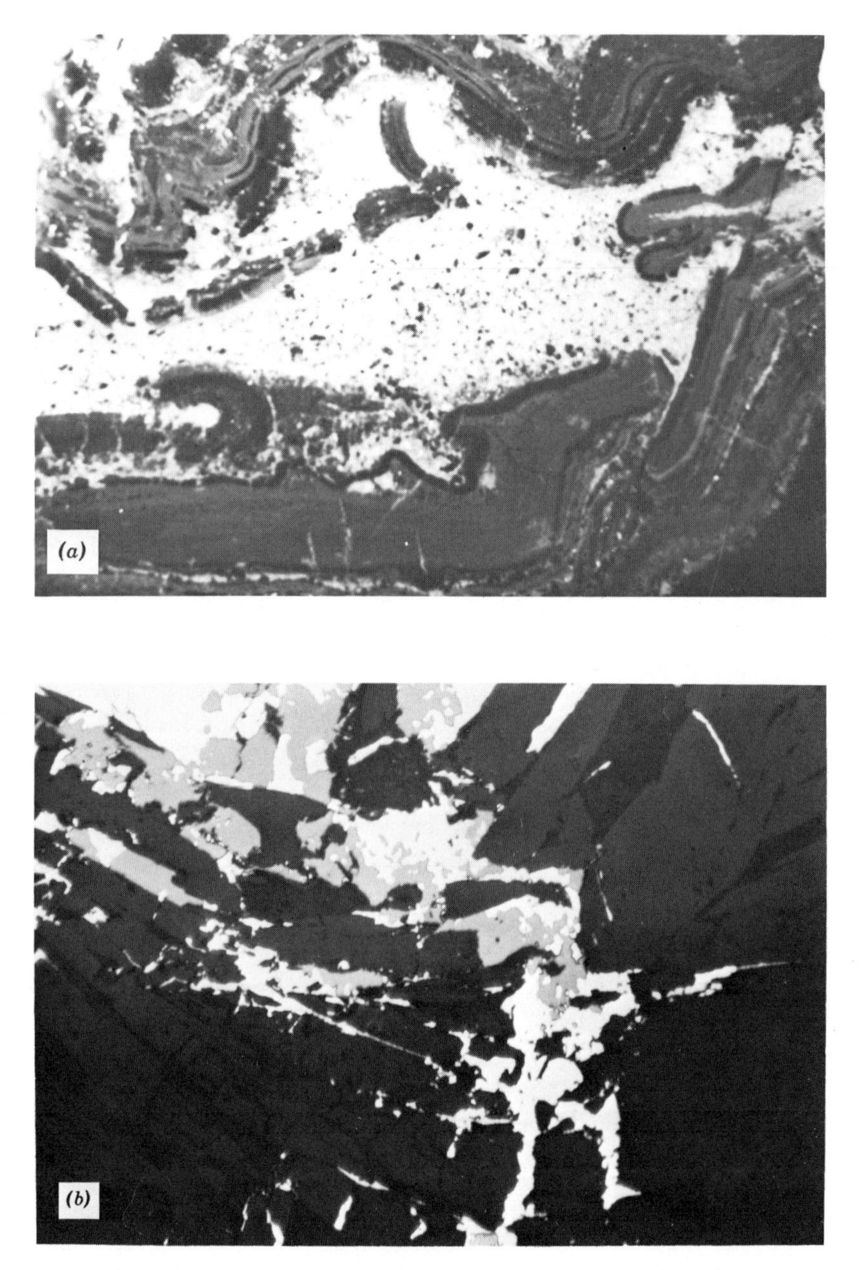

FIGURE 10.23 Textures observed in metamorphosed massive sulfides: *(a)* severe distortion and disagreggation of primary banding in interlayered sphalerite (medium gray) and galena (white), Mt. Isa, Australia (width of field = 8 cm); *(b)* chalcopyrite (white) and sphalerite (light gray) injected into fractured amphiboles, Great Gossan Lead, Virginia (width of field = 700 μm) (From Henry, et. al., *Econ. Geol.* **74**, 651, 1979; used

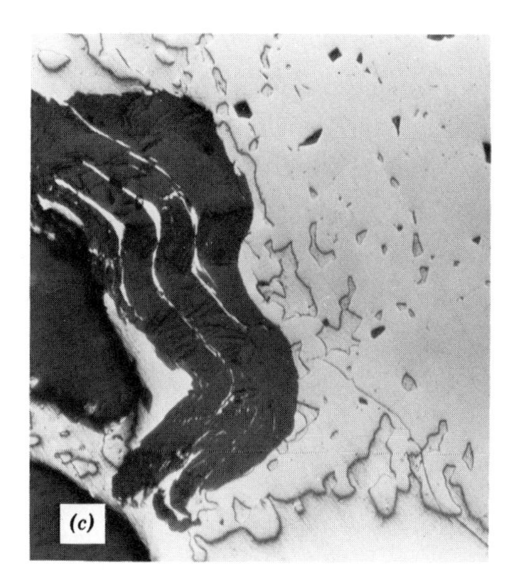

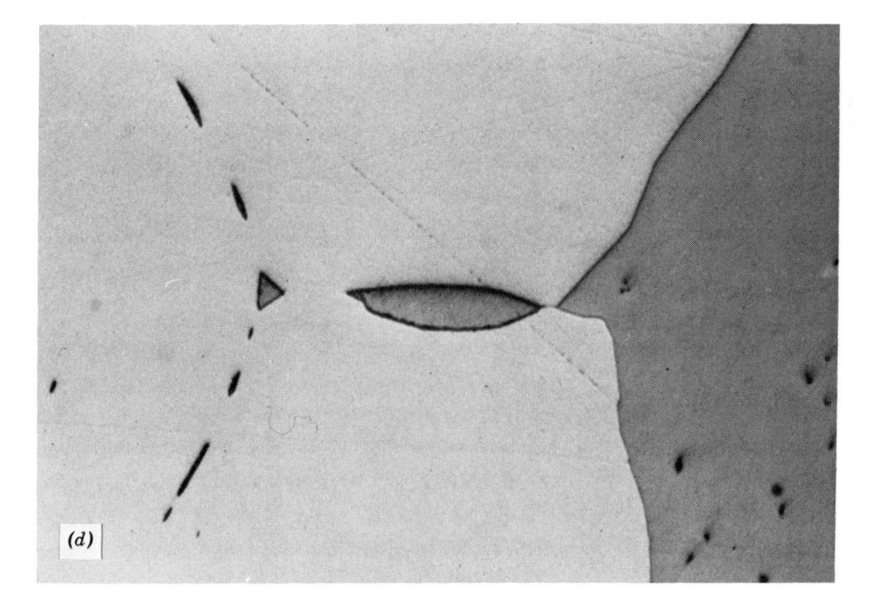

with permission); *(c)* pyrrhotite injected along cleavages in a deformed biotite, Great Gossan Lead, Virginia (width of field = 400 μm); *(d)* lenslike inclusions of galena defining the grain boundaries of recrystallized pyrite, Mineral District, Virginia (width of field = 520 μm).

cm across but may, in extreme cases, reach 10–20 cm across (e.g., Ducktown, Tennessee).

Mineralogic changes in sulfide ores depend on the grade of metamorphism. At lower grades, the more refractory sulfides (pyrite, sphalerite, arsenopyrite) tend to retain their original compositions and structures while softer sulfides (chalcopyrite, pyrrhotite, and galena) readily recrystallize. However, at moderate to high grades, pyrite often begins to lose sulfur and be converted to pyrrhotite, and both pyrite and pyrrhotite may undergo oxidation to magnetite. Chalcopyrite commonly exhibits the development of laths of cubanite and very fine (~ 1 μm) wormlike inclusions of mackinawite. Sphalerite and tetrahedrite, which retain original zoning at low grades of metamorphism, are homogenized at higher grades and tend to be brought into equilibrium with adjacent iron sul-

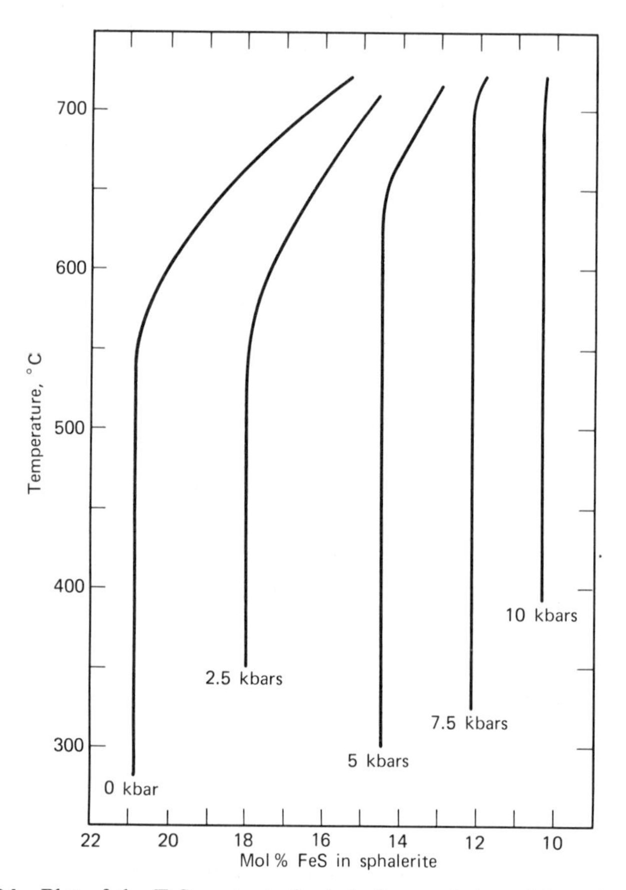

FIGURE 10.24 Plot of the FeS content of sphalerite coexisting with pyrite and hexagonal pyrrhotite at 0, 2.5, 5, 7.5, and 10 kbars at temperatures from 300 to 700°C. (After S. D. Scott, *Amer. Min.*, **61**, 662, 1976; used with permission.)

fides. The FeS content of sphalerite coexisting with pyrite and pyrrhotite, which varies as a function of pressure, has been calibrated as a geobarometer (Figure 10.24). To employ this geobarometer, the temperature of metamorphism must be determined by some independent means (e.g., fluid inclusions, trace element, or isotope partitioning). Sphalerites in metamorphosed ores commonly contain rows of small ($< 5\ \mu m$) inclusions of chalcopyrite that appear to be the remnants of "chalcopyrite disease," primary depositional intergrowths. During recrystallization, the small chalcopyrite inclusions tend to concentrate and coalesce along grain boundaries. Experimental studies (Hutchison and Scott, 1980) have shown that the solubility of CuS in sphalerite is very small below 500°C; hence chalcopyrite should have little effect on the application of the sphalerite geobarometer. Nevertheless, chalcopyrite-bearing sphalerites often yield anomalous pressure estimates, possibly because the presence of the copper promotes low temperature reequilibration, and should be avoided in geobarometric studies. Furthermore, Barton and Skinner (1979) have suggested that sphalerite reequilibrates by outward diffusion of FeS more readily when in contact with pyrrhotite than with the more refractory pyrite during postmetamorphic cooling. As a result, sphalerites coexisting with pyrrhotite commonly have lost some FeS after the peak of metamorphism and also indicate higher than actual metamorphic pressures. Accordingly, in applying the sphalerite geobarometer, it is important to:

1. Use only coexisting sphalerite, pyrite, and pyrrhotite (the coarser the better).
2. Avoid grains that contain or coexist with chalcopyrite.
3. Choose the most FeS-rich sphalerites as indicative of the pressure during metamorphism.

Arsenopyrite, although not abundant in these ores, may be a useful indicator of temperature if equilibrated with pyrite and pyrrhotite or other aS_2-buffering assemblage (Kretschmar and Scott, 1976) (Figure 8.16). As with all geothermometers, it must be applied with caution and other independent checks on temperature should be employed if possible.

Origin of the Textures

The textures of metamorphosed ores result mainly from the dynamic deformation and heating accompanying regional metamorphism (see Sections 7.6 and 7.7). In some localities, deformation is minimal and recrystallization is the dominant change. In zones of intense deformation, the mineralogical character of the ore may be a factor that contributes to the obliteration of premetamorphic features and to the development of chaotic textures. This is especially true of ores rich in pyrrhotite, chalcopyrite, and galena, all of which suffer dramatic loss of shearing strength as temperature rises (Figure 10.25). The extreme flow of such

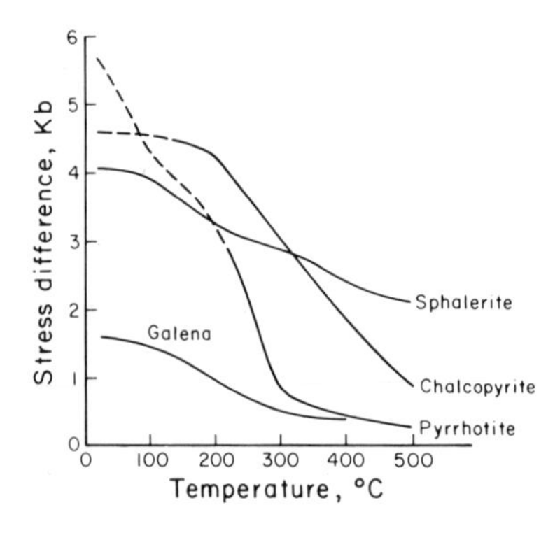

FIGURE 10.25 Shearing strength of some common sulfides as a function of temperature. (After W. C. Kelly and B. R. Clark, *Econ. Geol.* **70,** 431, 1975; reproduced with permission of the publisher.)

ores often results in disaggregation of primary banding, the tectonic incorporation of wall rock fragments in "ball textures" (Vokes, 1973), and total reorientation—*durchbewegung*—of any surviving original features, thus rendering paragenetic interpretation difficult or impossible.

Selected References

Barton, P. B. and Skinner, B. J. (1979) Sulfide mineral stabilities. In H. L. Barnes, ed., *Geochemistry of Hydrothermal Ore Deposits,* 2nd ed., Wiley-Interscience, New York, pp. 278–403.

Hutchison, M. N. and Scott, S. D. (1980) Sphalerite geobarometry in the Cu-Fe-Zn-S system. *Econ. Geol.* (in press).

Kelly, W. C. and Clark, B. R. (1975) Sulfide deformation studies: III Experimental deformation of chalcopyrite to 2000 bars and 500°C. *Econ. Geol.* **70**, 431–453.

Kretschmar, U. and Scott, S. D. (1976) Phase relations involving arsenopyrite in the system Fe-As-S and their application. *Can. Mineral.* **14**, 364–386.

Lawrence, L. J. (1973) Polymetamorphism of the sulphide ores of Broken Hill, N.S.W. Australia. *Mineral. Deposita* **8**, 211–236.

McDonald, J. A. (1967) Metamorphism and its effects on sulphide assemblages. *Mineral. Deposita* **2**, 200–220.

Mookherjee, A. (1976) Ores and metamorphism: Temporal and genetic relationships. In K. H. Wolf, ed., *Handbook of Stratabound and Stratiform Ore Deposits.* Elsevier, Amsterdam-Oxford-New York, Vol. 4, pp. 203–260.

Rickard, D. T. and Zweifel, H. (1975) Genesis of Precambrian sulfide ores, Skellefte District, Sweden. *Econ. Geol.* **70**, 255–274.

Scott, S. D. (1973) Experimental calibration of the sphalerite geobarometer. *Econ. Geol.* **68**, 466–474.

Scott, S. D., Both, R. A., and Kissin, S. A. (1977) Sulfide petrology of the Broken Hill region, New South Wales. *Econ. Geol.* **72**, 1410–1425.

Shadlun, T. N. (1971) Metamorphic textures and structures of sulphide ores, in Y. Takeuchi, ed., *Proc. IMA-IAGOD Meeting 70, Soc. Min. Geol. Jap.* Special Issue No. 3, 241–250.

Vokes, F. M. (1969) A review of the metamorphism of sulphide deposits. *Earth Sci. Rev.* **5**, 99–143.

Vokes, F. M. (1973) "Ball texture" in sulphide ores. *Geol. Foren. Stockholm Forh.* **195**, 403–406.

10.10 SKARN DEPOSITS

Mineralogy

The mineralogy of skarn deposits varies widely, hence generalizations should be regarded with caution. This discussion is confined to skarns that are important as sources of iron, molybdenum, tungsten, copper, lead, zinc, and tin and makes no attempt to treat the more unusual occurrences.

Major: (but highly variable from one deposit to another)—magnetite, molybdenite, sphalerite, galena, chalcopyrite, wolframite, scheelite-powellite (fluorescent under UV light).

Minor: pyrrhotite, cassiterite, hematite, gold, silver-bismuth (-selenium) sulfosalts.

Associated Minerals: quartz, various garnets, amphiboles, pyroxenes, calc-silicates, olivines, talc, anhydrite (some phases fluorescent under UV light).

Mode of Occurrence

Skarns (tactites) are composed dominantly of coarse-grained, commonly zoned calc-silicates, silicates and aluminosilicates, and associated sulfides and iron oxides. They form in high-temperature contact metamorphic halos at the junction of intrusions and carbonate-rich rocks or, more rarely, Al- and Si-rich rocks. The occurrence of ore minerals in skarns ranges from massive iron oxides or sulfides in some deposits to disseminated grains and veinlets of sulfides, molybdates and tungstates in others. Reaction skarns are narrow rims, often rich in Mn-silicates and carbonates, formed between an intrusion and carbonate-rich host rocks. Replacement skarns (ore skarns) are large areas of silicate replacement of carbonate rocks resulting from the passage of mineralizing solutions. These often contain appreciable amounts of Fe, Cu, Zn, W, and Mo.

Examples

Eagle Mountain (Fe), Darwin (Pb + Zn + Ag), Bishop (W + Mo + Cu), California; Twin Buttes (Cu), Christmas (Cu), Arizona; Hanover (Pb + Zn), Magdalena (Pb + Zn), New Mexico; Cotopaxi (Cu + Pb + Zn), Colorado; Cornwall and Morgantown (Fe), Pennsylvania; Iron Springs (Fe), Utah; Lost

Creek (W), Montana; Gaspé Copper (Cu), Murdockville (Cu), Quebec; Kamioka (Zn), Nakatatsa (Zn), Kamaishi (Fe + Cu), Chichibu (Fe + Cu + Zn), Mitate (Sn), Japan; Renison Bell (Sn), Tasmania; King Island, Australia.

Mineral Associations and Textures

Skarn deposits are typified by compositional banding, an abundance of garnets and calcsilicate minerals and a wide variation in grain-size. The garnets and calc-silicates are often poikiloblastic with enclosed pyroxenes and ore minerals. The thickness of compositional bands and the size of mineralized areas vary from a few millimeters to hundreds of meters, depending on the nature of the intrusion and its fluid content and the type of host rock. In tungsten-bearing skarns such as that at Bishop, California, the ore minerals occur as tiny inclusions, grain coatings, narrow veinlets, and occasionally as irregular polycrystalline aggregates up to 10 cm across. Sulfides occurring as disseminated grains (Figure 10.26*a*) and vein fillings are apparently late in the paragenetic sequence and frequently display replacement textures.

In massive magnetite replacement bodies (Figure 10.26*b*) such as those at Cornwall and Morgantown (Pennsylvania) and Iron Springs (Utah), the ores commonly display a laminated texture of alternating fine- to coarse-grained anhedral magnetite with greenish chlorite- and carbonate-rich bands seen even at the microscopic scale. Pyrite, commonly nickel- and cobalt- bearing, is present as irregular lenses and euhedral crystals. Irregular polycrystalline aggregates of chalcopyrite frequently display rims of secondary phases such as bornite, digenite, and covellite.

Replacement zinc-lead ores such as those at Hanover, New Mexico, and Darwin, California, consist of fine- to coarse-grained anhedral sphalerite, galena, pyrite, and lesser amounts of other sulfides interspersed with calc-silicates, garnets, and feldspars. Rose and Burt (1979) have noted that ores often tend to be restricted to particular zones within the skarn, apparently as a result of: (1) ground preparation; (2) skarn and ore forming solutions using the same "plumbing" systems; (3) coprecipitation of some skarn and ore minerals.

Formation

Skarns form as zoned sequences along the contact of acid igneous intrusives with carbonates or more rarely Al + Si rich rocks (shales, gneisses) through the diffusion of hot reactive fluids. Burt (1974) has pointed out that the mineral zoning in many skarns can be explained by simple diffusion models that assume simultaneous development of all major zones as a result of chemical potential gradients set up between dissimilar host rocks. Several types of diagrams, such as that presented in Figure 10.27, have been used to define the physico-chemical conditions under which the various zones have developed. Fluid inclusion and iso-

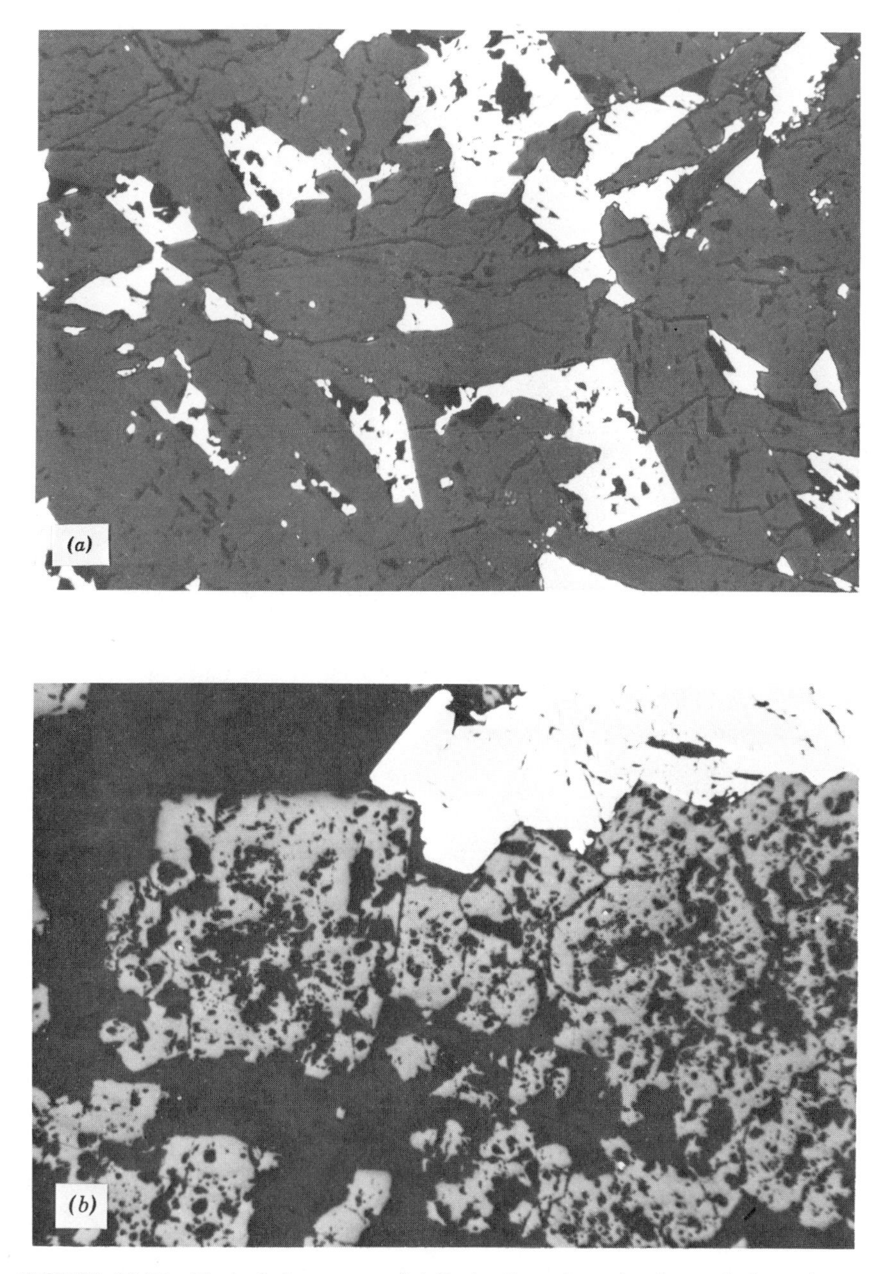

FIGURE 10.26 Typical skarn ores: *(a)* finely disseminated galena, chalcopyrite, and sphalerite in the skarn ore, Cotopaxi, Colorado (width of field = 2000 μm); *(b)* coarse-grained subhedral magnetite with pyrite in the replacement ores, Morgantown, Pennsylvania (width of field = 2000 μm).

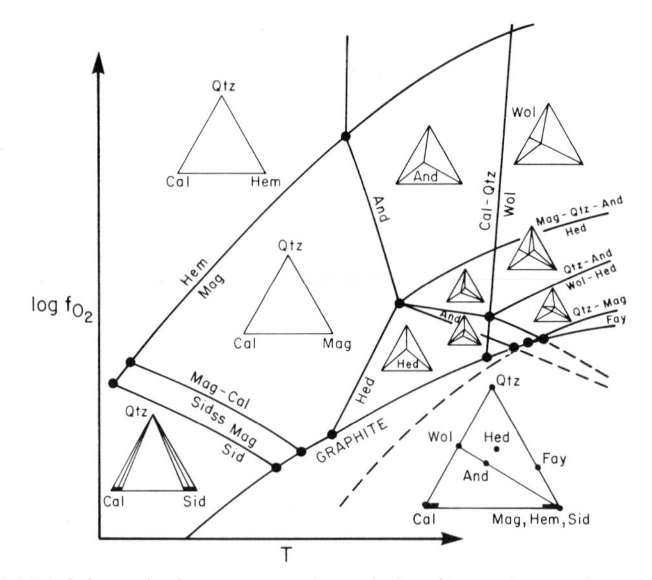

FIGURE 10.27 Schematic, low-pressure, isobaric log fO_2-T diagram for phases in equilibrium with vapor in the system Ca-Fe-Si-C-O. Abbreviations: cal, calcite; qtz, quartz; hem, hematite; sid, siderite; wol, wollastonite; and, andradite; hed, hedenbergite; fay, fayalite; mag, magnetite. ("Some phase equilibria in the system Ca-Fe-Si-C-O," Ann. Rept. Geophysical Lab.; after D. Burt, *Carnegie Institute Washington Yearbook*, Vol. 70, 1971, p. 181, reproduced with permission.)

tope studies, including those of skarn ores closely associated with porphyry-type deposits, indicate temperatures of formation from 225 to more than 600°C and a considerable degree of interaction of hypersaline magmatic fluids with convecting ground water. Rose and Burt (1979) have summarized the genesis of a typical skarn deposit as occurring in the manner outlined as follows:

1 Shallow intrusion occurs of granitic (more rarely mafic) magma at 900–700°C into carbonate sediments.

2 Contact metamorphism at 700–500°C takes place with some reaction with, and recrystallization of, carbonates to form calc-silicates.

3 Metasomatism and iron-rich skarn formation at 600–400°C occurs as a result of introduced magmatic and meteoric waters. The fluid properties change with time, becoming progressively enriched in sulfur and metals. The formation of skarn proceeds outward into the carbonate wall-rocks (exoskarn) and, from the calcium acquired by the fluid, into the solidifying intrusion (endoskarn). Diffusion gradients result in the formation of a series of skarn alteration zones but as temperatures drop, skarn destruction may begin and the

formation of replacement bodies of magnetite, siderite, silica, or sulfides occurs.

4 Superposition of oxides and sulfides at 500–300°C occurs with the formation of scheelite and magnetite commonly preceding sulfides.

5 Late hydrothermal alteration at 400–200°C causes skarn destruction and the breakdown of garnet (to calcite, quartz, hematite, pyrite, epidote, chlorite), clinopyroxene (to calcite, fluorite, quartz, oxides, sulfides, etc), and wollastonite (to calcite, quartz, fluorite).

Skarn deposits are commonly related to porphyry-type deposits (Section 9.5) and occur in carbonate beds adjacent to the intrusions. Skarns are also sometimes spatially and genetically related to the greisen (tin-tungsten-molybdenum-beryllium-bismuth-lithium-fluorine) type of mineralization that occurs locally adjacent to acid intrusives.

Selected References

Atkinson, W. W. and Einaudi, M. T. (1978) Skarn formation and mineralization in the contact aureole at Carr Fork, Bingham, Utah. *Econ. Geol.* **73**, 1326–1365.

Burt, D. M. (1974) *Metasomatic Zoning in Ca-Fe-Si Exoskarns, Geochemical Transport and Kinetics.* Carnegie Institute Wash. Pub. 634, 287–293.

Burt, D. M. (1974) Skarns in the United States—A review of recent research. *IAGOD Working Group on Skarns, 1974 Meeting*, Varna, Bulgaria.

Collins, B. I. (1977) Formation of scheelite-bearing and scheelite-barren skarns at Lost Creek, Pioneer Mountain, Montana. *Econ. Geol.* **72**, 1505–1523.

Perry, D. V. (1969) Skarn genesis at the Christmas Mine, Gila County, Arizona. *Econ. Geol.* **64**, 255–270.

Rose, A. W. and Burt, D. M. (1979) Hydrothermal alteration. In H. L. Barnes, ed., *Geochemistry of Hydrothermal Ore Deposits*, 2nd ed. Wiley-Interscience, New York, pp. 173–235.

Shimazaki, H. (1980) Characteristics of skarn deposits and related acid magmatism in Japan. *Econ. Geol.* **75**, 173–183.

Shoji, T. (1975) Role of temperature and CO_2 pressure in the formation of skarn and its bearing on mineralization. *Econ. Geol.* **70**, 739–749.

Zharikov, V. A. (1970) Skarns. *Int. Geol. Rev.* **12**, 541–559, 619–647, 760–775.

10.11 EXTRATERRESTRIAL MATERIALS: METEORITES AND LUNAR ROCKS

Mineralogy

Major: troilite, kamacite, taenite, copper, schreibersite, ilmenite, chromite, cubanite.

Minor: graphite, cohenite, mackinawite, pentlandite, magnetite, daubréelite, alabandite, sphalerite, rutile, armalcolite [$(Fe,Mg)Ti_2O_5$] (lunar only).

Secondary: goethite, lepidocrocite, maghemite, magnetite, pentlandite, pyrite.

Mode of Occurrence

Opaque minerals are present in nearly all meteorites but the proportion is variable, ranging from 100% in some irons to only a few percent in some chondrites and achondrites where the opaque minerals are interstitial between olivine, orthopyroxene, and minor plagioclase. The lunar rocks and the soils contain many of the same opaque minerals, although as much as 75% of the opaques in the soils are considered to be of meteoritic origin.

Examples

Meteorites are subdivided into four major groups, which in decreasing order of abundance are: chondrites (primarily silicates with visible chondrules), irons (nearly all opaques), achondrites (primarily silicates without chondrules), and stony-irons (roughly equal amounts of silicates and opaques). The many specimens studied appear to be fairly representative of the meteorite material in the solar system. The degree to which our lunar samples reflect the composition of the moon's surface is uncertain; the similarity of specimens from distant localities is encouraging but the number of samples is very limited.

Opaque Minerals in Meteorites

Although the relative amounts of opaque minerals in meteorites range from nearly 100% in irons to only a few percent in some chondrites and achondrites, the ore minerals which occur most abundantly in virtually all meteorites are kamacite, taenite, "plessite" (a fine intergrowth of kamacite and taenite), and troilite. Kamacite is α-iron, which contains a maximum of about 6% nickel; taenite is γ-iron, which usually contains between about 27 and 60% nickel. The iron meteorites with less than about 6% Ni, the hexahedrites, normally consist of large cubic (cube = hexahedron) crystals of kamacite. Cleavages and twinning are brought out as fine lines (Neumann lines or bands) by etching polished surfaces. Accessory minerals include grains of schreibersite, troilite, daubréelite, and graphite. With increasing nickel content (6–14%), the hexahedrites grade into octahedrites, the most common of iron meteorites. The octahedrites are so named because they show broad bands of kamacite bordered by taenite lamellae parallel to octahedral planes—the Widmanstätten structure (Figure 10.28). The iron-nickel phases are recongnizable by their high reflectance, hardness, fine polish and isotropism; kamacite is generally light bluish-gray whereas taenite is

FIGURE 10.28 Widmanstatten structure in one of the Henbury, Australia, meteorites shown after etching. (Photograph courtesy of American Meteorite Laboratory, Denver, Colorado.)

white with a slight yellowish tint. Plessite is present in the angular interstices between bands and accessory minerals present in minor amounts include schreibersite, troilite, copper, cohenite, mackinawite, graphite, chromite, daubréelite, sphalerite, and alabandite.

In the stony meteorites, troilite, recognized by its pinkish brown color, moderate anisotropism, and lower reflectance and hardness than the alloys, is frequently as abundant as the iron phases (Figure 10.29*a*). It usually occurs as single phase and occasionally occurs as twinned anhedral grains, but intergrowths with pentlandite, daubréelite, and mackinawite are common.

Chromite is relatively common in small amounts and varies from euhedral crystals to myrmekitic intergrowths with silicates; ilmenite is occasionally present as exsolution lamellae. Other minor phases commonly associated with iron phases include: copper, schreibersite, oldhamite (Ca,Mn)S, cohenite, and graphite (as a breakdown product of cohenite). Minor phases commonly associated with the troilite include: daubréelite (as lamellae in troilite), pentlandite, mackinawite, niningerite (Mg,Fe,Mn)S, sphalerite, and chalcopyrite.

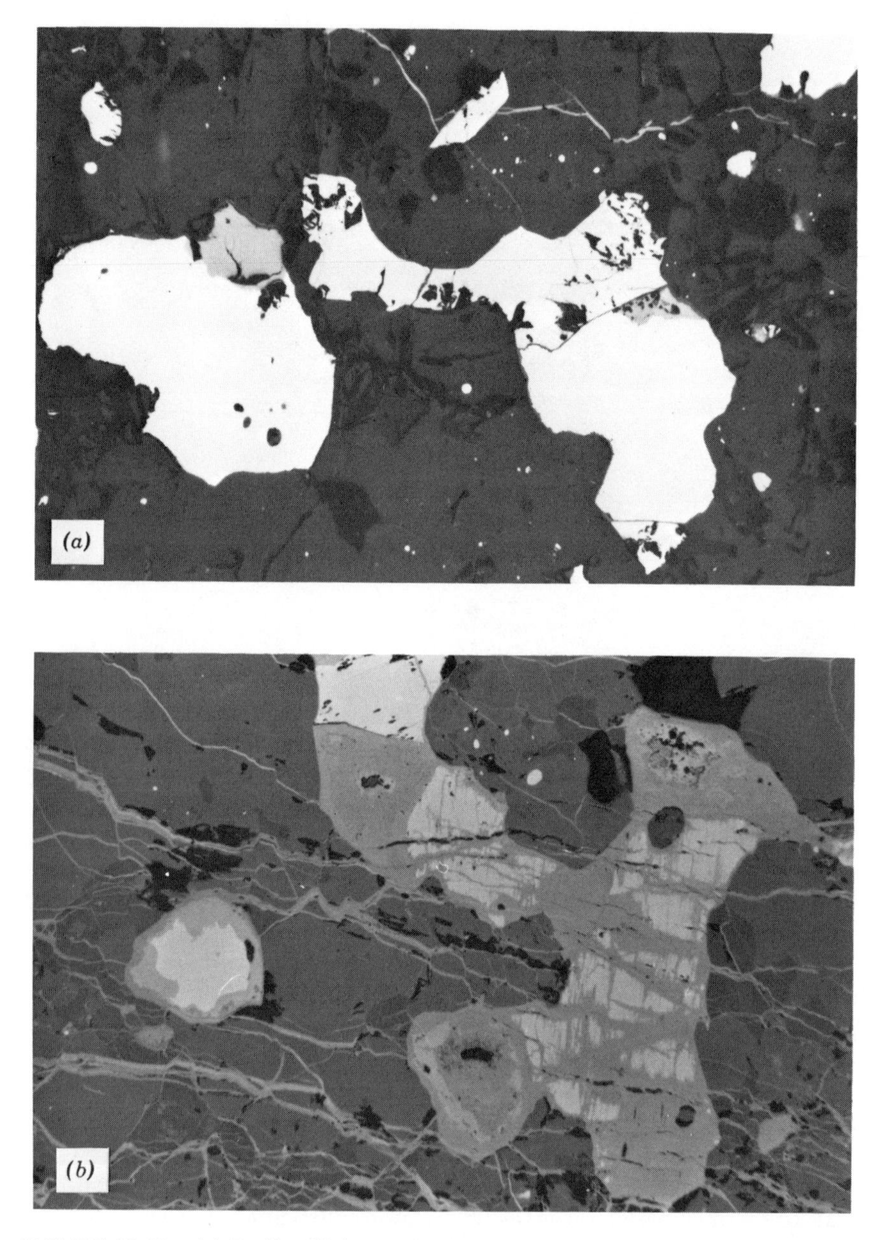

FIGURE 10.29 *(a)* Troilite (light gray) with native iron (white). Ashmore Meteorite, Texas. (Width of field = 520 μm.) *(b)* Goethite (medium gray) produced by alteration of troilite (light gray). Ashmore Meteorite, Texas. (Width of field = 520 μm.)

Secondary Minerals Resulting from Weathering of Meteorites

Meteorites, when exposed to the earth's atmosphere, undergo fairly rapid weathering with the formation of secondary phases similar to those seen in terrestrial gossans. Kamacite, taenite, troilite, and schreibersite weather rapidly; chromite, ilmenite, and magnetite weather more slowly and daubréelite very slowly. The iron-bearing minerals are converted into a "limonite" (actually goethite and lepidocrocite) shell with interspersed thin lenses of magnetite and maghemite (Figure 10.29*b*). Troilite is converted either to magnetite or locally to marcasite and pyrite. Secondary pentlandite forms when nickel released from the weathering of taenite and kamacite reacts with troilite.

Opaque Minerals in Lunar Rocks

The intense investigation of the retrieved lunar specimens has revealed a suite of opaque minerals very similar to that observed in meteorites. In fact, as much as 75% of the lunar opaques (and perhaps nearly 100% of those larger than 125 μm) are estimated to be of an initial meteoritic origin from the impact of meteorites on the lunar surface. However, the distinction between original lunar and original meteoritic phases is not always clear. The four most abundant ore minerals in presumed primary lunar rocks, are ilmenite, kamacite, taenite, and troilite (Figure 10.30). The best, but cautiously used, criterion for distinguishing lunar kamacite and taenite from that of meteoritic origin is the presence of much higher (0.8–3.0+%) cobalt contents and commonly lower nickel ($<4\%$) contents of the former. The meteoritic kamacite and taenite are similar to that directly observed in meteorites expect that much of it has lost its original Widmanstätten structure as a result of shock or thermal metamorphism. Presumed primary lunar iron occurs as dendrites, thin veinlets, needles, and tiny (<5 μm) globules on silicates or on troilite. Much of this iron is interpreted as having formed as a reduction breakdown product of fayalite-rich olivines or primary iron-containing oxides.

Troilite is disseminated throughout the crystalline lunar rocks as subrounded interstitial grains, generally less than 50 μm across. It also occurs as thin veinlets in ilmenite and as spherules on the walls of small cavities. Its common association with iron suggests that it may have formed through the crystallization of an Fe-FeS melt which was immiscible in the silicate liquid. Optically, the lunar troilite is identical to the meteoritic material, but the former commonly contains less nickel and phosphorous than the latter.

Certain lunar rocks are relatively rich in titanium, which is present in a variety of oxide spinels but largely occurs as ilmenite, the most abundant lunar opaque mineral constituting as much as 20% by volume of some rocks. The ilmenite is present in various forms:

1 Blocky euhedral to subhedral crystals (<100 μm) (Figure 10.30).

2 Thin, rhombohedral platelets parallel to (0001).

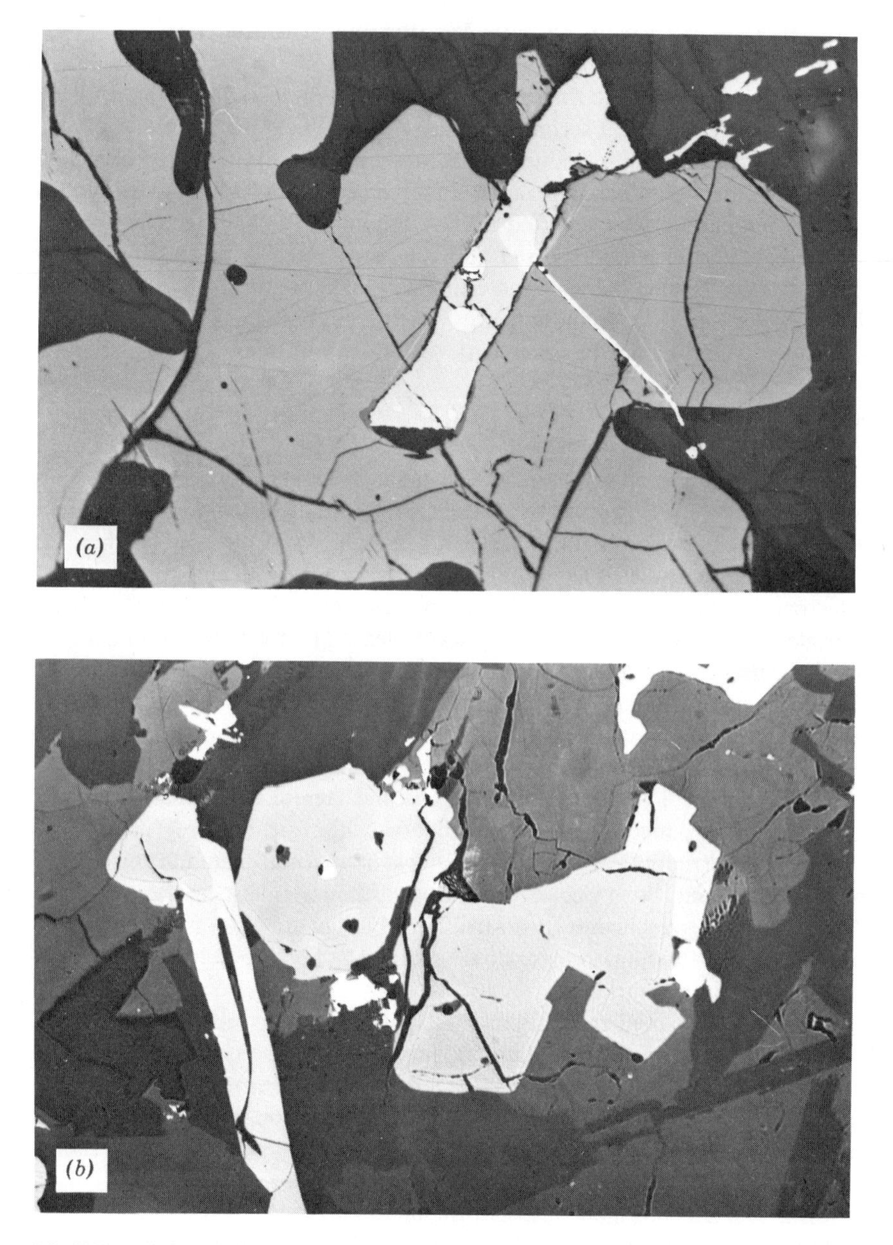

FIGURE 10.30 Opaque minerals in lunar rocks: *(a)* native iron (white) in troilite surrounded by ilmenite (width of field = 520 μm); *(b)* ilmenite rimmed by ulvöspinel with associated chromite and native iron (width of field = 520 μm). (NASA samples 70017, 224; 12002, 396, respectively.)

3 Coarse skeletal crystals (<0.5 mm) intergrown with pyroxene, troilite, and iron. The blocky grains sometimes have cores or rims of armalcolite or chromian ulvöspinel (Figure 10.30). Lunar ilmenite, like terrestrial ilmenite, is tan colored, distinctly anisotropic, occasionally twinned, and may be translucent. Rutile is occasionally present as inclusions and lamellae within the ilmenite.

The oxide spinels are represented in the lunar rocks. They are rather variable in composition but can be described as members of two groups: (1) aluminous members of the chromite-ulvöspinel series; (2) chromian members of the hercynite-spinel series.

The former are more numerous than the latter and occur in all lunar rock types as euhedral to subhedral grains in troilite, olivine, and pyroxene. Commonly, chromite cores have ulvöspinel rims. Reduction has frequently resulted in the formation of oriented laths of ilmenite and iron or a rim of ilmenite in which there are needles of iron. The hercynite-spinel minerals are rarer but are present as small euhedral to subhedral crystals (<200 μm) in lunar basalts and peridotites. These minerals are reddish to pale brown in reflected light but red to orange in thin section.

Armalcolite, $(Fe,Mg)Ti_2O_5$, a lunar mineral related to terrestrial pseudobrookite and named for the three astronauts who first brought back lunar samples, occurs as small subhedral to euhedral grains coexisting with iron in fine-grained basalts. Frequently, it is rimmed or partly replaced by ilmenite.

Other opaque mineral phases that occur only rarely in lunar samples and that are commonly associated with the troilite as rims, along fractures, and as exsolution lamellae include copper, mackinawite, pentlandite, bornite, chalcopyrite, cubanite, sphalerite (15–25% Fe), and niningerite. Schreibersite and cohenite have been identified but are probably of meteoritic origin.

Selected References

El Goresy, A. (1976) Oxide minerals in lunar rocks. In D. Rumble, ed., *Oxide Minerals*. Mineral. Soc. Am. Short Course Notes, Vol. 3, p. EG-1–46.

El Goresy, A. (1976) Oxide minerals in meteorites. In D. Rumble, ed., *Oxide Minerals*. Mineral. Soc. Am. Short Course Notes, Vol. 3, p. EG–47–72.

Frondel, J. W. (1975) *Lunar Mineralogy*. Wiley, New York.

Levinson, A. A. and Taylor, S. R. (1971) *Moon Rocks and Minerals*. Pergamon, New York.

Mason, B. H. (1962) *Meteorites*. Wiley, New York.

Mason, B. H. and Melson, W. G. (1970) *The Lunar Rocks*. Wiley, New York.

Ramdohr, P. (1973) *The Opaque Minerals in Stony Meteorites*. Elsevier, Amsterdam.

11

Applications of Ore Microscopy in Mineral Technology

11.1 INTRODUCTION

The extraction of specific valuable minerals from their naturally occurring ores is variously termed "ore dressing," "mineral dressing," and "mineral benefication." For most metalliferous ores produced by mining operations, this extraction process is an important intermediate step in the transformation of natural ore to pure metal. Although a few mined ores contain sufficient metal concentrations to require no beneficiation (e.g., some iron ores), most contain relatively small amounts of the valuable metal, from perhaps a few percent in the case of base metals to a few parts per million in the case of precious metals. As Chapters 7, 9, and 10 of this book have amply illustrated, the minerals containing valuable metals are commonly intergrown with economically unimportant (gangue) minerals on a microscopic scale.

Most mineral beneficiation operations involve two principal stages. The first of these is reduction in size of the particles of mined ore (which may initially be blocks up to several meters in diameter) to a size as close as possible to that of the individual metal-bearing mineral particles. This process of *comminution* achieves the *liberation* of valuable minerals from the gangue and, in the case of complex ores, liberation of different valuable minerals from one another. Since the size reduction required to achieve liberation is commonly down to a few hundreds of microns or even less in diameter, extensive crushing followed by grinding (or *milling*) of the ores is required. The second stage in beneficiation is that of *mineral separation* in which the valuable minerals are removed as a *concentrate* (or *concentrates*) and the remaining, commonly valueless materials (the *tailings*), are discarded. This separation is commonly achieved by exploiting differences in the physical, chemical, or surface properties between ore and gangue minerals. For example, the fact that many metalliferous ores are more dense than associated gangue minerals can be exploited using heavy media for separation or other methods of gravity concentration such as mineral jigs or shaking tables. The fact that certain ore minerals are strongly attracted by magnetic fields (e.g., magnetite, monoclinic pyrrhotite) or exhibit metallic or semiconducting electrical properties can be exploited in certain magnetic and electrical methods of separation. However, the most widely employed method of separation is *froth flotation* in which the surface chemistry of fine ore particles suspended in an aqueous solution is modified by addition of conditioning and activating reagents to be selectively attracted to fine air bubbles that are passed through this suspension or *pulp.* These air bubbles, with the associated mineral particles, are trapped in a froth that forms on the surface of the pulp and can be skimmed off to effect the separation.

The technical details of the various comminution and separation methods are outside the scope of this book and can be obtained from such works as Pryor (1965), Gaudin (1957), and Wills (1980). However, in the study of the mined ores and the products of various stages of the comminution and separation processes, ore microscopy has a very important industrial application. It facilitates the identification of the valuable minerals and of minerals that may prove troublesome during beneficiation or during later stages of extraction. It also provides information on the sizes of particles, the nature of their intergrowth, and the nature of the boundaries ("locking") between them. The efficiency of comminution and separation techniques can be monitored at any stage by the examination of mounted and polished products under the ore microscope. Thus from the initial assessment of the commercial exploitability of a prospective ore through the planning of a processing plant, the setting up of a pilot plant, and the first efficient operation of the full scale beneficiation scheme, a vital role is played by the ore microscopist.

Certain ores, rather than undergoing the complete processes of comminution and physical particle separation described previously, may have the valuable metals removed from them by chemical dissolution. For example, gold may be dissolved by cyanide solutions or copper in the form of copper sulfides may be dissolved (*leached*) by acid solutions. When crushing and grinding of the ores is required to expose the minerals to the action of the solutions, ore microscopy is again important in planning and monitoring efficient cyanidation or acid leaching. The technologies of such processes lie more in the general field of metallurgy than of mineral beneficiation, although the term *mineral technology* can be taken to embrace all of them. In this chapter, the applications of ore microscopy in mineral technology will be considered. Although the products of the roasting and smelting of ores that follows beneficiation are sometimes substances with no natural (mineral) equivalents, the techniques of ore microscopy remain applicable.

Further information on the subject discussed in this chapter may be found in publications by Schwartz (1938), Edwards (1954), Gaudin (1957), Amstutz (1961), Rehwald (1966), Ramdohr (1969), and Hagni (1978). The range of textural information required in mineral beneficiation and obtainable primarily from ore microscopy is summarized in Table 11.1.

TABLE 11.1 Information Available from Mineralogic Studies

Compositional or mineralogic data
- Subdivided into
  - Metallic ore minerals (and/or)
  - Nonmetallic ore minerals
  - Non-ore metallic (pyrite, etc.)
  - Gangue minerals
- With special reference to (selection of examples)
  - Specific gravity
  - Solubility
  - Radioactivity
  - Magnetic properties
  - Cleavability (sliming properties, sheeting and coating properties, such as sericite, clays, talc, covellite, etc.)
  - New phases in artificial products (slags, mattes, speisses, sinters, etc.)
  - State of oxidation
  - Objectionable minerals (minerals with P, S, As in certain iron ores, or Bi in lead ores, etc.)
  - Chemical composition of minerals (other elements contained in solid solution, like Fe in sphalerite, Ag in tetrahedrite, etc.)
  - On the basis of the aforementioned information the best method of concentrating can be chosen.
  - Compositional changes to be expected in the wall rock, in adjacent zones (oxidation, enrichment, leaching, etc.), or at depth, which will bear on the milling operations, as mining proceeds

Geometric data (textures and structures)
- Of
  - Metallic ore minerals (and/or)
  - Nonmetallic ore minerals
  - Non-ore metallic minerals
  - Gangue Minerals
- With Special information on
  - Locking types (including such data as tarnish, coating, veining, etc.) porosity, pitting, etc.

Quantitative data
  - Amounts of metallic ore minerals (and/or)
  - Amounts of nonmetallic ore minerals (and/or)
  - Amounts of non-ore metallic minerals (and/or)
  - Amounts of gangue minerals
- With quantitative information on the qualitative and geometric properties listed above, for example,
  - Relative grain size or particle size
  - Relative size of locking
  - Relative amounts of locking (as a whole)
  - Relative proportions of individual minerals in the locked particles (middlings)
  - Metal value percentage in ore minerals
  - Chemical analyses of samples (tailings, ores, concentrates, etc.), estimated or computed on the basis of the particle counting data

11.2 MINERAL IDENTIFICATION IN MINERAL BENEFICIATION

The techniques described in earlier sections of this book (Chapters 3, 5, and 6) can all, of course, be applied in the identification of opaque minerals in both untreated ores and products of various stages of comminution or separation. The first concern in the untreated ore is identification of the phase(s) that carry the valuable metal(s), since the initial information available is commonly only a bulk chemical analysis of the ore. That analysis provides neither information on the mineral phases present nor on their sizes and textural relations; it is possible for different mineral associations to yield very similar bulk analyses.

As outlined by Schwartz (1938) and by Ramdohr (1969), the precise identification and characterization of the ore minerals can save a great deal of work in the establishment of an efficient beneficiation system. Examination of the untreated ore will enable the assessment of the feasibility of using density, magnetic, or electrical methods of separation since such properties are well characterized for most minerals. However, fine intergrowths of dense ore minerals with gangue phases can result in ranges in specific gravity and loss of valuable metals or dilution of concentrate.* Similar problems can arise from fine intergrowths of "magnetic" and "nonmagnetic" phases (e.g., removal of ferrimagnetic magnetite and pyrrhotite from the nickel-bearing pentlandite in the Sudbury ores may result in nickel losses due to fine pentlandite flames in the pyrrhotite). The flotation properties of most ore minerals have also been extensively studied so that identification is an important first step in application of this separation method. However, flotation behavior can be very adversely affected by oxide coatings or tarnishing of ore mineral grains; such coatings may be detected under the microscope and either removed by acids prior to flotation or subjected to flotation using different reagents. Inefficient separation by flotation also occurs when the particles consist of grains of more than one mineral phase that are "locked" (bound in some manner) together; the result is either loss of ore mineral or contamination by the attached grains. Special problems may also arise with ores that contain complex minerals (e.g., minerals of the tetrahedrite group, although dominantly copper sufosalts, may contain high zinc, mercury, or silver that will appear in the concentrate). As well as assessment of the problems of mineral separation following mineral identification, the efficiency of separation can be monitored by examination of products at the various stages of benefication. In this regard, it is important that tailings as well as concentrates are thoroughly studied. Identification and characterization of the ore minerals is also important for subsequent metallurgical processing; for example, titanium is more difficult to extract from ilmenite than from rutile.

*When certain methods of separation are used, such material may appear in a third, intermediate fraction that is between concentrate and tailings in composition and is termed the "middlings."

The identification and characterization of the gangue minerals, which may include worthless opaque phases, is also very important. These materials may have economic potential and their behavior during ore processing must be assessed. Particularly important is the identification (and subsequent removal in some cases) of impurities which may adversely affect the efficiency of later concentration or refining processes or lower the quality of the final product. An example of the former is the presence of iron sulfides, stibnite, or copper sulfides in gold ores that are to be treated by dissolution in cyanide solution; such materials also react with the cyanide solution resulting both in its consumption and contamination. An example of the latter is the presence of phosphorus-bearing minerals in iron ores that reduce ore quality below that required for steelmaking. Problems can also arise from the presence of fine layer silicates such as kaolin, talc, or sericite when flotation is used for ore concentration. These minerals also tend to float and thus reduce the grade of the concentrate. Even the presence of inert gangue phases such as quartz may be important in assessing efficient comminution. When such hard materials are associated with soft sulfides (e.g., galena), the gangue is ground to a given size more slowly so there is a danger of overgrinding the galena producing fine materials (*slimes*), which are difficult to recover. Also of great importance is the identification of waste materials likely to have an adverse effect on the environment if allowed to disperse into the air or rivers (e.g., asbestos minerals).

Following identification of ore and gangue phases, the next important step is quantitative determination of their relative amounts in the untreated ores and in the ores after comminution and at the various stages of mineral separation. Such determinations must be statistically sound and it is most important to study sufficient samples of ores and mill products to ensure that the material examined is representative. The quantitative determination of mineral ratios from polished sections can be undertaken using point counting or various methods of image analysis by electronic scanning equipment. The detailed applications of such methods and their reliability have been discussed by Petruk (1976) and Jones (1977).

The combination of the various methods of quantitative mineralogical analysis with bulk chemical analysis and chemical analysis of various mineral fractions (and, where possible, of individual mineral grains) makes possible the determination of relative amounts of ore and gangue minerals, percentages of each ore mineral, and average composition of each ore mineral. In the untreated ores, it is also of great importance to know the sizes and size distributions of the various ore minerals and to monitor the distribution of ore minerals in various size fractions during comminution. Measurement may be accomplished under the microscope (e.g., using a micrometer or grating eyepiece) or through the use of various electronic devices (see Petruk, 1976 and Jones, 1977).

The question of the sizes and size distributions of ore (and gangue) particles is closely linked to the important role of studies of ore textures in relation to beneficiation problems, which will now be discussed.

11.3 ORE TEXTURES IN MINERAL BENEFICIATION

Since the first stage in the beneficiation of ores is comminution in order to liberate the particles of valuable minerals from each other and from the gangue, knowledge of the sizes and intergrowth relationships of ore mineral grains is of great importance. Only through careful examination of the ores in polished section can the optimum grain size for effective liberation be determined. Insufficient grinding may result in loss of valuable minerals in the tailings; overgrinding wastes energy and may produce slimes that are difficult to treat later in the processing. The efficiency of the grinding methods employed at the pilot stage must also be monitored by the examination of polished grain mounts of their products.

The great variety of intergrowths that may occur between ore and gangue minerals and between different ore minerals that may eventually require separation, has been well illustrated already (see Chapters 7, 9, and 10). In considering the problems of liberation of the ore minerals, a fairly simple classification based on the geometry of the intergrowths and locking textures, without any genetic implications, is useful. Such a classification of textures has been suggested by Amstutz (1961). This classification forms the basis of Figure 11.1, which also incorporates some information on the liberation characteristics of the ore types illustrated.

As well as the type and scale of the intergrowths, the nature of the boundaries between intergrown particles is important. This will show whether or not the rupturing of larger particles during grinding is likely to occur at grain boundaries. For similar reasons, information regarding fractures and fissures in the ore minerals as well as the porosity of the material are important data derived from examination under the microscope. In addition to the influence such textural features have on the comminution process, they have importance in the flotation and leaching methods of ore treatment.

The extent to which liberation has been achieved at each stage in grinding can, of course, be assessed by quantitative determination under the ore microscope. This may be seen in the photomicrographs of ground ores in Figure 11.2. The results of such examination may well be presented as in Table 11.2 (after Gaudin, 1957). Here, a lead-zinc ore pulp has been separated into six successively finer size fractions by elutriation (a method that depends on size and density of particles, hence the six fractions contain different size ranges of the major phases). Examination of each fraction under the ore microscope makes possible estimates of free and mixed grains of galena, sphalerite, and gangue and hence of the percentage liberation of each phase at each size range.

FIGURE 11.1 Geometric classification of ore mineral textures and their liberation characteristics.

Texture and Nature of Interlocking	Diagram	Liberation Properties in Relatively Large Particles, Examples of Ores
Equigranular. Straight, rectilinear, cuspate margins. Simple locking.	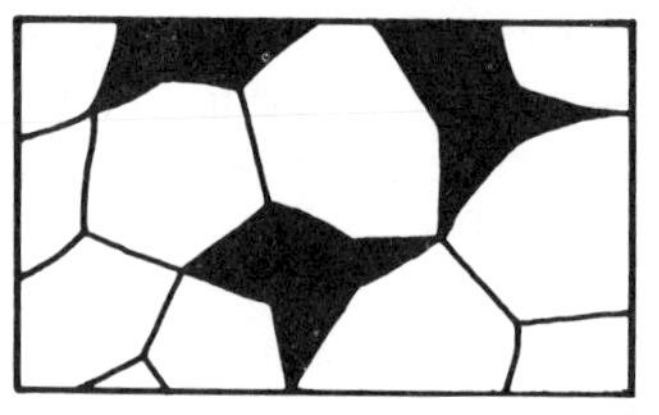	Fairly easy liberation. Common occurrence especially in orthomagmatic and highly metamorphosed and recrystallized ores. Also in ores showing successive depositional sequence.
Mutually curving boundaries with negligible interpenetration. Simple locking.		Fairly easy liberation. Common occurrence in simultaneously crystallized ores where interfacial free energies are similar.
Mottled, spotty, careous, with partial penetration. Relatively simple locking.	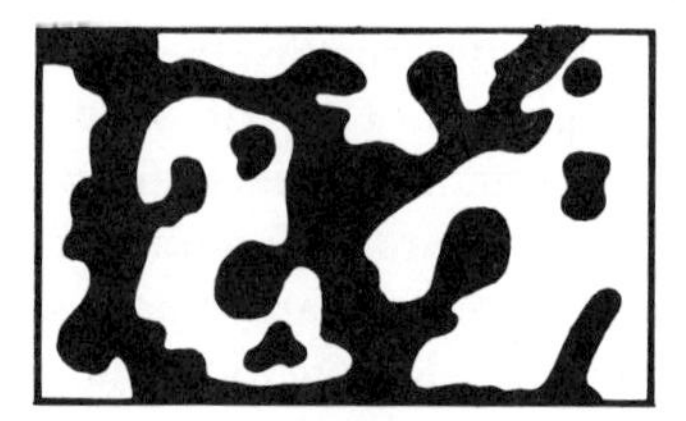	Fairly easy liberation. Common occurrence in ores where interreplacement processes have been active.
Graphic, myrmekitic, visceral locking. Deep micropenetration.		Complete liberation difficult or impossible. Not common as a major texture in ores. Produced by exsolution and replacement. Examples: galena/sphalerite and chalcocite/bornite.

FIGURE 11.1 Geometric classification of ore mineral textures and their liberation characteristics. *(Continued)*

Texture and Nature of Interlocking	Diagram	Liberation Properties in Relatively Large Particles, Examples of Ores
Disseminated, droplike, emulsion, eutectoidal locking. Finely dispersed phases.	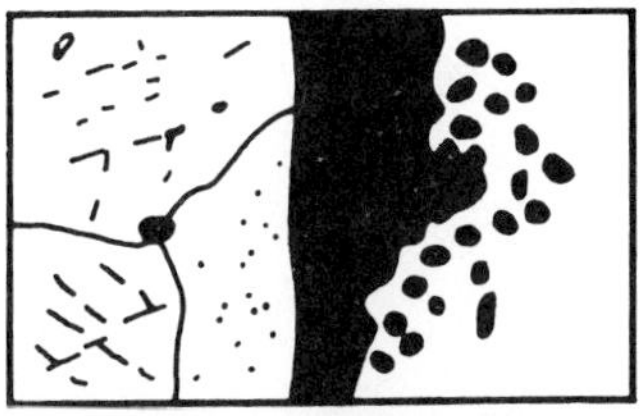	Complete liberation difficult or impossible; chemical treatment often required. Common occurrence by exsolution (left) Au/arsenopyrite chalcopyrite/sphalerite; by replacement (right) pyrite/sphalerite.
Intergranular rim; coating mantled, enveloped, atoll-like locking.	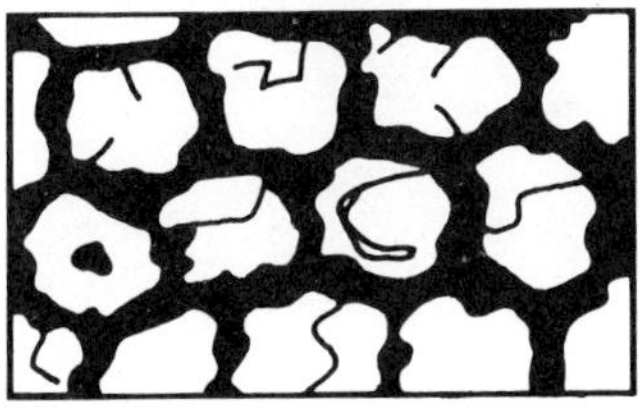	Liberation may be difficult if free grain is continously enveloped by layer. Not uncommon; often formed by replacement reaction. Examples: hematite film on gold; chalcocite or covellite on pyrite, galena, or sphalerite.
Concentric, spherulitic, scalloped, colloform-layered locking.		Liberation fairly difficult or difficult; common occurrence in Fe, Mn, and Al ores. Also U (pitchblende) intergrained with sulfide. Usually associated with colloidal precipitation.
Planar, lamellar, sandwich-type locking. Lamellae may vary in size.		Liberation fairly easy to variable. Produced by exsolution (examples: cubanite/chalcopyrite, ilmenite/magnetite). Also by replacement. Examples: magnetite and hematite.
Reticulate (netlike) boxwork. Finely interpenetrating locking.	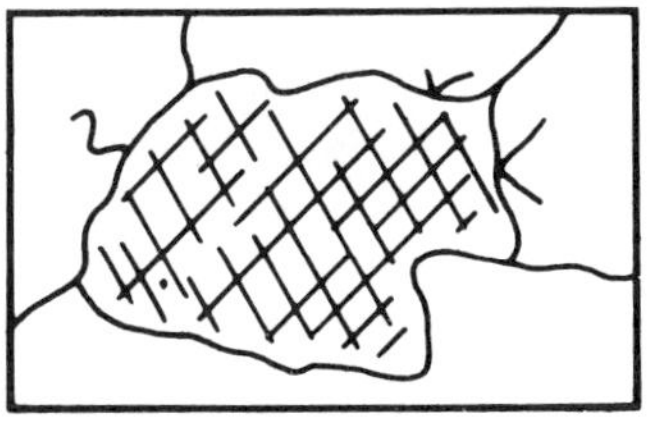	Liberation variable to difficult. Common occurrence by replacement. Examples: bornite/chalcopyrite, anglesite/covellite/galena. Also by exsolution. Examples: hematite/ilmenite/magnetite.

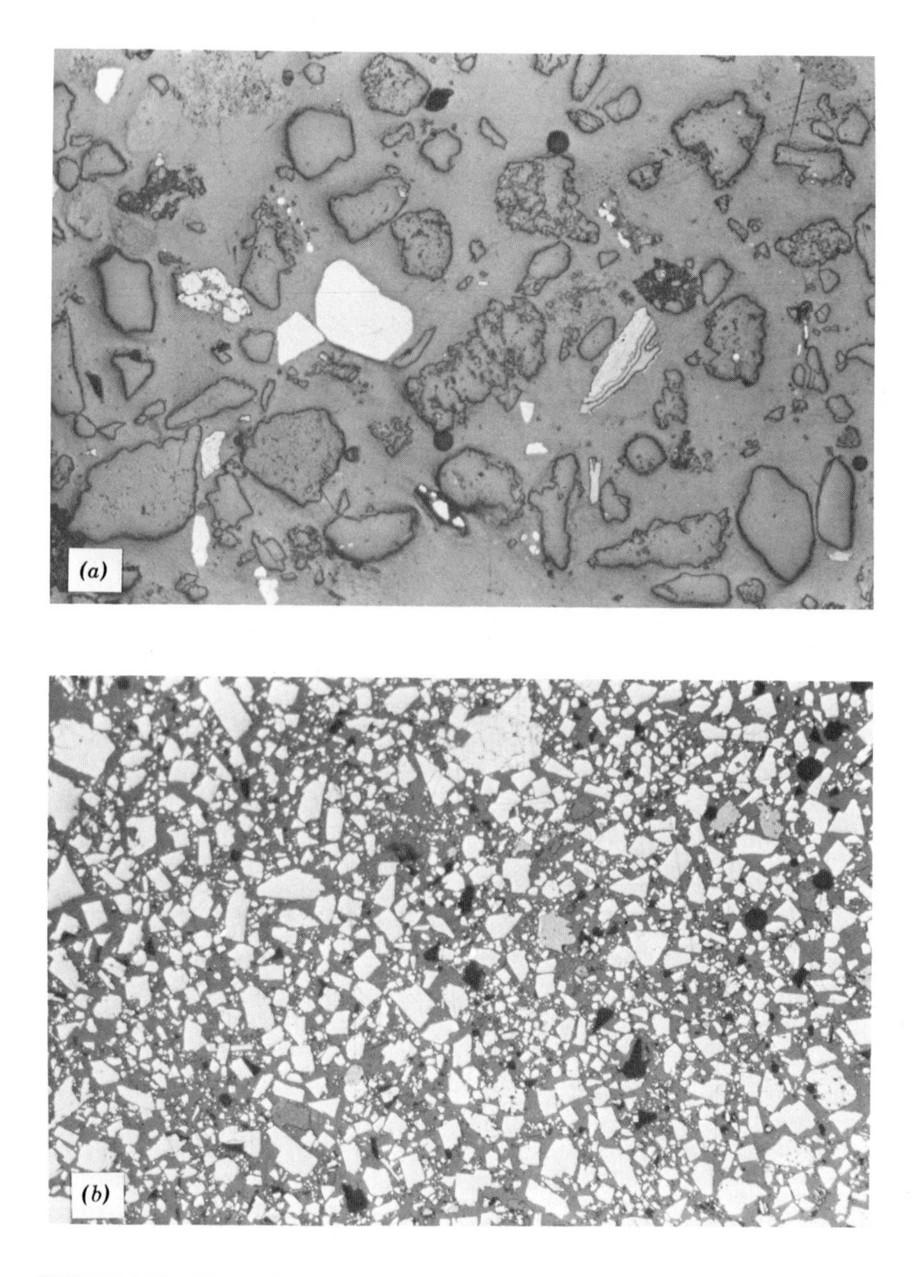

FIGURE 11.2 Photomicrographs of grain mounts of ground ores: *(a)* ground ore prior to concentration (width of field = 520 μm); *(b)* lead concentrate from Viburnum Trend, southeast Missouri—dominantly galena fragments with minor sphalerite (darker gray). (Width of field = 520 μm.)

TABLE 11.2 Microscope Determination of Liberation of Minerals in a Complex Lead-Zinc Ore* Elutriated After Grinding

	Size Range					
	A	B	C	D	E	F
Size, μm						
Galena	105/75	75/52	52/37	37/26	26/18	−18
Sphalerite	150/105	105/75	75/52	52/37	37/26	−26
Gangue	210/150	150/105	105/75	75/52	52/37	−37
Percentage by weight (from quantitative microscope determination)	100.0	100.0	100.0	100.0	100.0	100.0
Free galena	7.1	8.6	8.6	8.4	8.1	9.3†
Free sphalerite	10.3	11.6	11.3	10.7	10.3	10.4†
Free gangue	59.2	68.0	75.6	79.9	81.5	80.3
Mixed galena-sphalerite (average ½ each)	16.3	8.3	3.2	0.9	0.1	
Mixed galena-gangue (average ¼ galena)	3.7	1.9	0.7	0.1	0.0	
Mixed sphalerite-gangue (average ¼ sphalerite)	2.4	1.2	0.5	0.0	0.0	
Mixed galena-sphalerite-gangue (average ⅕ each sulfide)	1.0	0.4	0.1	0.0	0.0	
Percentage of each size in relation to total weight	1.5	7.3	14.6	16.2	9.7	50.7
Percentage liberation at each size						
Galena	43	65	83	94	99	100†
Sphalerite	53	71	87	96	99	100†
Gangue	92	97	99	100	100	100†
Average liberation for whole pulp: galena 91.3%, sphalerite 92.8%, gangue 99%						

*Ore contains 8.4% lead and 7.7% zinc.
†Calculated.

11.4 EXAMPLES OF APPLICATIONS OF ORE MICROSCOPY IN MINERAL BENEFICIATION

A number of contrasting examples can be used to illustrate this application of ore microscopy.

11.4.1 Gold Ores

Economic occurrences of gold generally consist of very small amounts of dispersed native gold or gold-silver alloys. Even in the well-known ores of the Wit-

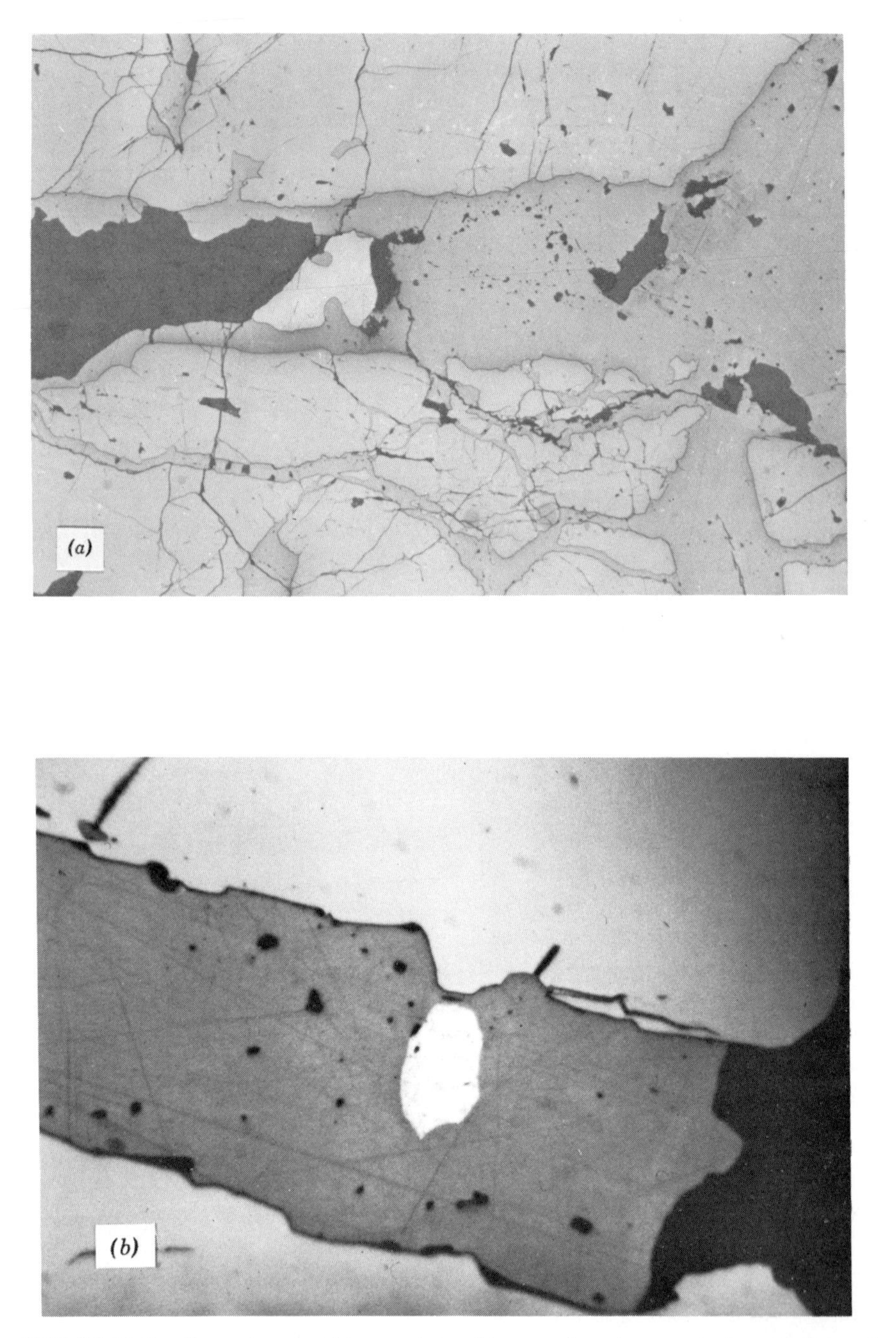

FIGURE 11.3 Examples of gold ores: *(a)* gold occurring along grain boundaries and fractures in pyrite, Witwatersrand, South Africa (width of field = 2000 μm); *(b)* gold occurring within chalcopyrite, Witwatersrand, South Africa (width of field = 210 μm); *(c)* gold grain coated with magnetite and goethite, Alaska (width of field = 2000 μm).

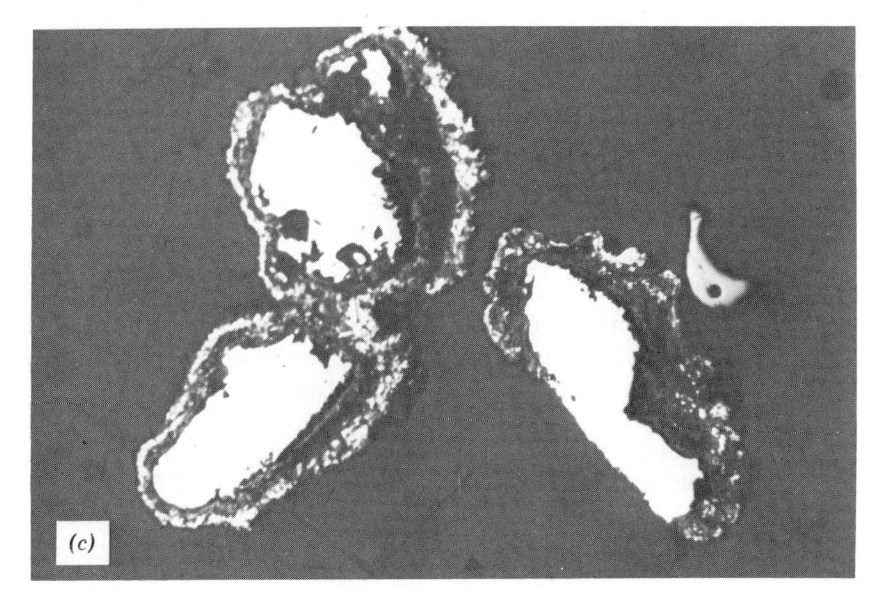

FIGURE 11.3 (*Continued*)

watersrand in South Africa, the average concentration of gold is only about 16 ppm (0.4 troy oz/ton). The ores containing native gold may contain large amounts of quartz and minor (uneconomic) sulfides; little quartz but large amounts of valueless sulfides (pyrite, pyrrhotite, arsenopyrite); or valuable base metal sulfides of antimony, arsenic, copper, lead, or zinc. Although some coarser gold particles may be separated from gangue by utilizing the high density of gold (cf. the "panning" of early prospectors), much gold is removed from ore by dissolution in a cyanide solution (*cyanidation*) or in mercury amalgam. For density separation, the gold particles must be liberated from gangue; for cyanidation or amalgamation, the gold must be sufficiently exposed to permit attack by the cyanide solution (or mercury). This is shown by two of the examples of gold ores illustrated in Figures 11.3*a* and 11.3*b*. In the first case, gold occurs along grain boundaries and fractures, and in the second, as minute particles within the sulfide. An equivalent amount of grinding will produce a much greater "effective liberation" of gold in the first ore because of the tendency to break along fractures and boundaries. This ore can be subjected to cyanidation after (or during) grinding whereas the second ore may need roasting to release the gold before it can be successfully cyanided. Many ores will, of course, contain a certain amount of both types of intergrowth; the relative amounts and size ranges can then be determined by microscopic examination. If gold particles occur largely as inclusions in a particular phase (e.g., pyrite), it may be possible to concentrate this phase by flotation and subject only this to fine grinding or roasting to liberate the gold.

Where the cyanidation process is to be employed in gold extraction, it is particularly important that microscopic studies are made to determine the presence of deleterious minerals. The dissolution process depends on an adequate supply of oxygen; pyrrhotite, marcasite, and some pyrites consume oxygen, thus inhibiting the process. Other minerals, notably stibnite, copper sulfides, and some arsenopyrites and pyrites may dissolve in the cyanide solution resulting in excessive consumption of cyanide and even reprecipitation of the gold in extreme cases. In cases where the concentration of deleterious minerals is such as to seriously affect the efficiency of the process, it may be necessary to remove them by flotation prior to cyanidation. As with cyanidation, certain minerals can adversely affect amalgamation (e.g., stibnite, enargite, realgar, tetrahedrite, pyrrhotite, arsenopyrite, and pyrite react with amalgam) and may have to be removed.

Other problems in the processing of gold ores may result from the presence of a coating on the surface of the gold particles (commonly of iron oxide as shown in Figure 11.3*c*). This can result in losses of gold during separation (especially if a magnetic process is employed to remove oxide impurities that may then carry gold with them) and can prevent dissolution of the gold unless removed by grinding. Other coatings inhibiting dissolution may form during processing. With an ore as valuable as gold, an important aspect of microscopic studies for efficient extraction is the examination of tailings. If losses are occurring in the tailings, the reasons for such losses can then be determined.

11.4.2 Copper Ores

Copper is obtained largely from sulfide ores and chalcopyrite is the single most important copper ore mineral. The chemistry and metallurgy of chalcopyrite have recently been reviewed by Habashi (1978). Other copper-iron (bornite, cubanite, talnakite, mooihoekite) and copper (covellite, chalcocite, digenite, djurleite) sulfides are often associated with chalcopyrite and may be locally important. The copper contents of each of these phases is different and a careful quantitative determination of the mineralogy is an important step in the assessment of ore grade. Many of these phases (particularly bornite, chalcocite, and covellite) may result from the alteration of chalcopyrite in processes of secondary enrichment (see Figure 7.10). Examples of major copper deposits are the *porphyry coppers* (see Section 9.5), large deposits that are often mined at average copper concentrations less than 0.5%. Such deposits contain large amounts of pyrite as well as chalcopyrite; some copper may occur as sulfosalts (tetrahedrite, enargite), and silver and gold may also occur in small but economically important quantities. Quartz, feldspars, biotite, chlorite, sericite, anhydrite, clay minerals, and other layer silicates are the dominant gangue minerals.

As Gaudin (1957) has pointed out, the sulfide copper ores are particularly well suited to flotation recovery methods. Where the ore is largely chalcopyrite and pyrite, liberation of chalcopyrite can usually be achieved by normal grinding methods. Flotation can then be carried out to selectively concentrate the chalcopyrite. The preparation of a copper concentrate may be more difficult if chalcopyrite, pyrite, and other copper sulfides are intimately intergrown. Another problem may arise in recovering the gold or silver, which may well follow pyrite into a "tailings fraction." Solutions to all of these problems require careful study of mined ores and mill products by ore microscopy.

Where the ores have been partly oxidized, some of the copper may be present as easily soluble oxides, basic sulfates, or carbonates from which copper can be easily extracted by acid leaching. Such methods are also being more widely employed as a method of extraction of copper from sulfides in *dump leaching* of very low grade ores (Fletcher, 1970). Here again, microscopic study of textures to estimate the efficiency of the leaching process is important.

Another major source of copper is the volcanogenic massive sulfide deposit (Section 10.8). The chalcopyrite in these ores occurs as anhedral interstitial grains and to a variable but significant extent as very small ($< 5\mu m$) blebs and veinlets within sphalerite (Figure 7.15). This finely dispersed chalcopyrite can create separation problems and may result in appreciable copper reporting in the zinc concentrate. It may also be desirable to remove, during beneficiation, certain phases that cause problems during smelting (e.g., arsenopyrite in order to reduce arsenic emissions). Such problems are readily anticipated if detailed ore microscopic work has first been undertaken.

11.4.3 Chromium Ores

The only significant ore mineral of chromium is chromite (ideal composition $FeCr_2O_4$), which occurs in ultramafic and mafic igneous rocks (see Section 9.2). Although chromite often occurs in a silicate matrix (see Figures 9.1 and 9.2) as single phase euhedral-subhedral grains that can be readily separated by grinding and gravity concentration, deficiencies in the grade of a concentrate may result from intimate intergrowths with gangue minerals or variation in chromite composition. The former is often a result of severe fracturing of the chromite and infilling of the fractures with serpentine; the latter results from the substitution of Fe^{3+} and Al^{3+} for Cr^{3+} in the chromite, the composition of which may even vary within a single grain, producing a zonal distribution of chromium concentration. Variations in magnetic properties of chromites as a function of iron content may permit selective magnetic concentration of chromium-rich fractions in some cases. The efficient gravity concentration of low-grade chromite ore from Cyprus described by Mousoulos and Papadopoulos (1976) involves a combination of heavy media, jigs, and shaking tables.

11.4.4 Iron-Titanium Oxide Ores

Some iron-titanium oxide ores present particular milling problems because of the intimate association of the mineral phases. Figure 11.4 illustrates two ex-

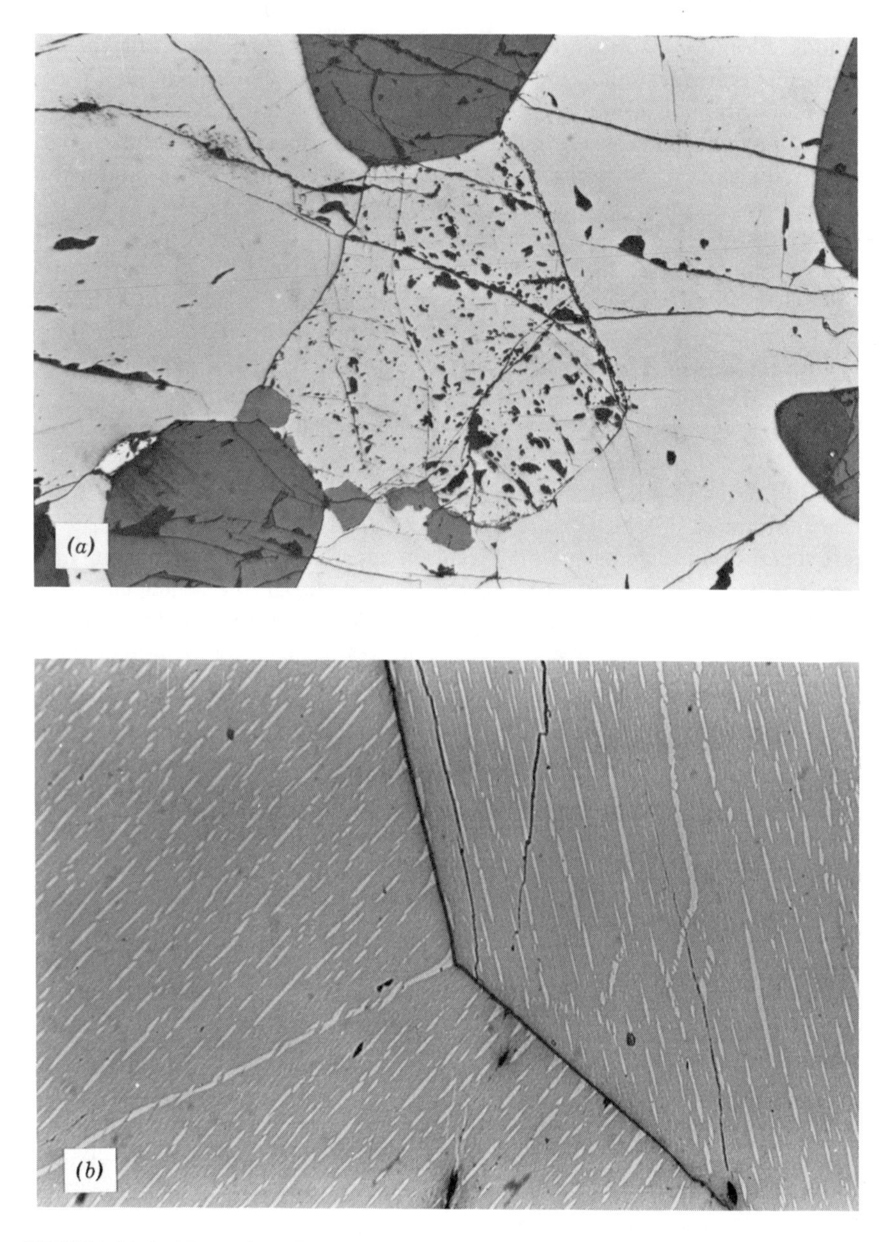

FIGURE 11.4 Examples of iron-titanium oxide ores: *(a)* coarse magnetite grain (dark gray, pitted) flanked by coarse ilmenite grains, Storgangen, Norway (width of field = 2000 μm); *(b)* fine lamellae of hematite within ilmenite, Blaafjell, Norway (width of field = 520 μm).

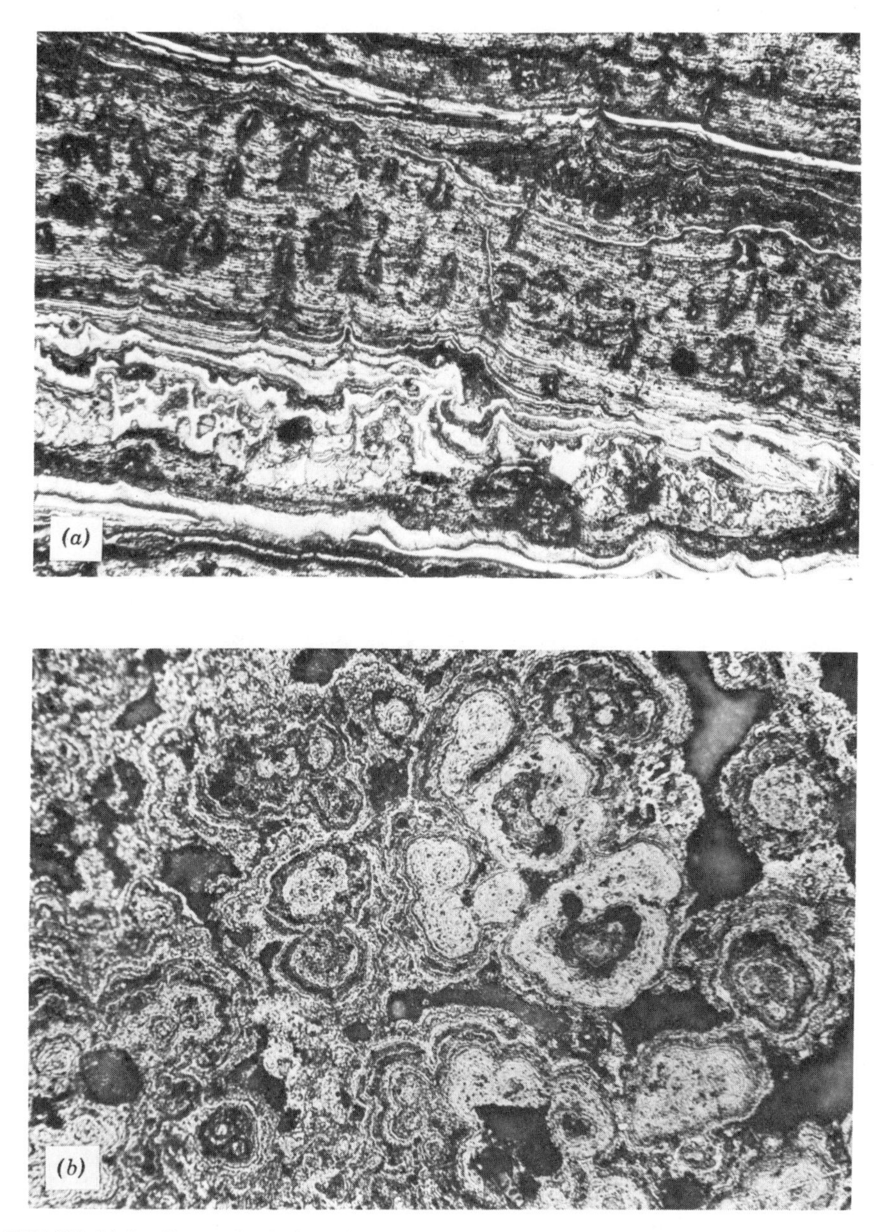

FIGURE 11.5 Textural relationships between mineral phases in manganese nodules sectioned in different orientations: *(a)* Blake Plateau, Atlantic Ocean (width of field = 2000 μm); *(b)* Pacific Ocean (width of field = 520 μm).

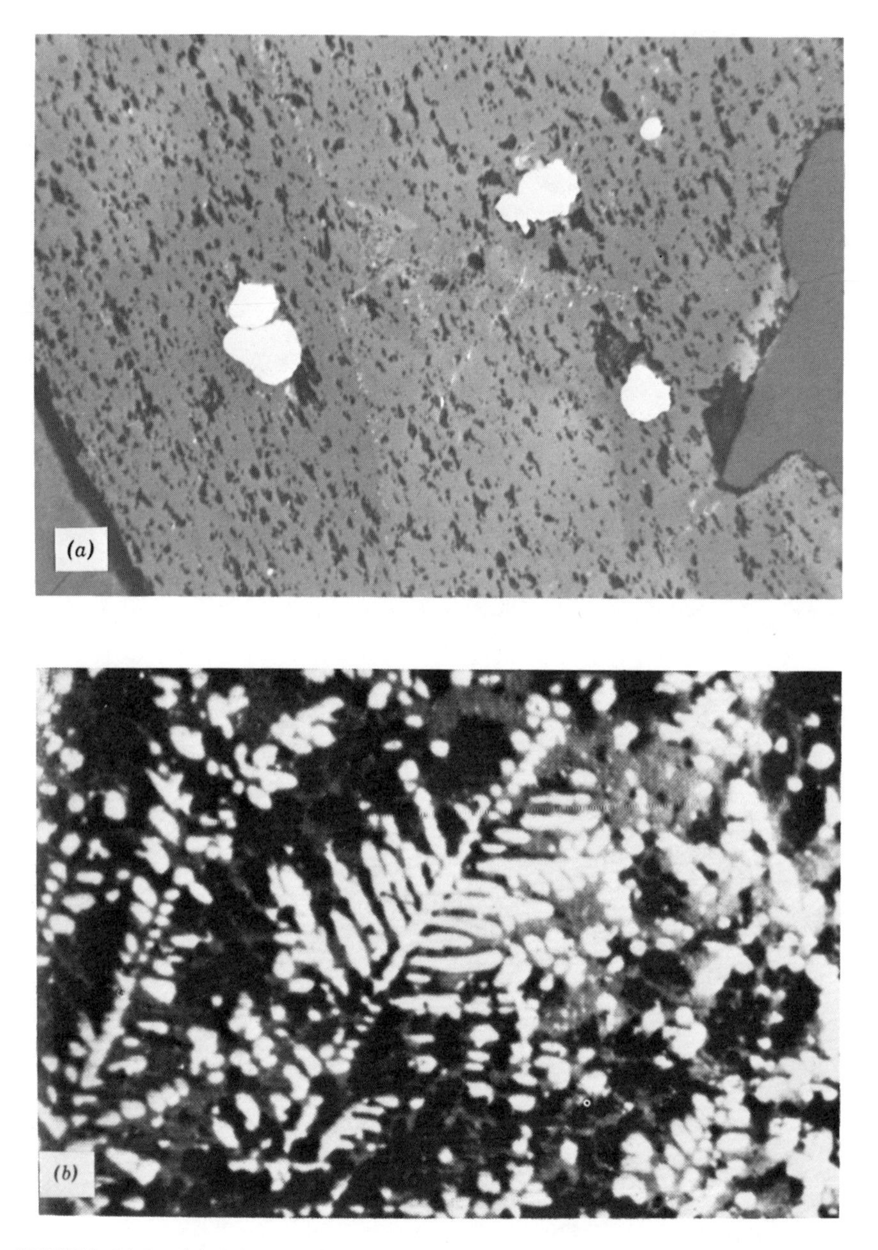

FIGURE 11.6 *(a)* Spheres of iron occurring with slag from steel-making furnace. (Width of field = 520 μm.) (Sample courtesy of Bethlehem Steel Corporation.) *(b)* Dendritic crystals of magnetite in a matrix of glass; sinter from Aswan iron ore. (Width of field = 200 μm.) (Reproduced from E. Z. Basta et. al., *Trans. I.M.M.* **78**, C3, 1969; with permission.) *(c)* Copper-nickel matte with small euhedral crystals of Fe, Ni disulfide in a matrix of Cu_2S (dark gray) and NiS (light gray). A large subhedral grain of metallic nickel is visible in the upper left-hand corner. (Width of field = 2000 μm.)

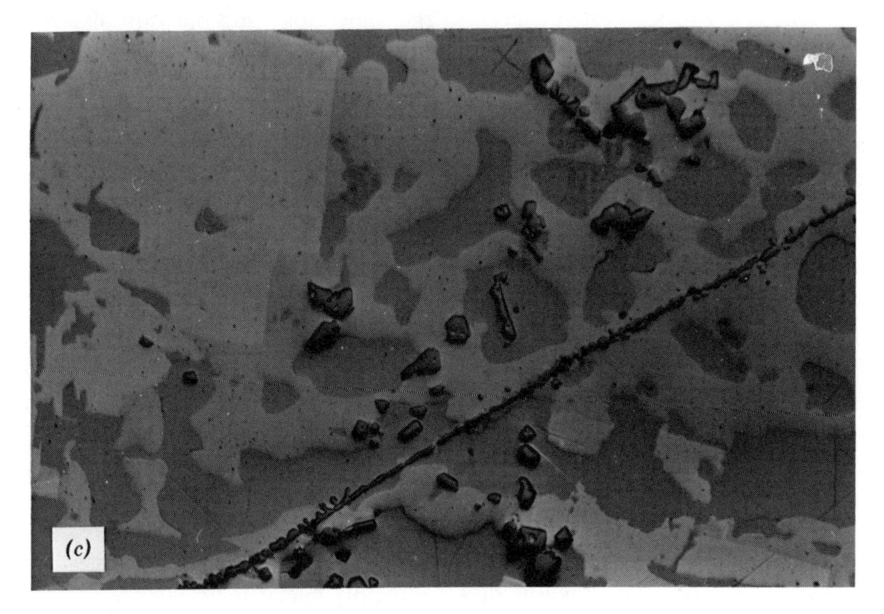

FIGURE 11.6 (*Continued*)

treme example of ores in which, in Figure 11.4*a*, the separation is readily achieved and, in Figure 11.4*b*, the separation is difficult to achieve. In the former case, ilmenite and magnetite coexist in a coarse equigranular aggregate and can readily be separated magnetically after grinding. In the latter case, however, the intimate intergrowth of ilmenite with hematite, down to a submicroscopic scale, makes a clean mechanical separation of the phases impossible.

11.4.5 Manganese Nodules

The manganese nodules of the deep ocean floors are an important potential source of not only manganese but also of other base metals; nodule deposits being considered for mining assay as high as 2.3% Cu, 1.9% Ni, 0.2% Co, and 36% Mn (dry weight) (Fuerstenau and Han, 1977). Study of the nature and distribution of the phases containing these valuable metals in the nodules is an important step towards their economic exploitation. Although their detailed mineralogy is complex, the predominant minerals in the nodules are manganese (IV) oxides related to the terrestrial minerals todorokite, birnessite, and δ-MnO_2; also present is crystallographically poorly ordered goethite ("incipient goethite") (Burns and Burns, 1977; 1979). The nickel, copper, and cobalt are taken up by todorokite in postdepositional processes. The textural relationships between mineral phases in a nodule are shown in Figures 11.5*a* and 11.5*b*.

Brooke and Prosser (1969) examined the mineralogy and porosity of several such nodules and investigated the problem of selective extraction of copper and nickel. Tests showed that selective leaching of copper and nickel using dilute sulfuric acid is a possible method of extraction. A whole range of possible extraction methods (acid leaching, ammonia leaching, smelting, chlorination, and segregation roasting) have also been reviewed by Fuerstenau and Han (1977).

11.5 THE STUDY OF SINTER AND SMELTER PRODUCTS

The compositions and textural relationships of the products of sintering and smelting (mattes, slags, etc.) can be studied using the techniques outlined in Chapters 1–6. Although these are not minerals, many do closely resemble ore minerals in composition and sometimes in texture and optical properties. Their history of crystallization may also be followed using textural interpretations similar to those outlined for ores. A few very brief examples can be given to illustrate this vast subject.

Basta et al. (1969) have studied the mineralogy of fluxed sinters of fines from the Aswan iron ore. These oolitic ores are mainly hematite with some goethite, quartz, carbonate, chlorite; some minor sulfur, phosphorus, and manganese are also present. Sintering was at ~ 1200°C with limestone, pyrite cinders, and coke as a fuel. Studies of such sinters show iron, iron oxides (magnetite, hematite, wüstite), and calcium ferrites as spherules, dendrites, and crystals in a matrix of silicates—both crystalline (gehlenite, olivine, wollastonite) and glassy (Figure 11.6*a* and 11.6*b*). Minor sulfides may also be present (pyrrhotite, chalcopyrite). The effect of varying the amounts of limestone added is related to mineralogical variations that are in turn related to such properties as strength and reducibility of the sinters.

The reduction of ilmenite has been studied by Jones (1974). At temperatures less than 1000°C, natural ilmenites are reduced by carbon monoxide to metallic iron and reduced rutiles ($Ti_nO_{2n-1}, n \geq 4$) but minor magnesium or manganese prevent the reaction from going to completion. These problems are lessened at higher temperatures, and at 1200°C the products of reduction are metallic iron and anosovite ($Fe_{3-x}Ti_xO_5$) solid solution. These reaction products were characterized at various stages of reduction by ore microscopy (and electron probe microanalysis).

The mattes produced in copper smelters vary considerably in composition but include many of the phases familiar from mineralogical studies in the Cu-Fe-S and Cu-Ni-S systems. Figure 11.6*c* illustrates the nature of a copper-nickel matte containing 40% Cu and 40% Ni. The matte contains an early crystallized iron-nickel alloy phase, dispersed small grains of an Fe-Ni disulfide and a matrix of NiS (millerite), and a copper sulfide solid solution. This last phase has exsolved on cooling to give a basketweave texture of two copper sulfide phases.

11.6 CONCLUDING REMARKS

This very brief outline of the applications of ore microscopy in mineral technology serves only to illustrate the great importance of this area of application. The increasing world demand for metals necessitates the economic extraction of metals from ores of lower and lower grade, which requires efficient, skilled beneficiation. Thus careful study under the reflected-light microscope of the ore and of the products of various comminution and separation processes is necessary. If, following initial discovery, exploratory drilling to determine overall size, grade, and geological setting, mineralogical analysis to establish feasibility of metal extraction, and pilot testing of the extraction processes, it is decided to proceed with a mining operation, the ore microscopist is still needed to monitor variations in the mineralogy of the ore and the effects of such variation on processing. Variations laterally or vertically in the orebody may call for changes in beneficiation procedures or mixing of ores mined from different areas; the microscopic study of the ores may also be of great value as a guide to mine exploration and development. In all these applications, the ore mineralogy has to be related to the geological and engineering problems of mining on the one hand and to the problems of beneficiation and metallurgical treatment on the other.

BIBLIOGRAPHY

Amstutz, G. C. (1961) Microscopy applied to mineral dressing. *Colo. School Mines* **56**, 443–484.

Basta, E. Z., El Sharkowi, M. A., and Salem, M. W. (1969) Mineralogy of some fluxed sinters produced from the Aswan iron ore. *Trans. Inst. Min. Metall.* **78**, C1–C13.

Brooke, J. N. and Prosser, A. P. (1969) Manganese nodules as a source of copper and nickel-mineralogical assessment and extraction. *Trans. Inst. Min. Metall.* **78**, C64–C73.

Burns, R. G. and Burns, V. M. (1977) Mineralogy. In G. P. Glasby, ed., *Marine Manganese Deposits.* Elsevier Oceanography Series, 15, Elsevier, Amsterdam.

Burns, R. G. and Burns, V. M. (1979) Manganese oxides. In *Marine Minerals.* Min. Soc. Am. Short Course Notes, Vol. 6, pp. 1–46.

Edwards, A. B. (1954) *Textures of the Ore Minerals and Their Significance.* Aust. Inst. Min. Metall., Melborne.

Fletcher, A. W. (1970) Metal winning from low grade ore by bacterial leaching. *Trans. Inst. Min. Metall.* **79**, C247–C252.

Fuerstenau, D. W. and Han, K. N. (1977) Extractive metallurgy. In G. P. Glasby, ed., *Marine Manganese Deposits.* Elsevier Oceanography Series, 15, Elsevier, Amsterdam.

Gaudin, A. M. (1957) *Flotation.* McGraw-Hill, New York.

Habashi, F. (1978) *Chalcopyrite: Its Chemistry and Metallurgy.* McGraw-Hill, New York.

Hagni, R. D. (1978) Ore microscopy applied to beneficiation. *Min. Eng.* **30**, 1137–1147.

Jones, D. G. (1974) Optical microscopy and electron-probe microanalysis study of ilmenite reduction. *Trans. Inst. Min. Metall.* **83**, C1–C9.

Jones, M. P. (1977) Automatic image analysis. In J. Zussman, ed., *Physical Methods in Determinative Mineralogy.* Academic, London-New York.

Mousoulos, L. and Papadopoulos, M. Z. (1976) Gravity concentration of Troodos chromites. Cyprus. *Trans. Inst. Min. Metall.* **85**, C73–C77.

Petruk, W. (1976) The application of quantitative mineralogic analysis of ores to ore dressing. *Can. Min. Metal. Bull.*

Pryor, E. J. (1965) *Mineral Processing.* Applied Science, London.

Ramdohr, P. (1968) *The Ore Minerals and Their Intergrowths.* Pergamon, Oxford.

Rehwald, G. (1965) The application of ore microscopy in beneficiation of ores of the precious metals and of the nonferrous metals. In H. Freund, ed., *Applied Ore Microscopy.* Macmillan, New York.

Schwartz, G. M. (1938) Review of the application of microscopic study to metallurgical problems. *Econ. Geol.* **33**, 440–453.

Wills, B. A. (1980) *Mineral Processing Technology.* Pergamon, Oxford.

Appendix 1

Table of Diagnostic Properties of the Common Ore Minerals

This appendix contains data to help in the microscopic identification of the most commonly encountered opaque minerals (approximately 100 minerals are included). The data presented are as follows:

1. The mineral name.
2. The chemical formula; this is generally in its simplest form (e.g., the end member of a solid solution series) although major substitutions are shown.
3. The crystal system.
4. A description of the color of the mineral (the symbol "→ galena, bluish" indicates that the mineral described appears bluish against galena).
5. A description of any observable bireflectance and reflection pleochroism.
6. A description of the presence, intensity, and character of any anisotropism.
7. A description of the character of any observable internal reflections.
8. The quantitative reflectance values ($R\%$) in air at 546 and 589 nm wavelength. These data are consistent with the COM Data File (Henry, 1977); however, the COM data are for a single sample and do not necessarily reflect the ranges of values that have been reported for many minerals.
9. Quantitative color values quoted using the C.I.E. System and giving chromaticity coordinates (x and y) and the luminance ($Y\%$) relative to the standard C-illuminant.
10. Quantitative indentation microhardness (Vickers hardness number) at a load of 100 g (VHN_{100}) unless another load is specified. For some minerals information is given on indentation characteristics as follows: p, perfect; f, fractured; sf, slightly fractured; cc, concave; cv, convex.
11. Polishing hardness (PH) given as less than, equal to, or greater than other common ore minerals.

12 Mode of occurrence and other characteristic properties; this is general information on crystal morphology, cleavage, twinning, characteristic alteration effects, and commonly associated minerals.

The data presented in the tables have mainly been derived from the following sources, which should be consulted for further details and information on other minerals:

Uytenbogaart, W. and Burke, E. A. J. (1971) *Tables for Microscopic Identification of Ore Minerals.* Elsevier, Amsterdam.

Ramdohr, P. (1969) *The Ore Minerals and Their Intergrowths.* Pergamon, Oxford.

Schouten, C. (1962) *Determinative Tables for Ore Microscopy.* Elsevier, Amsterdam.

Henry, N. F. M., ed. (1977) Commission on Ore Microscopy: IMA/COM Quantitative Data File (first issue). Applied Mineralogy Group, Mineralogical Society, London.

Inserted before the main part of the appendix is an "identification scheme" as an aid to determining an unknown mineral. This simplistic scheme should be used as no more than a preliminary guide to the possible identity of a phase.

IDENTIFICATION SCHEME*

Distinctly Colored

Blue	Isotropic (or weakly anisotropic)	Chalcocite, digenite
	Anisotropic	Covellite
Yellow	Isotropic (or weakly anisotropic)	Gold, chalcopyrite
	Anisotropic	Chalcopyrite, millerite, delafossite, cubanite, mackinawite, valleriite
Red-brown to brown	Isotropic (or weakly anisotropic)	Bornite, copper, bravoite
	Anisotropic	Idaite, valleriite, delafossite, mawsonite
Pink, purple, violet	Isotropic (or weakly anisotropic)	Bornite, copper, bravoite, violarite
	Anisotropic	Breithauptite

Distinctly Colored Internal Reflections (In minerals which are not distinctly colored)

Blue	Anatase, azurite
Yellow	Sphalerite, orpiment, rutile, cassiterite
Red to brown	Cinnabar, proustite, pyrargyrite, tennantite, sphalerite, cuprite, chromite, orpiment, wolframite

Weakly Colored (if at all)*

Blue	Isotropic	Tetrahedrite
	Anisotropic with internal reflections	Hematite, cuprite, cinnabar, hausmannite, proustite, pyrargyrite
	Anisotropic without internal reflections	Psilomelane
Green	Isotropic (or weakly anisotropic)	Tetrahedrite, acanthite
	Anisotropic	Stannite, polybasite
Yellow	Isotropic	Pyrite, pentlandite
	Anisotropic	Marcasite, niccolite
Red-brown to brown	Isotropic	Magnetite, ulvöspinel
	Anisotropic	Pyrrhotite, ilmenite, enargite
Pink, purple, violet	Isotropic	Cobaltite, linnaeite
	Anisotropic	Niccolite, famatinite

Not Colored to Any Degree*

$R\% \geq 51.7$ (pyrite)

Isotropic	Hardness high	(Pyrite) gersdorffite, skutterudite
	Hardness medium-low	Silver, platinum, allargentum
Anisotropic	Hardness high	(Marcasite) rammelsbergite, pararammelsbergite, safflorite, loellingite, arsenopyrite

	Hardness medium-low	Bismuth, antimony, arsenic, dyscrasite, tetradymite, sylavnite
R% 51.7 (pyrite) to 43.1 (galena)		
Isotropic	Hardness high	Siegenite, ullmannite
	Hardness medium-low	Galena, freibergite, alabandite
Anisotropic	Internal reflections	Pyrargyrite
	No internal reflections	Bismuthinite, stibnite, cosalite, kobellite
R% 43.1 (galena) to 20.0 (magnetite)		
Isotropic	No internal reflections	Carrollite, tetrahedrite, maghemite, bixbyite (magnetite)
	Internal reflections	Realgar, tennantite, pearcite
Anisotropic	Internal reflections	Hematite, enargite, miargyrite, pyrargyrite, boulangerite, chalcostibite, orpiment, realgar, chalcophanite
	No Internal reflections	Molybdenite, pyrolusite, berthierite, boulangerite, chalcostibite, jamesonite, tenorite, stephanite, stromeyerite, mawsonite, pyrolusite
R% $\geq$ 20.0 (magnetite)		
Isotropic	No internal reflections	Chromite, coffinite
	Internal reflections	Brannerite, sphalerite
Anisotropic	Internal reflections	Columbite-tantalite, manganite, chalcophanite, scheelite, cassiterite, lepidocrocite, zincite, uraninite, manganite, wolframite, goethite, rutile
	No internal reflections	Graphite, braunite

*Categories defined are intended only as a rough guide to identification—the following tables should be used to confirm any possible identification.

ALPHABETICAL LISTING OF ORE MINERALS WITH DIAGNOSTIC PROPERTIES

Note: Information is reported as follows:

1. Name 2. Formula 3. Crystal System	C—Color B/P—Bireflectance/Pleochroism A—Anisotropy IR—Internal Reflections	*R*—Reflectance at 546 and 589 nm in Air QC—Quantitative Color Coordinates	VHN—Vickers Microhardness at 100 g Load PH—Polishing Hardness	Mode of Occurrence; Other Characteristic Properties
Acanthite Ag_2S Monoclinic	C—gray, with a greenish tint → Galena, darker, greenish gray → Silver, dark greenish gray B/P—very weak A—distinct if well polished IR—not present	*R*—30.3–31.3 29.0–29.8	VHN—23–26 (p) PH—less than most minerals	Occurs as euhedral cubic crystals pseudomorphous after argentite (stable > 176°C) and as anhedral polycrystalline aggregates. Difficult to polish without scratches because of softness, but twinning often visible when well polished. Occurs as irregular inclusions in galena; often associated with pyrite, galena, sphalerite, tetrahedrite, covellite, proustite, pyrargyrite, polybasite. The high temperature polymorph, argentite, always inverts to acanthite on cooling but its former existence may be evidenced by cubic morphology.

Note: Information is reported as follows:

1. Name 2. Formula 3. Crystal System	C—Color B/P—Bireflectance/Pleochroism A—Anisotropy IR—Internal Reflections	*R*—Reflectance at 546 and 589 nm in Air QC—Quantitative Color Coordinates	VHN—Vickers Microhardness at 100 g Load PH—Polishing Hardness	Mode of Occurrence; Other Characteristic Properties
Alabandite MnS Cubic	C—gray → Sphalerite, distinctly lighter B/P—not present A—isotropic. Sometimes with weak anomalous A IR—common, dark green to brown	*R*—22.8 22.3 QC—O.301 0.305 22.8	VHN—240–251 (p) PH ~sphalerite	Occurs as euhedral crystals and as anhedral aggregates; resembles sphalerite. Cleavage, lamellar twinning, and zonal textures may be visible. Occurs with pyrite, chalcopyrite, pyrrhotite, pyrolusite, Mn-sphalerite, Mn-carbonate.
Allargentum $Ag_{1-X}Sb_X$ Hexagonal	C—white, slightly grayish →silver, grayish B/P—not present A—weak IR—not present	*R*— ~70	VHN— PH > silver	Occurs as lamellar intergrowths in silver, especially that from Cobalt, Ontario. Originally identified as dyscrasite, which is very similar but is Ag_3Sb.

Allemontite A mixture of As or Sb with AsSb	C-white B/P—weak A—distinct IR—not present	*R*—50–70	VHN—85–100 PH ~antimony	Occurs as a myrmekitic intergrowth which may be on such a fine scale it is only discernible as two phases under high power magnification. Two phases are often more visible after slight oxidation or etching. Occurs with stibnite in Co-Ni-Ag-Bi-As ores and pegmatites.
Antimony Sb Trigonal	C—white → Arsenic, slightly more white → Galena, brighter white → Silver, less bright → Dyscrasite, similar B/P—weak A—distinct; yellowish gray, brownish, bluish gray IR—not present	*R*—71.1–73.0 70.0–72.1	VHN—84–98 (p) PH > stibnite < arsenic	Occurs as fine- to coarse-grained aggregates, rarely euhedral. Cleavage and twinning (often polysynthetic) commonly visible. Occurs with stibnite, pyrite, arsenopyrite, Co-Ni arsenides, and with stibarsen as fine graphic to myrmekitic intergrowths known as "allemontite."
Argentite—see Acanthite				
Arsenic As	C—white; tarnishes rapidly → Antimony, slightly darker	*R*—51.3–56.4 50.4–55.8	VHN—83–149 (sf)	Occurs as fine- to coarse-grained anhedral aggregates

Note: Information is reported as follows:

1. Name 2. Formula 3. Crystal System	C—Color B/P—Bireflectance/Pleochroism A—Anisotropy IR—Internal Reflections	*R*—Reflectance at 546 and 589 nm in Air QC—Quantitative Color Coordinates	VHN—Vickers Microhardness at 100 g Load PH—Polishing Hardness	Mode of Occurrence; Other Characteristic Properties
Trigonal	gray →Skutterudite and safflorite, slightly darker gray →Galena, white with a creamy tint B/P—weak in air; distinct in oil; grayish white to yellow or bluish gray A—distinct; gray to yellowish gray IR—not present		PH > bismuth, silver	and commonly as colloform bands. Twinning and a basal cleavage are often visible. Occurs with rammelsbergite, skutterudite, proustite, arsenopyrite, pyrite, and stibarsen as fine graphic to myrmekitic intergrowths of "allemontite." The very rapid (a few hours) tarnish is diagnostic.
Arsenopyrite FeAsS Monoclinic	C—white → Pyrite, white → Loellingite, safflorite, creamy-white → Antimony, grayish white → Galena, sphalerite, white	*R*—51.85–52.2 51.8–53.2 QC—a 0.315 0.320 52.5	VHN—715–1354 1081 on (001) (sf) PH > skutterudite, magnetite < pyrite, cobaltite	Commonly observed as euhedral to subhedral crystals with characteristic rhomb shape when a minor phase; also as anhedral granular masses when abundant. Lamellar

	with pale yellow tint B/P—weak A—strong; blue, green IR—not present	b 0.318 0.325 51.8 c 0.310 0.316 51.8		twinning common. Occurs with pyrite, loellingite, glaucodot, pyrrhotite, chalcopyrite, sphalerite, galena, cobaltite, gold, molybdenite. Good polish, white color, anisotropism, and crystal form are characteristic.

Berthierite
$FeSb_2S_4$
Orthorhombic
	C—white-gray with a pink or brown tint
	B/P—strong and characteristic
	//a brownish pink
	//b grayish white
	//c white
		R—42.0–36.6
		41.0–36.5
			VHN—102–213	Occurs as euhedral needlelike crystals and as subhedral aggregates, with stibnite, chalcopyrite, pyrite, arsenopyrite, pyrrhotite, gudmundite, sphalerite, galena.

	A—very strong, blue, gray white; brown, pink IR—not present	QC—0.301 0.308 41.9	PH ~ stibnite < sphalerite	
Bismuth Bi Trigonal	C—white to creamy white; pinkish cream → Silver, creamy → Arsenic, pinkish creamy → Sulfosalts, pinkish creamy B/P—weak but distinct, creamy to pinkish	*R*—66.7 68.8	VHN—15–18(p) PH < all associated minerals	Occurs as irregular masses or inclusions of anhedral crystals. Twinning is common and may be induced by grinding or scratching. Occurs with sulfosalts, pyrite, pyrrhotite, sphalerite, chalcopyrite,

Note: Information is reported as follows:

1. Name 2. Formula 3. Crystal System	C—Color B/P—Bireflectance/Pleochroism A—Anisotropy IR—Internal Reflections	*R*—Reflectance at 546 and 589 nm in Air QC—Quantitative Color Coordinates	VHN—Vickers Microhardness at 100 g Load PH—Polishing Hardness	Mode of Occurrence; Other Characteristic Properties
	A—distinct to strong IR—not present			bismuthinite, cassiterite, molybdenite, wolframite, arsenopyrite, Co-Ni arsenides, silver, galena.

Bismuthinite
Bi_2S_3
Orthorhombic
	C—white; in oil with bluish gray tint
	→ Bismuth, darker, bluish gray
	→ Chalcopyrite, bluish gray
	→ Galena, lighter, creamy white
	B/P—weak to distinct
	//a bluish gray white
	//b gray white
	//c creamy white
	A—very strong, especially in oil; gray, yellow, violet,
		R—38.5–45.4
		38.1–45.0
			VHN—110–136 (sf)
			PH > bismuth
			< chalcopyrite
				Occurs as subhedral lathlike crystals; less commonly as granular masses. Cleavage // (010) common. Stress-induced twinning and undulose extinction often seen. Occurs with bismuth, pyrite, pyrrhotite, arsenopyrite, chalcopyrite, sphalerite, stannite, cassiterite, wolframite, molybdenite.

	straight extinction; large crystals often undulose IR—not present			
Bixbyite $(Mn,Fe)_2O_3$ Cubic	C—gray with cream to yellow tint → Braunite, jacobsite, hausmannite, lighter, yellowish → Hematite, brownish B/P—usually absent; sometimes very weak in oil A—isotropic; sometimes weakly anomalous IR—not present	*R*—22.2 22.0 QC—0.308 0.316 22.1	VHN—946–1402 (p) PH > hausmannite ~braunite	Occurs as euhedral crystals and as granular aggregates. Cleavage (111), lamellar twinning, and zonal growth may be visible. Occurs with hematite, braunite, pyrolusite, hausmannite.
Bornite Cu_5FeS_4 Tetragonal	C—pinkish brown to orange; tarnishes purplish, violet or iridescent B/P—slight bireflectance may be visible on grain boundaries A—very weak IR—not present	*R*—21.3 24.5 QC—0.350 0.338 22.0	VHN—95–105 (sf) PH > galena, chalcocite < chalcopyrite	Occurs as irregular polycrystalline aggregates and as coatings on, or lamellae intergrown with, chalcopyrite. Cleavage may be visible; twinning infrequent and difficult to see. Lamellar exsolution and replacement textures with chalcopyrite, enargite, digenite are common; alters on grain boundaries and fractures to covellite. Occurs with pyrite, chalcopyrite, enargite, digen-

Note: Information is reported as follows:

1. Name 2. Formula 3. Crystal System	C—Color B/P—Bireflectance/Pleochroism A—Anisotropy IR—Internal Reflections	*R*—Reflectance at 546 and 589 nm in Air QC—Quantitative Color Coordinates	VHN—Vickers Microhardness at 100 g Load PH—Polishing Hardness	Mode of Occurrence; Other Characteristic Properties
				ite, covellite, linnaeite, sphalerite, galena, magnetite, tetrahedrite, hematite.
Boulangerite $Pb_5Sb_4S_{11}$ Monoclinic	C—white with bluish gray tint → Galena, darker greenish gray → Stibnite, slightly lighter → Jamesonite, darker B/P—distinct, gray white to green gray A—distinct, tan, brown, bluish gray IR—rare, red	*R*—41.8–37.4 40.7–36.5 QC—0.303 0.312 41.4	VHN—92–125(sf) PH < galena	Usually occurs as granular or fibrous aggregates with galena, sphalerite, chalcopyrite, tetrahedrite, or other Pb-Sb sulfosalts.
Brannerite $(U,Ca,Ce)(Ti,Fe)_2O_6$	C—gray B/P—not present	*R*—15.0–15.1 14.7–14.8	VHN—690(p)	Occurs as euhedral prismatic to needlelike crystals and as

Monoclinic (metamict)	A—not present IR—coarse crystals: brownish gray; fine grained material: blue gray to bluish white, dark brown to yellowish			subhedral aggregates. Often forms as replacement (sometimes as a pseudomorph) after uraninite and rutile. Usually contains included laths of pyrrhotite and anatase and may have a "dusting" of small radiogenic galena crystals. Occurs with uraninite, rutile, pitchblende, pyrite, coffinite, galena, sphalerite, tetrahedrite, pyrrhotite, anatase, magnetite.
Braunite $(Mn,Fe,Si)_2O_3$ Tetragonal	C—gray with brownish tint → Magnetite, less brown → Pyrolusite, psilomelane, darker → Manganite, hausmanite, similar but weaker bireflectance → Bixbyite, jacobsite, more gray B/P—weak but distinct, gray A—weak but distinct, gray to blue; often undulose IR—rare, dark brown to deep red	*R*—18.6–19.5 18.1–18.9 QC—0.299 0.304 18.6	VHN—1027–1225 (at 50g) PH > magnetite < bixbyite	Occurs as anhedral granular masses and as subhedral to euhedral crystals. Zonal textures reported. Associated with jacobsite, bixbyite, hematite, pyrolusite, magnetite.

Note: Information is reported as follows:

1. Name 2. Formula 3. Crystal System	C—Color B/P—Bireflectance/Pleochroism A—Anisotropy IR—Internal Reflections	*R*—Reflectance at 546 and 589 nm in Air QC—Quantitative Color Coordinates	VHN—Vickers Microhardness at 100 g Load PH—Polishing Hardness	Mode of Occurrence; Other Characteristic Properties
Bravoite $(Fe,Ni,Co)S_2$ Cubic	C—composition dependent; Fe-rich: creamy to pinkish; Co- and Ni-rich: pinkish to brownish to violet B/P—not present A—not present IR—not present	*R*—31.0–53.9 (lowest for Co and Ni-rich)	VHN—668–1535 PH < pyrite > sphalerite	Zonal texture very characteristic, the darker zones being richer in Ni and Co. Commonly occurs as isolated cube or octahedral crystals but may be associated with chalcopyrite, sphalerite, galena, linnaeite, siegenite, tetrahedrite, maucherite, safflorite, bismuth, niccolite.
Breithauptite NiSb Hexagonal	C—pink with violet tint → Niccolite, darker, violet tint B/P—strong, pinkish to pinkish violet A—very strong, bluish green, bluish gray, violet red	*R*—36.9–48.2 43.7–53.0	VHN—412–584 PH < niccolite, rammelsbergite, safflorite	Occurs as subhedral to euhedral grains, often with zonal structure. Occurs with niccolite, silver, safflorite, galena, chromite, pentlandite, pyrrhotite, Ag-sulfosalts. Color and very strong anisotropism are

	IR—not present			diagnostic; only similar mineral is niccolite. Violarite appears similar but does not show the zonal texture.
Carrollite $CuCo_2S_4$ Cubic	C—creamy white, sometimes with a slight pinkish tint B/P—not present A—not present IR—not present	*R*—45	VHN—525–542 PH > chalcopyrite < pyrite	Occurs as anhedral granular masses to subhedral and euhedral octahedra. Usually associated with copper minerals, chalcopyrite, bornite, chalcocite, digenite, cobalt-pyrite, pyrrhotite, siegenite.
Cassiterite SnO_2 Tetragonal	C—brownish gray → Stannite, wolframite, ilmenite, rutile, magnetite, brownish gray B/P—distinct, gray to brownish gray A—distinct, gray; in oil, masked by internal reflections IR—abundant, yellow to yellow-brown	*R*—11.5–12.4 11.3–12.2	VHN—1168–1332(p) PH very high < pyrite	Occurs as compact anhedral masses and as subhedral to euhedral crystals which are often well zoned. Commonly twinned; cleavage may be visible. Occurs with pyrite, arsenopyrite, stannite, wolframite, sphalerite, galena, rutile, hematite, magnetite, bismuth, bismuthinite, pyrrhotite. Resembles sphalerite but is anisotropic and usually exhibits lighter internal reflections.

Note: Information is reported as follows:

1. Name 2. Formula 3. Crystal System	C—Color B/P—Bireflectance/Pleochroism A—Anisotropy IR—Internal Reflections	*R*—Reflectance at 546 and 589 nm in Air QC—Quantitative Color Coordinates	VHN—Vickers Microhardness at 100 g Load PH—Polishing Hardness	Mode of Occurrence; Other Characteristic Properties
Chalcocite Cu_2S Monoclinic	C—bluish white → Galena, pyrite, bornite, copper, bluish gray to bluish white → Covellite, white B/P—very weak A—weak to distinct, emerald green to light pinkish IR—not present	*R*—33.1–33.4 31.5–31.8 QC—a 0.295 0.304 33.2 b 0.295 0.304 33.1 c 0.295 0.303 32.9	VHN—84–87(p) on (001) PH > acanthite ~ digenite < bornite	Occurs as anhedral polycrystalline aggregates and vein fillings with iron and copper-iron sulfides such as pyrite, chalcopyrite, bornite, digenite. Also associated with enargite, tetrahedrite-tennantite, sphalerite, galena, stannite. Often in exsolution intergrowth with bornite or low-temperature copper sulfides. Often appears isotropic, especially in supergene fine-grained aggregates.
Chalcophanite (Zn,Fe,Mn) $Mn_2O_5 \bullet nH_2O$ Triclinic	C, B/P—very strong and characteristic bireflectance especially in oil, white to gray	*R*—9.9–24.9 9.7–23.6	VHN—188–253(f) //cleavage	Occurs as aggregates of tabular and radiating crystals and as colloform bands in secondary Mn-ores. Perfect basal

	A—very strong, white to gray IR—absent except when Zn-rich which have deep red internal reflections			cleavage usually visible in crystals. Common as vein filling in other Mn-oxides such as psilomelane, pyrolusite, hausmannite.
Chalcopyrite $CuFeS_2$ Tetragonal	C—yellow to brassy yellow → Pyrite, more yellow → Gold, distinct greenish tint B/P—weak A—weak, but distinct, gray-blue to yellow-green IR—not present	*R*—44.6–45.0 46.5–47.2 QC—0.349 0.369 44.1	VHN—187–203 (basal section) 181–192 (vertical section) PH ~ galena < sphalerite	Occurs as medium to coarse-grained anhedral aggregates; rarely as well-developed tetrahedra. Commonly twinned; often contains laths of cubanite, "stars" of sphalerite, or "worms" of pyrrhotite or mackinawite. Basketweave exsolution with bornite common. Associated with pyrite, pyrrhotite, bornite, digenite, cubanite, sphalerite, galena, magnetite, pentlandite, tetrahedrite, many other minerals. Often alters along cracks and grain boundaries to covellite.
Chalcostibite $CuSbS_2$ Orthorhombic	C—white with pinkish gray tint → Silver, galena, grayish → Sphalerite, pinkish	*R*—36.9–42.2 35.2–39.6	VHN—283–309(sf)	Occurs as anhedral grains; rarely as euhedral prismatic crystals. Cleavage (001) and triangular pits may be visible.

Note: Information is reported as follows:

1. Name 2. Formula 3. Crystal System	C—Color B/P—Bireflectance/Pleochroism A—Anisotropy IR—Internal Reflections	*R*—Reflectance at 546 and 589 nm in Air QC—Quantitative Color Coordinates	VHN—Vickers Microhardness at 100 g Load PH—Polishing Hardness	Mode of Occurrence; Other Characteristic Properties
	B/P—distinct in oil, creamy to brown A—distinct; pinkish to greenish or bluish gray IR—rare, pale red		PH > silver < chalcopyrite, sphalerite	May be intergrown with enargite; occurs with pyrite, sphalerite, chalcopyrite, silver, galena, chalcocite, covellite, jamesonite, arsenopyrite, tetrahedrite, cinnabar.
Chromite $(Fe,Mg)(Cr,Al)_2O_4$ Cubic	C—dark gray to brownish gray → Magnetite, sphalerite, darker → Ilmenite, less brown-red B/P—not present A—usually absent but many show weak anisotropism IR—common, red brown; absent in Fe-rich samples	*R*—12.3 12.1 QC—0.304 0.309 12.2	VHN—1332(p) PH > magnetite, < hematite	Usually occurs as subhedral (rounded) to euhedral crystals or coarsely crystalline aggregates; cataclastic affects common. Zonal textures with lighter (Fe-enriched) rims are very common. "Exsolution" of hematite, ilmenite, magnetite, rutile, ulvöspinel uncommon but observed. Associated with

				magnetite, ilmenite, platinum, pentlandite, pyrrhotite, millerite.
Cinnabar HgS Trigonal	C—white with bluish gray tint → Galena, darker, bluish B/P—distinct in oil A—distinct; in oil often masked by internal reflections IR—intense and abundant, red	*R*—24.6–29.6 23.8–28.2 QC—0.297 0.302 24.5	VHN—82–156 PH > antimony < galena, pyrite	Occurs as subhedral to euhedral crystals and as polycrystalline aggregates of euhedral grains. Associated with metacinnabar (an isotropic polymorph), pyrite, marcasite, stibnite, chalcopyrite, tetrahedrite, bornite, gold, realgar, orpiment, galena, enargite, cassiterite. Resembles proustite and pyrargyrite in polished section.
Cobaltite (Co,Fe)AsS Orthorhombic	C—white with pink or violet tint → Arsenopyrite, pinkish → Pyrite, whiter B/P—weak, white to pinkish A—weak to distinct in oil, blue gray to brown IR—not present	*R*—50.5 51.9 QC—0.320 0.326 50.7	VHN—935–1131 PH > skutterudite, arsenopyrite < pyrite	Commonly occurs as euhedral crystals and as polycrystalline aggregates. Twinning, zoning, and cleavage may be visible. Occurs with niccolite, silver, gold, chalcopyrite, arsenopyrite, bismuth, uraninite, Ni-Co arsenides. The weak anisotropism will distinguish this from niccolite or breithauptite.

Note: Information is reported as follows:

1. Name 2. Formula 3. Crystal System	C—Color B/P—Bireflectance/Pleochroism A—Anisotropy IR—Internal Reflections	*R*—Reflectance at 546 and 589 nm in Air QC—Quantitative Color Coordinates	VHN—Vickers Microhardness at 100 g Load PH—Polishing Hardness	Mode of Occurrence; Other Characteristic Properties
Coffinite $U(SiO_4)_{1-x}(OH)_{4x}$ Tetragonal	C—gray B/P—very weak A—very weak to absent IR—Air: rare and weak Oil: pronounced, brownish	*R*—7.9–8.0 7.8–7.9	VHN—230–302(p) PH ~pitchblende	Occurs as euhedral tetragonal crystals, as fine aggregates and as colloform bands. Botryoidal encrustations and intergranular films, especially near organic matter, are common. Associated with pyrite, sphalerite, uraninite, pitchblende, bismuth, loellingite, rammelsbergite.
Cohenite Fe_3C Orthorhombic	C—creamy white → Pyrrhotite, lighter creamy → Iron, similar B/P—weak but distinct A—weak but distinct IR—not present		PH > iron	A meteoritic mineral, extremely rare on earth. Occurs as irregular grains with kamacite, schreibersite, graphite and troilite. Found in meteorites with 6 to 8 wt. %. Ni where it is a residual metastable phase.

				Twinning common in larger grains.
Columbite- Tantalite $(Fe,Mn)(Ta,Nb)_2O_6$ Orthorhombic	C—gray-white with brown tint → Magnetite, slightly less brown B/P—weak A—distinct, straight extinction IR—Fe-rich, deep red	*R*—15.3–17.3	VHN—240–1021	Occurs as euhedral crystals and anhedral aggregates. May be zoned and cleavage // (100) may be visible. May contain inclusions of cassiterite, galena, hematite, ilmenite, rutile, uraninite, wolframite and be contained within cassiterite. Occurs as oriented intergrowths with uraninite.
Copper Cu Cubic	C—pink, but tarnishes brownish → Silver, pink B/P—not present A—isotropic but fine scratches will appear anisotropic IR—not present	*R*—60.6 87.0	VHN—96–104(p) PH $>$ chalcocite $<$ cuprite	Occurs as coarse- to fine-grained aggregates; occasionally as dendritic or spearlike crystals. Lamellar twinning visible if etched. Zoning due to Ag or As not uncommon. Occurs with cuprite, chalcocite, enargite, bornite, pyrrhotite, iron, magnetite.
Cosalite $Pb_2Bi_2S_5$ Orthorhombic	C—white with pink or gray tint → Galena, yellowish to green tint	*R*—45.7–41.4 45.3–40.65	VHN—74–161	Occurs as granular masses, bundles of subhedral, elongate laths, and fibrous crystals. Twinning absent. Occurs with

Note: Information is reported as follows:

1. Name 2. Formula 3. Crystal System	C—Color B/P—Bireflectance/Pleochroism A—Anisotropy IR—Internal Reflections	*R*—Reflectance at 546 and 589 nm in Air QC—Quantitative Color Coordinates	VHN—Vickers Microhardness at 100 g Load PH—Polishing Hardness	Mode of Occurrence; Other Characteristic Properties
	B/P—weak to distinct A—weak to moderate; pinkish yellow, bluish, violet gray IR—not present	QC—0.304 0.308 45.8	PH > galena	other Bi and Sb sulfosalts, pyrite, pyrrhotite, chalcopyrite, gold, bismuth, sphalerite, arsenopyrite, tetrahedrite, wolframite, glaucodot.
Covellite CuS Hexagonal	C—indigo blue with violet tint to bluish white in air B/P—purple to violet red, to blue-gray in oil A—extreme, red-orange to brownish IR—not present	*R*—7.2–23.7 4.2–21.2 QC—0.224 0.226 6.8	VHN—128–138(sf) PH < chalcopyrite	Occurs as subhedral to anhedral masses, as laths and as platelike crystals. The brilliant blue color, and strong pleochroism and anisotropism are unmistakable, even when present as the tiny alteration laths commonly seen on copper and iron sulfides such as pyrite, chalcopyrite, bornite; also with enargite, digenite, tennantite, sphalerite. Blaubleibender

				(blue-remaining) covellite is similar except that it remains blue in oil; it occurs infrequently with covellite.
Cubanite $CuFe_2S_3$ Orthorhombic	C—creamy gray to yellowish brown → Pyrrhotite, more yellow, less pink → Chalcopyrite, more gray brown B/P—distinct, grayish to brownish A—strong, brownish to blue IR—not present	*R*—39.4–35.4 40.7–37.65 QC—0.331 0.341 39.4	VHN—247–287(sf) PH > chalcopyrite < pyrrhotite	Occurs most commonly as sharply bounded laths within coarse-grained chalcopyrite; also as irregular granular aggregates. Recognized by its distinct bireflectance and anisotropism. Also occurs with pyrrhotite, sphalerite, galena, mackinawite, pentlandite, magnetite, arsenopyrite.
Cuprite Cu_20 Cubic	C—Air: light bluish gray Oil: darker, more blue → Chalcopyrite, hematite, darker and greenish B/P—very weak A—strong anomalous anisotropism gray-blue to olive green IR—deep red, characteristic	*R*—26.6 24.6 QC—0.287 0.300 26.3	VHN—193–207(sf) PH > chalcopyrite copper, tenorite	Occurs as euhedral octahedra and in a fine-grained "earthy" form. Replaces copper sulfides and copper. Also occurs with goethite, tenorite, delafossite, pyrite, marcasite.

Note: Information is reported as follows:

1. Name 2. Formula 3. Crystal System	C—Color B/P—Bireflectance/Pleochroism A—Anisotropy IR—Internal Reflections	*R*—Reflectance at 546 and 589 nm in Air QC—Quantitative Color Coordinates	VHN—Vickers Microhardness at 100 g Load PH—Polishing Hardness	Mode of Occurrence; Other Characteristic Properties
Delafossite $CuFeO_2$ Trigonal	C, B/P—distinct bireflectance Air: yellow rose brown to rose brown Oil: pinkish gray to browngray → Enargite, tenorite, more yellow A—distinct to strong, bluish gray, straight extinction IR—not present	*R*—20–25	PH < cuprite, goethite	Occurs as masses of subparallel crystals and sheaflike bundles or as fine inclusions in goethite. Concentric and botryoidal textures common. Occurs with goethite, limonite, cuprite, tenorite, copper, pyrite, bornite, chalcocite, covellite, galena, tennantite.
Digenite Cu_9S_5 Cubic	C—grayish blue → Galena, bornite, blue → Chalcocite, darker blue B/P—not present A—isotropic; sometimes with	*R*—23.1 21.0 QC—0.282 0.292	VHN—67–76(p) PH ~chalcocite,	Occurs as irregular aggregates of anhedral grains that contain lamellar intergrowths with other copper sulfides or bornite. Also with chalcopyrite,

	weak anomalous anisotropism IR—not present	22.6	galena	pyrite, tetrahedrite, enargite; alters to covellite.
Dyscrasite Ag_3Sb Orthorhombic	C—white → Galena, creamy white → Silver, slightly grayer → Antimony, slightly creamy B/P—weak, white to creamy white A—weak to distinct IR—not present	*R*—62.1–63.2 62.8–64.0	VHN—146–160(sf) PH $>$ galena, silver $<$ chalcopyrite	Occurs as euhedral platelike to square crystals and as aggregates of anhedral crystals with arsenic, galena, cobaltite, pyrite. (The "dyscrasite" of Cobalt, Ontario, is actually allargentum.)

Enargite
Cu_3AsS_4
Orthorhombic
	C—pinkish gray to pinkish brown in air; darker in oil
	→ Bornite, pinkish white
	→ Chalcocite, galena, pinkish to grayish brown
	B/P—distinct in oil:
	//a grayish pink
	//b pinkish gray
	//c grayish violet
	A—strong, blue, green, red, orange
	IR—deep red may occur
		R—24.2–25.2
		23.8–25.7
			VHN—285–327
			PH $>$ galena, chalcocite, bornite
			$\sim$ tennantite
			$<$ sphalerite
				Occurs as anhedral to subhedral grains. Cleavage (110) often seen and usually untwinned. Occurs with pyrite, chalcopyrite, bornite, sphalerite, tennantite, galena, chalcocite, covellite, arsenopyrite.

Note: Information is reported as follows:

1. Name 2. Formula 3. Crystal System	C—Color B/P—Bireflectance/Pleochroism A—Anisotropy IR—Internal Reflections	*R*—Reflectance at 546 and 589 nm in Air QC—Quantitative Color Coordinates	VHN—Vickers Microhardness at 100 g Load PH—Polishing Hardness	Mode of Occurrence; Other Characteristic Properties
Famatinite Cu_3SbS_4 Tetragonal	C—pale pinkish orange → Enargite, lighter B/P—distinct to strong in oil, orange brown to grayish violet A—very strong, brown to gray green IR—not present	*R*—24–27.4	VHN—205–397 PH > bornite, chalcopyrite ~ enargite < sphalerite	Occurs as anhedral to euhedral grains. Polysynthetic twinning nearly always visible, and star-shaped patterns may occur. Occurs with enargite, chalcopyrite, tetrahedrite, bornite, sphalerite, chalcocite, pyrite, galena, proustite, pyrargyrite.
Freibergite Ag-tetrahedrite Cubic	C—gray, faint yellow brown tint in oil → Proustite, brownish → Galena, grayish brown → Sphalerite, lighter B/P—not present A—isotropic IR—brownish red when visible	*R*—29.4	VHN—252–375 PH > Ag-sulfosalts < galena, sphalerite	Occurs as irregular masses and inclusions of anhedral crystals with, and in, chalcopyrite, bornite, argentite, proustite, galena, silver, Co-Fe-Ni arsenides, enargite.

Galena PbS Cubic	C—white, sometimes with pink tint → Sphalerite, white → Tennantite, pinkish B/P—not present A—isotropic but weak anomalous anisotropism may be visible IR—not present	*R*—43.1 41.9 QC—0.300 0.304 43.0	VHN—59–65(p) PH > proustite ~ chalcopyrite < tetrahedrite	Occurs as anhedral masses to euhedral cubes. The perfect (100) cleavage usually visible and seen as triangular pits. Very common and occurs with wide variety of common minerals. Often contains inclusions of tetrahedrite, Pb-Bi or Pb-Sb sulfosalts, silver, chalcopyrite, sphalerite. May occur as inclusions in chalcopyrite, sphalerite.
Gersdorffite NiAsS Cubic	C—white with yellow or pink tint → Skutterdite, more yellow → Linnaeite, less pink → Niccolite, bluish B/P—not present A—isotropic; some anomalous anisotropism IR—not present	*R*—54.2 54.3 QC—0.311 0.317 54.2	VHN—782–835(sf) PH > linnaeite ~ loellingite < pyrite	Occurs as euhedral crystals that may show zonal growth. Cleavage (100) common. Occurs with pyrite, chalcopyrite, silver, niccolite, skutterudite, bismuth, cobaltite, bornite, uraninite. Sometimes as pseudo-eutectic intergrowths with niccolite, maucherite, pyrrhotite, chalcopyrite.
Glaucodot (Co,Fe)AsS Orthorhombic	C—white to light cream → Arsenopyrite, more bluish white	*R*—50.0–50.6 50.4–50.7	VHN—1097–1115 (sf) PH < arsenopyrite,	Usually occurs as subhedral to euhedral crystals, often with inclusions. Associated with co-

Note: Information is reported as follows:

1. Name 2. Formula 3. Crystal System	C—Color B/P—Bireflectance/Pleochroism A—Anisotropy IR—Internal Reflections	*R*—Reflectance at 546 and 589 nm in Air QC—Quantitative Color Coordinates	VHN—Vickers Microhardness at 100 g Load PH—Polishing Hardness	Mode of Occurrence; Other Characteristic Properties
	B/P—weak, weaker than arsenopyrite A—distinct, less than for arsenopyrite IR—not present		cobaltite	baltite, pyrite, arsenopyrite, safflorite, skutterudite, niccolite, galena, rammelsbergite. Polishes very well.
Goethite FeO•OH Orthorhombic	C—gray, with a bluish tint → Sphalerite, more bluish → Hematite, darker → Lepidocrocite, darker B/P—weak in air; distinct in oil but often masked by internal reflections A—distinct, gray-blue, gray-yellow, brownish IR—brownish yellow to reddish brown	*R*—17.5–15.5 16.6–15.0 QC—0.291 0.296 17.5	VHN—667 PH ~ lepidocrocite < magnetite, hematite	Common in porous colloform bands with radiating fibrous texture, or as porous pseudomorphs after pyrite. Nearly always secondary, as veins, fracture fillings, or botryoidal coatings. Occurs with hematite, pyrite, lepidocrocite, pyrite, pyrrhotite, manganese-oxides, sphalerite, galena, chalcopyrite. Brownish to yel-

				lowish internal reflections help to distinguish from lepidocrocite.
Gold Au Cubic	C—bright golden yellow → Chalcopyrite, no greenish tint B/P—not present A—isotropic but incomplete extinction IR—not present	*R*—71.5 83.4 QC—0.384 0.391 72.7	VHN—53–58(p) PH $>$ galena $<$ tetrahedrite, chalcopyrite	Occurs as isolated grains and veinlets in many sulfides, especially pyrite, arsenopyrite, chalcopyrite. Recognized by its "golden" color and very high reflectance; addition of silver to form electrum changes color to whitish and increases *R*%.
Graphite C Hexagonal	C,B/P—very strong, bireflectance from brownish gray to grayish black → Molybdenite, darker A—very strong, straw yellow to brown or violet gray IR—not present	*R*—17.4–6.8 18.1–7.0	VHN—12–16(f) (at 50g) PH $<$ almost all minerals	Occurs as small plates, laths, and bundles of blades. Basal cleavage visible and undulose extinction common. Present as isolated laths in many igneous and metamorphic rocks; also as inclusions in sphalerite, pyrite, magnetite, pyrrhotite. Much more common than molybdenite.
Hausmannite Mn_3O_4 Tetragonal	C—bluish to brownish gray → Jacobsite, grayer → Bixbyite, darker	*R*—19.6–17.6 18.9–17.5	VHN-536–566(p) PH $>$ manganite,	Occurs as coarse-grained equigranular anhedral crystals, often in veinlets. Irregular

Note: Information is reported as follows:

1. Name 2. Formula 3. Crystal System	C—Color B/P—Bireflectance/Pleochroism A—Anisotropy IR—Internal Reflections	*R*—Reflectance at 546 and 589 nm in Air QC—Quantitative Color Coordinates	VHN—Vickers Microhardness at 100 g Load PH—Polishing Hardness	Mode of Occurrence; Other Characteristic Properties
	→ Braunite, less brown B/P—very distinct in oil, bluish gray to brownish gray A—strong, yellow brown to bluish gray IR—blood red, especially in oil		pyrolusite <jacobsite, bixbyite, braunite	twinning common. Occurs with other Mn-Oxides and alters to pyrolusite and psilomelane.
Hematite α - Fe_2O_3 Hexagonal	C—gray white with bluish tint → Ilmenite, magnetite, white → Pyrite, bluish gray → Goethite, lepidocrocite, white B/P—weak A—distinct, gray blue, gray yellow	*R*—30.2–26.1 29.15–25.1 QC—0.299 0.309 29.8	VHN—1038 PH > magnetite < pyrite	Usually occurs as bladed or needlelike subparallel or radiating aggregates. Lamellar twinning common. Also common as exsolution lenses or lamellae in ilmenite or magnetite, or as a host to lamellae of the same. Occurs with magnetite, ilmenite, py-

	IR—deep red common			rite, chalcopyrite, bornite, rutile, cassiterite, sphalerite.
Idaite Cu_5FeS_6 $\rightarrow Cu_3FeS_4$ Tetragonal	C,B/P—strong bireflectance from reddish orange or red brown to yellowish gray A—extreme, green or gray green IR—not present	*R*—27–33.6	VHN—176–260 PH $>$ covellite	Occurs as hypogene tabular crystals that occur with covellite, pyrite, bornite and as supergene alterations of bornite where it occurs as lamellae and veinlets. Recognized by the orangish color and the strong greenish anisotropism. (A new mineral of composition close to "idaite" has been named nukundamite.)
Ilmenite $FeTiO_3$ Trigonal	C—brownish with a pink or violet tint $\rightarrow$ Magnetite, darker, brownish B/P—distinct, pinkish brown, dark brown A—strong, greenish gray to brownish gray IR—rare, dark brown	*R*—20.1–17.0 20.2–17.4	VHN—659–703 (cv) PH $>$ magnetite $<$ hematite	Occurs as subhedral to anhedral grains and as "exsolution" lamellae or lenses in hematite or magnetite. Lamellar twinning common. Common accessory in igneous and metamorphic rocks. Occurs with magnetite, hematite, rutile, pyrite, pyrrhotite, chromite, pentlandite, tantalite.
Iron Fe	C—white, slight bluish or yellowish	*R*—57.7 58.0	VHN—158 (av.)	Common as irregular patches and droplike grains in stony

Note: Information is reported as follows:

1. Name 2. Formula 3. Crystal System	C—Color B/P—Bireflectance/Pleochroism A—Anisotropy IR—Internal Reflections	*R*—Reflectance at 546 and 589 nm in Air QC—Quantitative Color Coordinates	VHN—Vickers Microhardness at 100 g Load PH—Polishing Hardness	Mode of Occurrence; Other Characteristic Properties
Cubic α-Fe = Kamacite γ-Fe = Taenite	→ Pentlandite, much whiter → Cohenite, slightly bluish B/P—not present A—isotropic IR—not present	QC—0.311 0.316 57.9	PH < troilite, magnetite, cohenite	meteorites and as a major phase in iron meteorites; extremely rare on earth. α-Fe contains $\lesssim$ 6% Ni and is slightly bluish; γ-Fe contains ~27–60% Ni and is slightly yellowish. (111) intergrowths of γ-Fe and α-Fe form Widmanstätten structures, which are brought out by etching. Fine exsolution of cohenite occurs in α-Fe. Other associated minerals include troilite, copper, schreibersite, ilmenite, chromite. Oxidizes to hematite, goethite, lepidocrocite.

Jacobsite $(Mn,Fe,Mg)(Fe,Mn)_2O_4$ Cubic	C—rose brown to brownish gray → Magnetite, braunite, olive green → Hausmannite, less gray → Bixbyite, olive gray B/P—not present A—isotropic, sometimes slight anomalous anisotropism IR—deep red, especially when Mn-rich	*R*—19.4 19.4 QC—0.310 0.317 19.4	VHN—665–707(p) PH ~ magnetite < braunite	Occurs as anhedral grains and rounded subhedral crystals. Occurs with and alters to other Fe-Mn minerals such as goethite, pyrolusite, hematite, psilomelane.
Jamesonite $Pb_4FeSb_6S_{14}$ Monoclinic	C—white → Galena, similar or slightly greenish → Stibnite, lighter B/P—strong, white to yellow green A—strong, gray, tan, brown, blue IR—reddish in Bi-jamesonite	*R*—37.4–42.9 36.7–41.5	VHN—113–117(p) PH < galena	Occurs as needle- or lathlike crystals or bundles. Cleavage // long dimension common; often twinned. Occurs with galena, pyrite, pyrargyrite, boulangerite, chalcopyrite, sphalerite, tetrahedrite, arsenopyrite.
Kamacite See Iron				
Kobellite $Pb_2(Bi,Sb)_2S_5$	C—white → Galena, slightly darker	*R*—40.0–46.1 39.1–45.7	VHN—100–117(sf) PH > bismuth	Occurs as granular to tabular aggregates with well-developed

Note: Information is reported as follows:

1. Name 2. Formula 3. Crystal System	C—Color B/P—Bireflectance/Pleochroism A—Anisotropy IR—Internal Reflections	*R*—Reflectance at 546 and 589 nm in Air QC—Quantitative Color Coordinates	VHN—Vickers Microhardness at 100 g Load PH—Polishing Hardness	Mode of Occurrence; Other Characteristic Properties
Orthorhombic	B/P—distinct, greenish white to violet gray A—distinct, gray to gray brown IR—not present	QC—0.302 0.309 47.1	< galena	(010) cleavage. Commonly twinned. Occurs with arsenopyrite, pyrite, pyrrhotite, chalcopyrite, bismuth, bismuthinite and as intergrowths with tetrahedrite.
Lepidocrocite γ-FeO•OH Orthorhombic	C—grayish white → Goethite, lighter and whiter → Hematite, greenish tint B/P—weak to distinct A—strong, gray IR—reddish, common	*R*—18.4–11.6 17.4–11.1 QC—0.291 0.297 18.3	VHN—402 PH < goethite	Occurs as weathering product of iron oxides and sulfides with (but less commonly than) goethite. Present as crusts, veinlets, and even as porous pseudomorphs.
Linnaeite Co_3S_4 Cubic	C—creamy white → Skutterudite, grayish white	*R*—49.5 49.6	VHN—450–613	Occurs as euhedral crystals and subhedral aggregates. May be intergrown in lamellar pat-

	→ Ullmannite, gersdorffite, creamy or yellowish B/P—not present A—isotropic IR—not present		PH > chalcopyrite, sphalerite < pyrite	tern with millerite, chalcopyrite, bornite, pyrrhotite, pyrite, bismuth, covellite, safflorite, niccolite.
Loellingite $FeAs_2$ Orthorhombic	C—white, with yellowish tint → Arsenopyrite, less yellow → Rammelsbergite, safflorite, similar B/P—weak but distinct, bluish white to yellowish white A—very strong, orange-yellow; red-brown, blue, green IR—not present	R—52.4–54.1 51.2–55.2	VHN—446–560(p) PH > chalcopyrite, sphalerite < arsenopyrite	Commonly occurs as interlocking to radiating aggregates of euhedral crystals; sometimes as skeletal crystals. Commonly twinned. Usually associated with other arsenides, dyscrasite, arsenic, arsenopyrite, uraninite, antimony, chalcopyrite, galena.
Mackinawite $Fe_{1+x}S$ Tetragonal	C—pinkish to reddish gray → Pyrrhotite, similar B/P—moderate to strong, pinkish gray to gray A—very strong, grayish white, bluish, brownish IR—not present	*R*—22–46	VHN—52–58 PH ~ pyrrhotite	Occurs as small wormlike grains and lamellae (more rarely as small plates) in pyrrhotite, chalcopyrite, cubanite, pentlandite. Probably much confused with valleriite which tends to have a more pronounced orange tint to its anisotropism. Most easily found as "bright" grains under nearly crossed nicols.

Note: Information is reported as follows:

1. Name 2. Formula 3. Crystal System	C—Color B/P—Bireflectance/Pleochroism A—Anisotropy IR—Internal Reflections	*R*—Reflectance at 546 and 589 nm in Air QC—Quantitative Color Coordinates	VHN—Vickers Microhardness at 100 g Load PH—Polishing Hardness	Mode of Occurrence; Other Characteristic Properties
Maghemite γ-Fe_2O_3 Cubic	C—bluish gray → Goethite, gray, lighter → Hematite, bluish gray → Magnetite, bluish B/P—not present A—isotropic IR—rare, brownish red	*R*—24.4 28.8 QC—0.293 0.304 24.1	VHN—412 (at 50g) PH > magnetite < hematite	Forms as a rare oxidation product of magnetite. Irregularly present in oxidizing magnetite as lamellae and porous patches.
Magnetite Fe_3O_4 Cubic	C—gray, with brownish tint → Hematite, darker brown → Ilmenite, less pink → Sphalerite, lighter B/P—not present A—isotropic, slight anomalous anisotropism IR—not present	*R*—20.0 20.3 QC—0.311 0.314 20.1	VHN—592(p) (av.) PH > pyrrhotite < ilmenite, hematite, pyrite	Occurs as euhedral, subhedral, and even skeletal crystals and as anhedral polycrystalline aggregates. Often contains exsolution or oxidation lamellae of hematite; lamellae of ilmenite and ulvöspinel also common. Associated with pyrrhotite, pyrite, pentlandite, chalcopyrite, bornite, sphalerite, galena.

				Alters to hematite, and goethite.
Manganite $MnO(OH)$ Monoclinic	C—Gray to brownish gray → Pyrolusite, darker gray B/P—weak, brownish gray A—strong, yellow, bluish gray, violet gray IR—blood red, common	*R*—14.8–20.7 14.3–19.9	VHN—698–772(p) PH < hausmannite, jacobsite	Occurs as prismatic to lamellar crystal aggregates often intergrown with pyrolusite and psilomelane. Cleavage on (010) and (110) may be visible. Commonly twinned. Occurs also with hausmannite, braunite, goethite.
Marcasite FeS_2 Orthorhombic	C—yellowish white with slight pinkish or greenish tint → Pyrite, whiter → Arsenopyrite, greenish yellow B/P—strong, brownish, yellowish green A—strong, blue, green yellow, purple gray IR—not present	*R*—48.2–55.8 48.4–54.6	VHN—1288–1681(f) PH ~ pyrite	Occurs as subhedral to lamellar intergrowths with pyrite as euhedral crystals. Also occurs as radiating colloform bands. Commonly twinned. Forms as hypogene crystals and as supergene veinlets in pyrrhotite and iron oxides. Often with pyrite but also occurs with most other common sulfides. Blue to yellowish anisotropism is diagnostic.
Maucherite $Ni_{11}As_8$ Tetragonal	C—white → Cobaltite, similar → Loellingite, brownish gray	*R*—47.8–48.5 50.0–50.7	VHN-715–743(p) PH > chalcopyrite, sphalerite	Commonly occurs as euhedral crystals and anhedral aggregates; may be twinned. May

Note: Information is reported as follows:

1. Name 2. Formula 3. Crystal System	C—Color B/P—Bireflectance/Pleochroism A—Anisotropy IR—Internal Reflections	*R*—Reflectance at 546 and 589 nm in Air QC—Quantitative Color Coordinates	VHN—Vickers Microhardness at 100 g Load PH—Polishing Hardness	Mode of Occurrence; Other Characteristic Properties
	→ Breithauptite, bluish gray B/P—not observed A—weak to distinct in oil, gray IR—not present		< safflorite, loellingite	be intergrown with niccolite or gersdorffite. Also occurs with chalcopyrite, cubanite, siegenite.
Mawsonite $Cu_7Fe_2SnS_{10}$ Tetragonal	C—brownish orange B/P—strong, orange to brown A—very strong, straw yellow to royal blue IR—not present	*R*—26.9–29.7 29.1–35.1 QC—0.339 0.340 27.3	VHN—166–210 PH > bornite	Occurs as irregular inclusions in or associated with bornite. Also associated with chalcopyrite, chalcocite, tetrahedrite, pyrite, galena, enargite, stannite.
Miargyrite $AgSbS_2$ Monoclinic	C—white in air; bluish tint in oil → Galena, darker with green gray tint → Freibergite, bluish	*R*—34.5–31.6 32.8–30.05	VHN—88–130	Occurs as granular anhedral aggregates (sometimes twinned) with sphalerite, galena, tetrahedrite, pyrargyrite, silver, polybasite, stephanite.

	→ Pyrargyrite, whiter B/P—moderate, white, bluish gray A—strong, blue gray to brownish but masked by internal reflections IR—deep red	QC—0.294 0.302 34.2	PH > pyrargyrite < stephanite, galena	
Millerite NiS Trigonal	C—yellow → Chalcopyrite, lighter, not greenish → linnaeite, pentlandite, yellower B/P—distinct in oil, yellow to blue or violet A—strong, lemon yellow to blue or violet IR—not present	*R*—50.2–56.6 51.9–59.05 QC—0.328 0.339 50.4	VHN—192–376 PH > chalcopyrite < pentlandite	Occurs as radiating aggregates and as anhedral granular masses. Also common as oriented intergrowths with linnaeite, violarite, pyrrhotite. Twinning and cleavage (1011) often visible. Usually associated with Ni-bearing sulfides, often as a replacement or alteration phase.
Molybdenite MoS_2 Trigonal	C,B/P—extreme bireflectance, white to gray with bluish tint → Graphite, lighter A—very strong, white with pinkish tint; dark blue if polars not completely crossed. IR—not present	*R*—38.5–19.5 38.8–19.0 QC—0.298 0.299 39.3	VHN—8–100 32–33(f) // cleavage PH < almost all minerals	Usually occurs as small, often deformed, plates and irregular inclusions; more rarely as rosettes or colloform bands. Cleavage (0001); twinning and undulatory extinction very common. Often in veins with pyrite, chalcopyrite, bornite, cassiterite, wolframite, bis-

Note: Information is reported as follows:

1. Name 2. Formula 3. Crystal System	C—Color B/P—Bireflectance/Pleochroism A—Anisotropy IR—Internal Reflections	*R*—Reflectance at 546 and 589 nm in Air QC—Quantitative Color Coordinates	VHN—Vickers Microhardness at 100 g Load PH—Polishing Hardness	Mode of Occurrence; Other Characteristic Properties
				muth, bismuthinite, but may occur in many sulfides. Softness, bireflectance, and anisotropism allow confusion only with graphite.
Niccolite NiAs Hexagonal	C,B/P—strong bireflectance, yellowish pink to brownish pink → Maucherite, skutterudite, bismuth, arsenic, more pink → Breithauptite, pinkish yellow A—very strong, yellow, greenish violet blue, blue gray IR—not present	*R*—51.6–47.2 56.0–53.3	VHN—363–372 PH > chalcopyrite ~ breithauptite < skutterudite, pyrite	Occurs as isolated subhedral and euhedral crystals, as anhedral aggregates, as concentric bands and as complex intergrowths (with pyrrhotite, chalcopyrite, maucherite). Commonly intergrown with arsenides. Often twinned and in radial aggregates.

Orpiment
As_2S_3
Monoclinic
	C—gray
	→ Realgar, slightly lighter
	→ Sphalerite, lighter
	B/P—strong
	Air: //a white; //b dull gray, reddish; //c dull gray–white
	Oil: //a gray–white; //b dark gray; //c gray white
	A—strong; in oil masked by internal reflections
	IR—abundant and intense; white to yellow
		R—27.5–23.0
		26.7–22.1
		
		QC—0.294 0.296
		27.6
			VHN—22–58
			
			PH > realgar
				Occurs as tabular interlocking anhedral masses and as needle- or lathlike crystals. Often formed on realgar; also with stibnite, arsenopyrite, arsenic, pyrite, enargite, sphalerite, loellingite.

Pararammelsbergite $NiAs_2$ Orthorhombic	C—whiter than associated Co-Ni-Fe arsenides B/P—very weak to distinct; yellowish to bluish white A—strong, but less than rammelsbergite and without blue IR—not present	*R*—56.3–57.8 56.8–57.8	VHN—762–792(sf) PH > niccolite < skutterudite	Occurs as tabular crystals with rectangular outlines and as mosaics of intergrown crystals. May be zoned but rarely twinned. Occurs with rammelsbergite, niccolite, skutterudite, gersdorffite, cobaltite, silver, pyrite, proustite.
Pearcite $Ag_{16}As_2S_{11}$ Monoclinic	C—Gray → Galena, darker → Pyrargyrite, darker brownish	*R*—31.9 31.05–31.2	VHN—180–192(sf)	Forms complete solid solution with polybasite. Occurs as platelike to equant grains with (or in) galena, tetrahedrite,

Note: Information is reported as follows:

1. Name 2. Formula 3. Crystal System	C—Color B/P—Bireflectance/Pleochroism A—Anisotropy IR—Internal Reflections	*R*—Reflectance at 546 and 589 nm in Air QC—Quantitative Color Coordinates	VHN—Vickers Microhardness at 100 g Load PH—Polishing Hardness	Mode of Occurrence; Other Characteristic Properties
	→ Tetrahedrite, similar B/P—air: weak Oil: distinct, green to gray with violet tint A—air: moderate Oil: strong, blue, gray, yellow green, brown IR—deep–red abundant		PH > argentite, ~ pyrargyrite, < stephanite	sphalerite, pyrite. Untwinned. Other associates include stephanite, pyrargyrite, stromeyerite, argentite, chalcopyrite. May be light etched.
Pentlandite $(Fe,Ni)_9S_8$ Cubic	C—light creamy to yellowish → Pyrrhotite, lighter → Linnaeite, darker, not pinkish B/P—not visible A—isotropic IR—not present	*R*—46.5 49.0 QC—0.332 0.339 46.9	VHN—268–285(sf) PH > chalcopyrite < pyrrhotite	Generally occurs as granular veinlets or as "flames" or lamellae in pyrrhotite; less commonly in chalcopyrite. Other associated minerals include magnetite, pyrite, cubanite, mackinawite. Alters to violarite and millerite along cracks and grain boundaries.

Platinum Pt Cubic	C—white B/P—not observed A—isotropic but incomplete extinction IR—not present	*R*—70.3 71.9 QC—0.318 0.324 70.7	VHN—122–129(p) PH > sphalerite < pyrrhotite	Occurs as isolated euhedral to subhedral crystals; sometimes zoned or with exsolution laths of iridium and osmium. Small grains of other platinum minerals may be present. Chromite, pyrrhotite, magnetite, pentlandite, chalcopyrite may be associated.
Polybasite $Ag_{16}Sb_2S_{11}$ Monoclinic	C—gray → Galena, darker → Pyrargyrite, darker brownish → Tetrahedrite, similar B/P—Air: weak Oil: distinct, green to gray with violet tint A—Air: moderate Oil: strong, blue gray, yellow-green, brown IR—deep red, abundant	*R*—32.5–30.7 31.4–30.0 QC—0.302 0.314 32.2	VHN— PH > argentite, ~ pyrargyrite < stephanite	Forms complete solid solution with pearcite. (See remarks for pearcite; polybasite occurrences are similar but are more likely in Sb-rich environments.)
Proustite Ag_3AsS_3 Trigonal	C—bluish gray → Pyrargyrite, darker B/P—distinct, yellowish,	*R*—28.1–27.4 26.4–25.8	VHN—103–137 (at 50g)	Forms complete solid solutions with pyrargyrite. Same characteristics as pyrargyrite

Note: Information is reported as follows:

1. Name 2. Formula 3. Crystal System	C—Color B/P—Bireflectance/Pleochroism A—Anisotropy IR—Internal Reflections	*R*—Reflectance at 546 and 589 nm in Air QC—Quantitative Color Coordinates	VHN—Vickers Microhardness at 100 g Load PH—Polishing Hardness	Mode of Occurrence; Other Characteristic Properties
	bluish gray A—strong, masked by internal reflection IR—always, scarlet-red	QC—0.287 0.291 28.1	PH ~ pyrargyrite	except found in more As-rich environments.
Psilomelane General name for massive hard manganese oxides	C—bluish gray to grayish white → Pyrolusite, darker → Braunite, manganite, jacobsite, hausmannite, bixbyite, lighter B/P—strong, white to bluish gray A—strong, white to gray IR—occasional, brown	*R*—15–30	VHN—203–813	Commonly occurs as botryoidal masses of very fine acicular crystals in concentric layers; often intergrown with pyrolusite and cryptomelane. Associated with other Mn-oxides.
Pyrargyrite Ag_3SbS_3	C—bluish gray → Proustite, slightly lighter	*R*—30.3–28.5 28.4–26.5	VHN—107–144 (at 50g)	Forms complete solution with proustite. Occurs as irregular

Trigonal	→ Galena, grayish blue B/P—distinct to strong A—strong, gray to dark gray; in oil, masked by internal reflections IR—intense red	QC—0.287 0.244 30.2	66–87 (// cleavage) PH > polybasite < galena	grains and aggregates. May be twinned and zoned. Often with galena, Sb-sulfosalts, pyrite, sphalerite, chalcopyrite, tetrahedrite, arsenopyrite, Ni-Co-Fe arsenides.
Pyrite FeS_2 Cubic	C—yellowish white → Marcasite, yellower → Arsenopyrite, creamy yellow → Chalcopyrite, lighter B/P—not present A—often weakly anisotropic, blue-green to orange-red IR—not present	*R*—51.7 53.5 QC—0.327 0.335 51.8	VHN—1505–1620(f) PH > arsenopyrite, marcasite < cassiterite	The most abundant sulfide; occurs as euhedral cubes and pyritohedra, anhedral crystalline masses, and colloform bands of very fine grains. Growth zoning, twinning, and anisotropy of hardness may be visible. Occurs in nearly all ore types and with most common minerals. Hardness, yellowish white color and abundance usually diagnostic.
Pyrolusite MnO_2 Tetragonal	C—creamy white → Magnetite, hematite, yellowish → Manganite, white B/P—distinct in oil, yellowish white to gray white A—very strong, yellowish, brownish, blue IR—not present	*R*—29.0–40.0 28.1–39.3	VHN—146–243(f) PH—very variable depending on grain size and orientation	Occurs as coarse-grained tabular crystals or as banded aggregates. Cleavage (110) and twinning may occur. Very fine-grained material may be intergrown with psilomelane, hematite, Fe-hydroxides. Also associated with manganite, braunite, magnetite, bixbyite.

Note: Information is reported as follows:

1. Name 2. Formula 3. Crystal System	C—Color B/P—Bireflectance/Pleochroism A—Anisotropy IR—Internal Reflections	*R*—Reflectance at 546 and 589 nm in Air QC—Quantitative Color Coordinates	VHN—Vickers Microhardness at 100 g Load PH—Polishing Hardness	Mode of Occurrence; Other Characteristic Properties
Pyrrhotite $Fe_{1-x}S$ Hexagonal ($\sim Fe_9S_{10}$) Monoclinic ($\sim Fe_7S_8$) FeS is troilite	C—creamy pinkish brown → Pentlandite, darker → Cubanite, more pinkish B/P—very distinct, creamy brown to reddish brown A—very strong, yellow gray, grayish blue IR—not present	*R*—34.0–39.2, 35.8–40.7 Hex 34.8–39.9, 36.9–41.6 Mono QC—0.330 0.334 35.3 (Mono)	VHN—Hex: 230–259(p) (anisotropic sections) 280–318 (p) (isotropic sections) Mono: 373–409(p) PH > chalcopyrite ~ pentlandite < pyrite	Usually occurs as anhedral granular masses. Not infrequently twinned, especially where stressed. Lamellar exsolution intergrowths of hexagonal and monoclinic forms are common; weathering of hexagonal pyrrhotite yields a rim of monoclinic pyrrhotite (usually slightly lighter in color). In Ni-ores exsolved lamellae and "flames" of pentlandite are common. Also often contains mackinawite lamellae. Occurs with most other common sulfides. Troilite occurs in meteorites usually as anhedral, equigranular masses with iron.

Rammelsbergite $NiAs_2$ Orthorhombic	C—white, more so than other Ni-Co-Fe arsenides B/P—very weak in air; distinct in oil, yellowish to bluish A—strong, pinkish, brownish, greenish, bluish IR—not present	*R*—53.2–56.3 53.5–56.1	VHN—585–803(sf) PH ~ skutterudite, < safflorite, loellingite	Occurs as fine-grained aggregates of interlocking crystals; often in zonal, spherulitic, radiating, and fibrous textures. Commonly with simple or complex twinning. May be intergrown with niccolite and Co-Ni-Fe arsenides; sometimes overgrowths on dendrites of silver or bismuth. Very similar to safflorite.
Realgar AsS Monoclinic	C—dull gray → Orpiment, slightly darker → Sphalerite, similar → Cinnabar, darker B/P—weak but distinct; gray with reddish to bluish tint A—strong; in oil masked by internal reflections IR—abundant and intense; yellowish red	*R*—22.1 20.9	VHN—47–60 PH < orpiment	Occurs as irregular platelike masses with orpiment. Also associated with stibnite, arsenopyrite, pyrite, arsenic, As-sulfosalts, tennantite, enargite, proustite.

Note: Information is reported as follows:

1. Name 2. Formula 3. Crystal System	C—Color B/P—Bireflectance/Pleochroism A—Anisotropy IR—Internal Reflections	*R*—Reflectance at 546 and 589 nm in Air QC—Quantitative Color Coordinates	VHN—Vickers Microhardness at 100 g Load PH—Polishing Hardness	Mode of Occurrence; Other Characteristic Properties
Rutile TiO_2 Tetragonal	C—gray, faint bluish tint → Magnetite, chromite, similar → Ilmenite, no brownish tint → Cassiterite, lighter B/P—distinct A—strong but masked by internal reflections IR—strong, abundant, white, yellowish, reddish brown	*R*—20.3 19.8	VHN—1132–1187(p) PH $>$ ilmenite $<$ hematite	Occurs as euhedral to subhedral needlelike to columnar crystals; frequently with hematite. Associated with Ti-hematite, Ti-magnetite, ilmenite, tantalite. Common in hydrothermally altered rocks.
Safflorite $(Co,Fe,Ni)As_2$ Orthorhombic	C—white with a bluish tint → Bismuth, bluish → Silver, grayish white B/P—very weak, bluish to gray A—strong IR—not present	*R*—55–60	VHN—285–464 PH $>$ skutterudite, $<$ loellingite	Occurs as radiating masses of anhedral to subhedral crystals in concentric layers with other arsenide minerals. Also present as euhedral crystals and as starlike triplets. Commonly twinned.

Scheelite $CaWO_4$	C—gray-white; darker in oil → Gangue, similar in air; lighter in oil B/P—not observed A—distinct but masked by internal reflections IR—common, white	*R*—9.9–10.0 9.7–9.9	VHN—387–409(f) PH < wolframite	Occurs as equant to lathlike polycrystalline aggregates, often as a partial replacement of wolframite. Also intergrown with Fe-oxides, huebnerite, ferberite, cassiterite. Fluoresces pale blue to yellow under ultraviolet light.
Schreibersite $(Fe,Ni)_3P$ Tetragonal	C—white in air; with brownish pink tint in oil → Cohenite, lighter → Iron, similar B/P—in oil distinct, pinkish brown to yellowish A—weak but distinct in oil IR—not present		VHN— PH > cohenite ~ iron	Occurs as oriented needle- and tablet-like inclusions in iron in meteorites.
Siegenite $(Co,Ni)_3S_4$ Cubic	C—creamy white with slight pink tinge → Cattierite, less pinkish B/P—not present A—isotropic IR—not present	*R*—45.4 47.2 QC—0.321 0.323 46.0	VHN—503–525(sf) PH ~ linnaeite	Occurs as euhedral and subhedral crystals and anhedral polycrystalline aggregates. Associated with Cu- and Cu-Fe sulfides, pyrite, vaesite, cattierite, uraninite.
Silver Ag Cubic	C—bright white with creamy tint; tarnishes rapidly → Antimony, arsenic,	*R*—94.2 95.0	VHN—55–63(p)	Occurs as irregular masses, veinlets, and inclusions and as dendrites within arsenides. In-

Note: Information is reported as follows:

1. Name 2. Formula 3. Crystal System	C—Color B/P—Bireflectance/Pleochroism A—Anisotropy IR—Internal Reflections	*R*—Reflectance at 546 and 589 nm in Air QC—Quantitative Color Coordinates	VHN—Vickers Microhardness at 100 g Load PH—Polishing Hardness	Mode of Occurrence; Other Characteristic Properties
	brighter and creamy B/P—not present A—isotropic; fine scratches often look anisotropic IR—not present	QC—0.314 0.321 94.2	PH > proustite, galena < tetrahedrite	complete extinction, tarnishes rapidly. Lamellar intergrowths with allargentum. Also with Ag-sulfosalts, Bi, argentite, galena, Cu-sulfides, Co-Fe-Ni arsenides.
Skutterudite $(Co,Ni)As_{2\text{-}3}$ Cubic	C—cream white to grayish white, often in zones → Cobaltite, white → Safflorite, yellowish B/P—not present A—isotropic; sometimes anomalous weak anisotropism IR—not present	*R*—54.4 53.8 QC—0.305 0.312 54.3	VHN—792–907(sf) PH ~ safflorite > linnaeite < arsenopyrite pyrite	Commonly and characteristically occurs as radial bladelike crystals with well-developed growth-zoning. Also as euhedral single crystals. May be intergrown with niccolite, bismuth, other Co-Fe-Ni arsenides; often present in Ag-Bi-U mineralization.

Sphalerite (Zn,Fe)S Cubic	C—gray, sometimes with brown tint → Magnetite, darker B/P—not present A—isotropic; sometimes weak anomalous anisotropism IR—common, yellow-brown to reddish brown	*R*—16.7 16.4 QC—0.303 0.309 16.6	VHN—218–227(p) PH > chalcopyrite, tetrahedrite < pyrrhotite, magnetite	Very common in many ore types. Occurs as irregular anhedral masses with pyrite, galena, chalcopyrite, pyrrhotite. Polishes well and is often featureless except for internal reflections. Also commonly contains rows of (or randomly dispersed) inclusions of chalcopyrite, pyrrhotite, galena, and, less commonly, stannite. Common growth-zoning of light and dark bands only visible in polished thin sections. Closely resembles magnetite except for internal reflections and absence of cleavage.
Stannite Cu_2FeSnS_4 Tetragonal	C—brownish olive-green → Tetrahedrite, darker brownish gray → Sphalerite, lighter, yellow-brown to olive green B/P—distinct, light brown to brown olive gray A—moderate, yellow-brown, olive green, violet gray IR—not present	*R*—27.3–26.0 27.3–26.1 QC—0.316 0.326 27.1	VHN—140–326 PH > chalcopyrite ~ tetrahedrite < sphalerite	Occurs as anhedral grains, granular aggregates and as oriented intergrowths with sphalerite, chalcopyrite, and tetrahedrite. Cleavage may be visible; compound twinning, sometimes in microcline pattern, common. In many ore types as a minor phase but common with bismuth and tungsten minerals.

Note: Information is reported as follows:

1. Name 2. Formula 3. Crystal System	C—Color B/P—Bireflectance/Pleochroism A—Anisotropy IR—Internal Reflections	*R*—Reflectance at 546 and 589 nm in Air QC—Quantitative Color Coordinates	VHN—Vickers Microhardness at 100 g Load PH—Polishing Hardness	Mode of Occurrence; Other Characteristic Properties
Stephanite Ag_5SbS_4 Orthorhombic	C—gray with pinkish violet tint → Galena, darker, pinkish → Polybasite, pyrargyrite, lighter B/P—weak but distinct, gray to pinkish gray A—strong in oil, violet to green IR—not present	*R*—30.4–28.1 29.7–27.5 QC—0.300 0.307 30.5	VHN—26–124 PH < tetrahedrite, > polybasite, pyrargyrite	Occurs as anhedral aggregates and euhedral columnar crystals. Compound twinning is common. Occurs with silver sulfosalts, Ni-Co-Fe arsenides, and common Cu-Fe sulfides.
Stibnite Sb_2S_3 Orthorhombic	C—white to grayish white → Bismuthinite, darker → Antimony, grayish B/P—strong, grayish white to white	*R*—31.1–48.1 30.6–45.2 QC—a 0.301 0.309	VHN—42–153 71–86 on (010) section (sf) PH > orpiment < chalcopyrite	Occurs as granular aggregates and lathlike crystals that often exhibit deformation textures, pressure twinning, and undulatory extinction. Associated

	A—very strong, often undulose, blue, gray, brown, pinkish brown IR—not present	41.8 b 0.306 0.317 30.6 c 0.294 0.305 47.3		with pyrite, pyrrhotite, sphalerite, chalcopyrite, and Sn, As, and Hg minerals.
Stromeyerite AgCuS Orthorhombic	C—gray with violet pinkish tint → Chalcocite, lavender gray B/P—weak but distinct in oil, gray brown to light gray with blue or pink tint A—strong, light violet, purple, brown, orange-yellow IR—not present	*R*—29.7–26.3 28.6–25.4 QC—0.286 0.285 30.1	VHN—30–32(sf) PH < galena, chalcocite	Occurs as a hypogene phase in granular aggregates and as a supergene phase in small veinlets. Often intergrown with other silver minerals, the common Cu-Fe and Fe sulfides, and sphalerite.
Sylvanite $(Au,Ag)Te_2$ Monoclinic	C—creamy white → Galena, lighter B/P—distinct, creamy white to brownish A—strong, light bluish gray to dark brown IR—not present	*R*—50.3–59.0 50.8–59.0	VHN—91–104(sf) PH > argentite < pyrargyrite	Occurs as skeletal blades. Well-developed cleavage and characteristic polysynthetic twins. Often intergrown with other gold-tellurides and associated with gold, galena, argentite, sphalerite, bornite, chalcopyrite, pyrite, Sb-, As- and Bi-sulfides.
Taenite See Iron				

Note: Information is reported as follows:

1. Name 2. Formula 3. Crystal System	C—Color B/P—Bireflectance/Pleochroism A—Anisotropy IR—Internal Reflections	*R*—Reflectance at 546 and 589 nm in Air QC—Quantitative Color Coordinates	VHN—Vickers Microhardness at 100 g Load PH—Polishing Hardness	Mode of Occurrence; Other Characteristic Properties
Tennantite $Cu_{12}As_4S_{13}$ Cubic (May contain Fe, Zn, Sb, etc.)	C—gray; sometimes with greenish or bluish tint → Galena, chalcocite, greenish → Pearcite, similar B/P—not present A—isotropic IR—common, reddish	*R*—30.0 29.8 QC—0.304 0.310 30.0	VHN—297–354(p) PH > galena ~ chalcopyrite < sphalerite	Forms complete solid solution with tetrahedrite. Occurrences the same as for tetrahedrite except in more As-rich environments.
Tenorite CuO Monoclinic	C—air: gray to gray white B/P—Oil: strongly pleochroic → Cuprite, brownish bluish → Chalcocite, brownish → Goethite, lighter, yellowish A—strong, blue to gray IR—not present	*R*—20.9–25.2	VHN—304–339 PH > chalcocite < goethite, cuprite	Occurs as aggregates of acicular crystals and as concentrically grown aggregates. May be twinned in lamellar fashion. Usually occurs with other oxides of Cu and Fe in weathering zone.

Tetradymite Bi_2Te_2S Trigonal	C—white with creamy tint → Chalcopyrite, lighter → Galena, yellowish B/P—weak A—distinct, bluish gray to yellow gray IR—not present	*R*—57.1–50.9 56.9–51.1 QC—0.311 0.318 56.9 0.312 0.319 50.8	VHN—25–76 PH > bismuth < galena	Occurs as tabular plates and granular aggregates. Basal cleavage common; twinning rare. Intergrowths with tellurobismuthinite, bismuth. Also occurs with common Cu-Fe and Fe sulfides, galena, gold, and Pb-Bi sulfosalts.
Tetrahedrite $Cu_{12}SbS_{13}$ Cubic (May contain Fe, Zn, Ag, As, Hg, etc.)	C—gray with olive or brownish tint → Galena, brownish or greenish → Chalcocite, blue-gray → Sphalerite, lighter B/P—not present A—isotropic IR—uncommon, increasingly common as As-content increases, reddish	*R*—30.5 30.0 QC—0.305 0.314 30.2 (note *R*% and color varies with composition)	VHN—285–322 (sf) PH > galena ~ chalcopyrite < sphalerite (note hardness varies with composition)	Forms complete solid solution with tennantite. Irregular masses of anhedral grains interstitial to common Cu-Fe-, Fe-sulfides, sphalerite, galena, arsenopyrite, and sulfosalts. Cleavages, twinning usually absent, but growth-zoning may be visible in thin section, especially in more As-rich members. Also occurs as rounded inclusions in galena and sphalerite.
Troilite See Pyrrhotite				
Ullmannite NiSbS	C—white with bluish tint → Gersdorffite, less yellow	*R*—46.4 45.7	VHN—536–592(sf)	Occurs as dispersed subhedral to euhedral crystals. Cleavage

Note: Information is reported as follows:

1. Name 2. Formula 3. Crystal System	C—Color B/P—Bireflectance/Pleochroism A—Anisotropy IR—Internal Reflections	*R*—Reflectance at 546 and 589 nm in Air QC—Quantitative Color Coordinates	VHN—Vickers Microhardness at 100 g Load PH—Polishing Hardness	Mode of Occurrence; Other Characteristic Properties
Cubic	→ Skutterudite, more yellow → Linnaeite, white B/P—not present A—isotropic IR—not present		PH > linnaeite ~ gersdorffite < pyrite	(100) may be visible and triangular cleavage pits occasionally seen. A minor phase in a variety of ores but usually associated with Cu-Fe sulfides and other Co-Fe-Ni antimonides and arsenides
Ulvöspinel Fe_2TiO_4 Cubic	C—brown to reddish brown → Magnetite, darker brown → //e of ilmenite, similar B/P—not present A—isotropic IR—not present	*R*—15.5 16.35	VHN— PH > magnetite	Usually observed as very fine dark isotropic exsolution lamellae in Ti-magnetite giving a "cloth-weave" texture. More rarely as octahedral crystals and as a matrix containing oriented cubes of magnetite. Associated with ilmenite and magnetite.

Uraninite UO_2, usually partly oxidized Cubic	C—brownish gray → Magnetite, less pink → Sphalerite, brownish B/P—not present A—isotropic IR—dark brown to reddish brown	*R*—13.6 13.6 QC—0.305 0.308 13.7	VHN—499–548(sf) (at 50g) PH > magnetite < pyrite	Occurs as growth-zoned crystals and as colloform, oolitic, and dendritic masses. (111) twinning common and (100) and (111) cleavage may occur. Often with pyrite, Cu-Fe sulfides, and other uranium minerals; may contain inclusions of gold.
Valleriite $(Fe,Cu)S_2 \bullet (Mg,Al)(OH)_2$ Hexagonal	C,B/P—very strong bireflectance and pleochroism, bronze to gray A—extreme, white to gray-bronze with satinlike texture IR—not present	*R*—14.2–22.0	VHN— PH > chalcopyrite ~ cubanite < pyrrhotite	Occurs as veinlets, interstitial fillings, and tiny inclusions in and around chalcopyrite, pyrrhotite, pentlandite, magnetite. Polishes poorly; has a characteristic bireflectance and pleochroism. The bronze anisotropy appears in a satinlike wavy pattern. Much confused with mackinawite, which tends to have a sharper extinction and less of an orange color or satinlike texture under crossed nicols.
Violarite $FeNi_2S_4$	C—brownish gray with violent tint	*R*—46.6	VHN—241–373	Most commonly occurs as a porous alteration product

Note: Information is reported as follows:

1. Name 2. Formula 3. Crystal System	C—Color B/P—Bireflectance/Pleochroism A—Anisotropy IR—Internal Reflections	*R*—Reflectance at 546 and 589 nm in Air QC—Quantitative Color Coordinates	VHN—Vickers Microhardness at 100 g Load PH—Polishing Hardness	Mode of Occurrence; Other Characteristic Properties
Cubic	→ Pentlandite, darker, violet tint → Pyrrhotite, lighter → Millerite, brownish violet B/P—not present A—isotropic IR—not present		PH > chalcopyrite, sphalerite ~ pentlandite < pyrrhotite	along grain boundaries and fractures of pentlandite, pyrrhotite, and millerite. Hypogene violarite occurs as equant anhedral grains with pyrite, millerite, pyrrhotite. Sometimes as fine lamellar intergrowths with millerite and chalcopyrite.
Wolframite $(Fe,Mn)WO_4$ Monoclinic	C—Air: gray to white Oil: gray with brown or yellow tint → Sphalerite, similar → Magnetite, darker → Cassiterite, lighter B/P—weak	*R*—15.0–16.2 14.7–15.9	VHN—312–342(sf) PH > magnetite, scheelite < pyrite, arsenopyrite	Occurs as euhedral platelets and as masses of interpenetrating laths. Cleavage distinct; twinning common. Often associated with scheelite, arsenopyrite, chalcopyrite, molybdenite, bismuth, bis-

	A—weak to distinct, yellow to gray IR—deep red, especially in oil			muthinite, gold, and cassiter-ite.
Zincite ZnO Hexagonal	C—pinkish brown B/P,A—masked by internal reflections IR—abundant, red to yellow-ish	*R*—11.8 11.4–11.5	VHN—190–219(sf) PH < franklinite, hausmannite	Occurs as rounded grains; cleavage (0001) may be visible. Forms oriented intergrowths with hausmannite. Associated with franklinite.

Appendix 2

Characteristics of Common Ore Minerals

Table A2.1 The Common Ore Minerals in Order of Increasing Minimum Reflectance in Air (at 546 nm)

Mineral	R%
Reflectance between 1 and 10%	
Graphite	6.8–17.4
Covellite	7.2–23.7
Coffinite	7.9–8.0
Scheelite	9.9–10.0
Chalcophanite	9.9–24.9
Reflectance between 10 and 20%	
Graphite	6.8–17.4
Covellite	7.2–23.7
Chalcophanite	9.9–24.9
Cassiterite	11.5–12.4
Lepidocrocite	11.6–18.4
Zincite	11.8
Chromite	12.3
Uraninite	13.6
Valleriite	14.2–22.0
Manganite	14.8–20.7
Brannerite	15.0–15.1
Wolframite group	15.0–16.2
Psilomelane	~15–30
Columbite-tantalite	15.3–17.3
Ulvöspinel	15.5
Goethite	15.5–17.5
Sphalerite	16.7
Ilmenite	17.0–20.1
Hausmannite	17.6–19.6
Braunite	18.6–19.5
Jacobsite	19.4
Molybdenite	19.5–38.5

Table A2.1 The Common Ore Minerals in Order of Increasing Minimum Reflectance in Air (at 546 nm)

Mineral	R%
Reflectance between 20 and 30%	
Covellite	7.2–23.7
Chalcophanite	9.9–24.9
Valleriite	14.2–22.0
Manganite	14.8–20.7
Psilomelane	~15–30
Molybdenite	19.5–38.5
Magnetite	20.0
Delafossite	~20–25
Tenorite	~20–25
Rutile	20.3
Bornite	21.3
Realgar	22.1
Mackinawite	~22–46
Bixbyite	22.2
Alabandite	22.8
Orpiment	23.0–27.5
Digenite	23.1
Famatinite	24–27.4
Enargite	24.2–25.2
Maghemite	24.4
Cinnabar	24.6–29.6
Stannite	26.0–27.3
Hematite	26.1–30.2
Stromeyerite	26.3–29.7
Cuprite	26.6
Mawsonite	26.9–29.7
Idaite	27–33.6
Proustite	27.4–28.1
Stephanite	28.1–30.4
Pyrargyrite	28.5–30.3
Pyrolusite	29.0–40.0
Freibergite	29.4
Tennantite	30.0
Reflectance between 30 and 40%	
Molybdenite	19.5–38.5
Mackinawite	~22–46
Hematite	26.1–30.2
Stephanite	28.1–30.4
Pyrargyrite	28.5–30.3
Pyrolusite	29.0–40.0
Tennantite	30.0
Acanthite	30.3–31.3
Tetrahedrite	30.5

Table A2.1 The Common Ore Minerals in Order of Increasing Minimum Reflectance in Air (at 546 nm)

Mineral	R%
Polybasite	30.7–32.5
Bravoite	~31–52
Stibnite	31.1–48.1
Miargyrite	31.6–34.5
Pearcite	31.9
Chalcocite	33.1–33.4
Pyrrhotite	34.0–39.9
Cubanite	35.4–39.4
Berthierite	36.6–42.0
Chalcostibite	36.9–42.2
Breithauptite	36.9–48.2
Boulangerite	37.4–41.8
Jamesonite	37.4–42.9
Bismuthinite	38.5–45.4
Kobellite	40.0–46.1
Reflectance between 40 and 50%	
Mackinawite	~22–46
Pyrolusite	29.0–40.0
Bravoite	~31–52
Stibnite	31.1–48.1
Berthierite	36.6–42.0
Chalcostibite	36.9–42.2
Breithauptite	36.9–48.2
Jamesonite	37.4–42.9
Boulangerite	37.4–41.8
Bismuthinite	38.5–45.4
Kobellite	40.0–46.1
Cosalite	41.4–45.7
Galena	43.1
Chalcopyrite	44.6–45.0
Carrollite	45
Siegenite	45.4
Ullmannite	46.4
Pentlandite	46.5
Violarite	46.6
Niccolite	47.2–51.6
Maucherite	47.8–48.5
Marcasite	48.2–55.8
Linnaeite	49.6
Reflectance between 50 and 60%	
Bravoite	~31–52
Niccolite	47.2–51.6
Marcasite	48.2–55.8

Table A2.1 The Common Ore Minerals in Order of Increasing Minimum Reflectance in Air (at 546 nm)

Mineral	R%
Glaucodot	50.0–50.6
Millerite	50.2–56.6
Sylvanite	50.3–59.0
Cobaltite	50.5
Tetradymite	50.0–57.1
Arsenic	51.3–56.4
Pyrite	51.7
Arsenopyrite	51.8–52.2
Loëllingite	52.4–54.1
Rammelsbergite	53.2–56.3
Gersdorffite	54.2
Skutterudite	54.4
Safflorite	~55–60
Pararammelsbergite	56.3–57.8
Iron	57.7
Reflectance over 60%	
Copper	60.6
Dyscrasite	62.1–63.2
Bismuth	66.7
Allargentum	~70
Platinum	70.3
Antimony	71.1–73.0
Gold	71.5
Silver	94.2

TABLE A2.2 The Common Ore Minerals in Order of Increasing Minimum Vickers Microhardness (at 100 g load)

Mineral	VHN
VHN from 1 to 100	
Molybdenite	8–100 (35//cleav)
Graphite	12–16
Bismuth	15–18
Orpiment	22–58
Acanthite	23–26
Stephanite	26–124
Stromeyerite	30–32
Stibnite	42–153 (71–86 on 010)

TABLE A2.2 The Common Ore Minerals in Order of Increasing Minimum Vickers Microhardness (at 100 g load)

Mineral	VHN
Realgar	47–60
Gold	53–58
Silver	55–63
Galena	59–65
Digenite	67–76
Cosalite	74–161
Mackinawite	~74–181
Cinnabar	82–156
Arsenic	83–149
Chalcocite	84–87
Antimony	84–98
Miargyrite	88–130
Sylvanite	91–104
Boulangerite	92–125
Bornite	95–105
Copper	96–104
VHN from 100 to 200	
Stephanite	26–124
Stibnite	42–153
Cosalitc	74–161
Mackinawite	~74–181
Cinnabar	82–156
Arsenic	83–149
Miargyrite	88–130
Sylvanite	91–104
Boulangerite	92–125
Bornite	95–105
Kobellite	100–117
Copper	96–104
Berthierite	102–213
Proustite	103–137
Pyrargyrite	107–144
Bismuthinite	110–136
Jamesonite	113–117
Platinum	122–129
Covellite	128–138
Stannite	140–326
Dyscrasite	146–160
Pyrolusite	146–243
Iron	158(av.)
Mawsonite	166–210
Idaite	176–260
Pearcite	180–192

TABLE A2.2 The Common Ore Minerals in Order of Increasing Minimum Vickers Microhardness (at 100 g load)

Mineral	VHN
Chalcopyrite	181–203
Chalcophanite	188–253
Zincite	190–219
Millerite	192–376
Cuprite	193–207
VHN from 200 to 300	
Berthierite	102–213
Stannite	140–326
Pyrolusite	146–243
Mawsonite	166–210
Idaite	176–260
Chalcopyrite	181–203
Chalcophanite	188–253
Zincite	190–219
Millerite	192–376
Cuprite	193–207
Psilomelane	203–813
Famatinite	205–397
Sphalerite	218–227
Coffinite	230–302
Pyrrhotite	230–409
Alabandite	240–251
Columbite-tantalite	240–1021
Violarite	241–373
Cubanite	247–287
Freibergite	252–375
Pentlandite	268–285
Chalcostibite	283–309
Enargite	285–327
Tetrahedrite	285–322
Safflorite	285–464
Tennantite	297–354
VHN from 300 to 400	
Stannite	140–326
Millerite	192–376
Psilomelane	203–813
Famatinite	205–397
Coffinite	230–302
Pyrrhotite	230–409
Columbite-tantalite	240–1021
Violarite	241–373
Freibergite	252–375
Chalcostibite	283–309

TABLE A2.2 The Common Ore Minerals in Order of Increasing Minimum Vickers Microhardness (at 100 g load)

Mineral	VHN
Enargite	285–327
Tetrahedrite	285–322
Safflorite	285–464
Tennantite	297–354
Wolframite group	312–342
Niccolite	363–372
Scheelite	387–409
VHN from 400 to 500	
Psilomelane	203–813
Pyrrhotite	230–409
Columbite-tantalite	240–1021
Safflorite	285–464
Scheelite	387–409
Lepidocrocite	402 (av.)
Maghemite	412 (av.)
Breithauptite	412–584
Loellingite	446–560
Linnaeite	450–613
Uraninite	499–548
VHN from 500 to 600	
Psilomelane	203–813
Columbite-tantalite	240–1021
Breithauptite	412–584
Loellingite	446–560
Linnaeite	450–613
Uraninite	499–548
Siegenite	503–525
Carrollite	525–542
Hausmannite	536–566
Ullmannite	536–592
Rammelsbergite	585–803
Magnetite	592 (av.)
VHN from 600 to 700	
Psilomelane	203–813
Columbite-tantalite	240–1021
Linnaeite	450–613
Rammelsbergite	585–803
Ilmenite	659–703
Jacobsite	665–707
Goethite	667 (av.)
Bravoite	668–1535
Brannerite	690 (av.)
Manganite	698–772

TABLE A2.2 The Common Ore Minerals in Order of Increasing Minimum Vickers Microhardness (at 100 g load)

Mineral	VHN
VHN from 700 to 800	
Psilomelane	203–813
Columbite-tantalite	240–1021
Rammelsbergite	585–803
Ilmenite	659–703
Jacobsite	665–707
Bravoite	668–1535
Manganite	698–772
Maucherite	715–743
Pararammelsbergite	762–792
Gersdorffite	782–835
Skutterudite	792–907
VHN from 800 to 900	
Psilomelane	203–813
Columbite-tantalite	240–1021
Rammelsbergite	585–803
Bravoite	668–1535
Gersdorffite	782–835
Skutterudite	792–907
VHN from 900 to 1000	
Columbite-tantalite	240–1021
Bravoite	668–1535
Arsenopyrite	715–1354
Skutterudite	792–907
Cobaltite	935–1131
Bixbyite	946–1402
VHN over 1000	
Columbite-tantalite	240–1021
Bravoite	668–1535
Arsenopyrite	715–1354
Cobaltite	935–1131
Bixbyite	946–1402
Braunite	1027–1225
Hematite	1038 (av.)
Glaucodot	1097–1115
Rutile	1132–1187
Cassiterite	1168–1332
Marcasite	1288–1681
Chromite	1332
Pyrite	1505–1620

Appendix 3

Ancillary Techniques

Two ancillary techniques—X-ray powder diffraction and electron probe microanalysis—are very commonly employed in conjunction with reflected-light microscopy in order to obtain more detailed information on the identities and compositions of phases. Discussion of the principles and detailed applications of these methods is inappropriate here, but the techniques are so important that this appendix has been included to suggest suitable references dealing with these topics in detail and to draw attention to some practical problems in applying them to polished sections.

X-RAY POWDER DIFFRACTION

The theory and applications of this technique are described in numerous texts and papers (e.g., Zussman, 1977; Azaroff and Buerger, 1958; Nuffield, 1966). All X-ray diffraction methods result from the diffraction of a beam of X-rays by a crystalline material. The precise method most often used in conjunction with reflected-light studies is X-ray powder photography using the Debye-Scherrer camera. Here the powdered sample can be a small "bead" only a fraction of a millimeter in diameter mounted on the tip of a fine glass fiber. When the sample is mounted in the camera, it is aligned so that a pencil beam of X-rays entering through a collimator strikes the sample and is diffracted to produce a pattern of lines on a strip of photographic film. The positions of these lines can then be measured and the information converted to the separation between layers of atoms in the crystal structure. The X-ray powder pattern provides in many cases a "fingerprint" identification of the material. In certain cases precise measurement of the pattern may also furnish compositional information [e.g., the iron content of a (Zn,Fe)S sphalerite, the Fe:S ratio of a pyrrhotite].

X-ray powder camera methods are particularly useful in ore microscopy because of the very small amount of material required. Small grains can be dug directly out of a polished section using a needle or special diamond pin "objective" and rolled up into a ball of collodion, which is then picked up on the end of a glass fiber and mounted in the camera.

ELECTRON PROBE MICROANALYSIS

The second technique also has been described in numerous texts and papers of which Long (1977), Reed (1975), and Smith (1976) serve as good examples for the mineralogist. A beam of electrons, generated in a potential field of 10–40 kV, strikes the specimen as a spot that may be as small as 1 μm in diameter; the spot can be maintained stationary or made to scan rapidly over an area up to several thousand square microns. The electron beam excites the emission of X-rays characteristic of the elements present from the sample, thus enabling the composition of the sample to be determined by analysis of the energies and intensities of the emitted X-rays. A quantitative analysis of a mineral for a particular element is obtained by comparing the intensity (i.e., count rate) of the characteristic X-rays from the sample with the intensity of radiation of the same energy generated by a standard of known composition. Consequently, a complete chemical analysis can be obtained for a spot that may be as little as a few microns in diameter. Concentrations from major elements down to 0.1 wt. % can be determined with an accuracy of a fraction of 1%. Most electron probes can rapidly scan areas of several thousand square microns and the backscattered electrons can be monitored and used to produce a magnified image (up to 100,000 ×) of the surface of the specimen. The characteristic X-rays emitted from this area of the surface can also be monitored and used to map out the distribution of elements within and between phases (see Figure A3.1).

A great advantage of the electron probe technique for the ore microscopist is that most instruments directly accept a standard size polished or polished thin section. The only other preparation usually needed is to apply (by sputtering under vacuum) a thin (∾200 Å) coat of carbon to the specimen surface in order to conduct away the charge. Therefore, samples can be studied under the ore microscope and drawings or photographs taken of interesting areas that can then be subjected to complete chemical analysis in the electron probe. The electron probe has revolutionized ore mineralogy by enabling the complete chemical analysis of very small grains, including many new minerals (e.g., platinum group metal minerals), the products of laboratory synthesis experiments aimed at establishing phase relations in ore mineral systems, and optically identifiable minerals that may contain economically important elements in solid solution (e.g., silver within the lattice of tetrahedrite).

A comprehensive treatment of the integration of the ore microscope with the electron microprobe has recently been prepared by Gasparrini (1980).

References

Azaroff, L. V. and Buerger, M. J. (1958) *The Powder Method in X-ray Crystallography.* McGraw-Hill, New York.

Gasparrini, C. (1980) The role of the ore microscope and electron microprobe in the mining industry. *CIM Bull. 73,* 73–85.

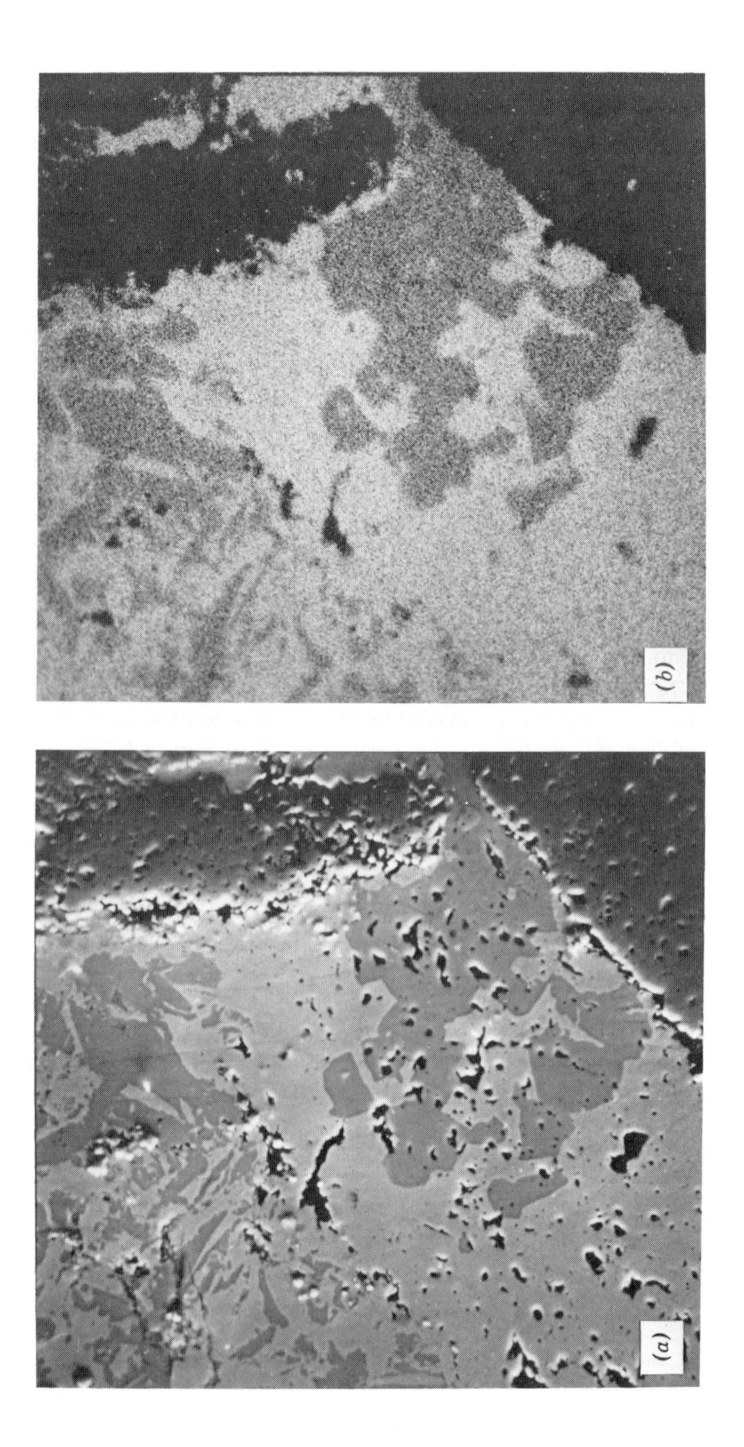
(b)
(a)

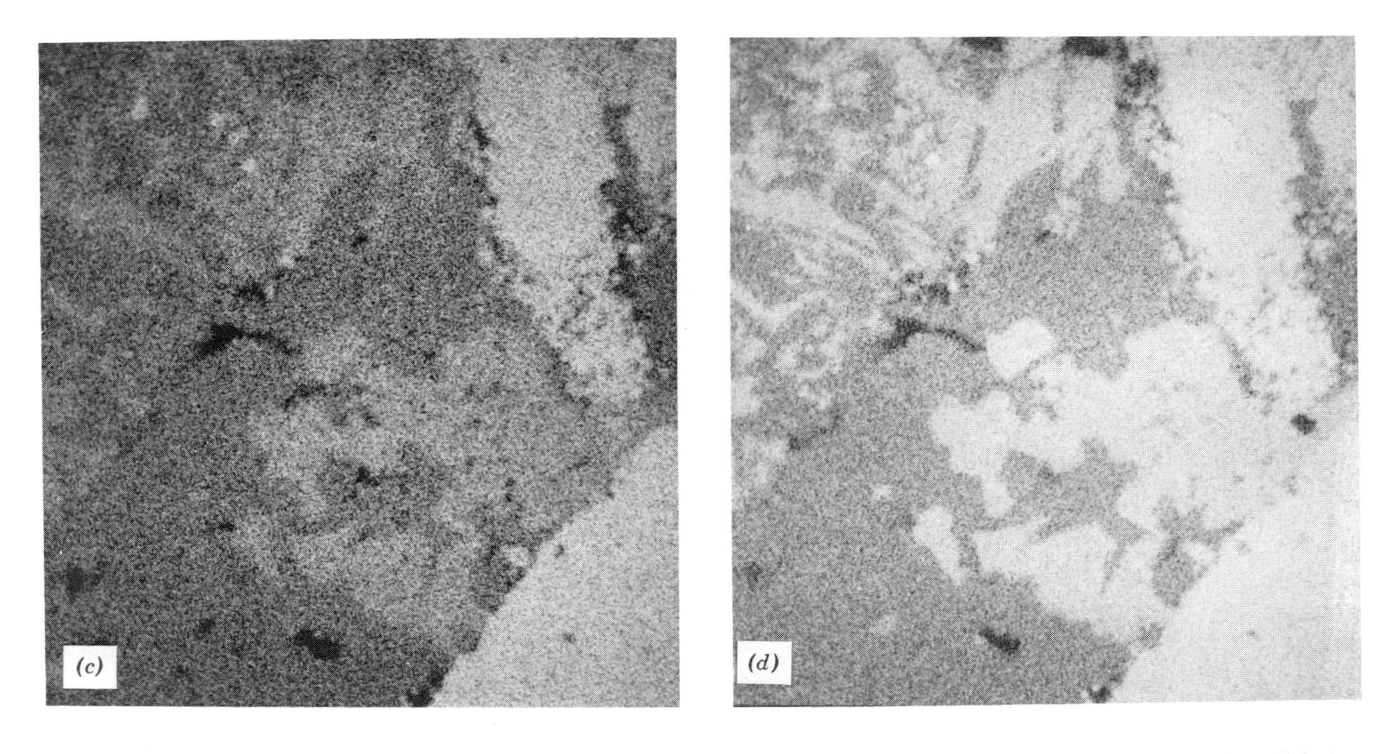

FIGURE A3.1 Select area photographs demonstrating the use of the electron microprobe in visually identifying minerals in polished surfaces: (**a**) back-scattered electron photograph of pyrite (upper right and lower right), chalcopyrite (dark gray) and bornite (light gray); (**b**) copper image (lighter color indicates more copper present); (**c**) sulfur image (lighter color indicates more sulfur present); (**d**) iron image (lighter color indicates more iron present). (Width of field = 1000 μm). (We are indebted to Todd Solberg for providing these photographs).

Long, J. V. P. (1977) Electron probe microanalysis. In J. Zussman, ed., *Physical Methods in Determinative Mineralogy*. Academic, New York.

Nuffield, E. W. (1966) *X-ray Diffraction Methods*. Wiley, New York.

Reed, S. J. B. (1975) *Electron Microprobe Analysis*. Cambridge University Press, Cambridge, England.

Smith, D. G. W., ed. (1976) *Microbeam Techniques*. Min. Assoc. Canada Short Course Handbook 1.

Zussman, J. (1977) X-ray diffraction. In J. Zussman, ed., *Physical Methods in Determinative Mineralogy*. Academic, New York.

Appendix 4

Plot of Common Ore Minerals in Terms of Reflectance and Vickers Hardness Number

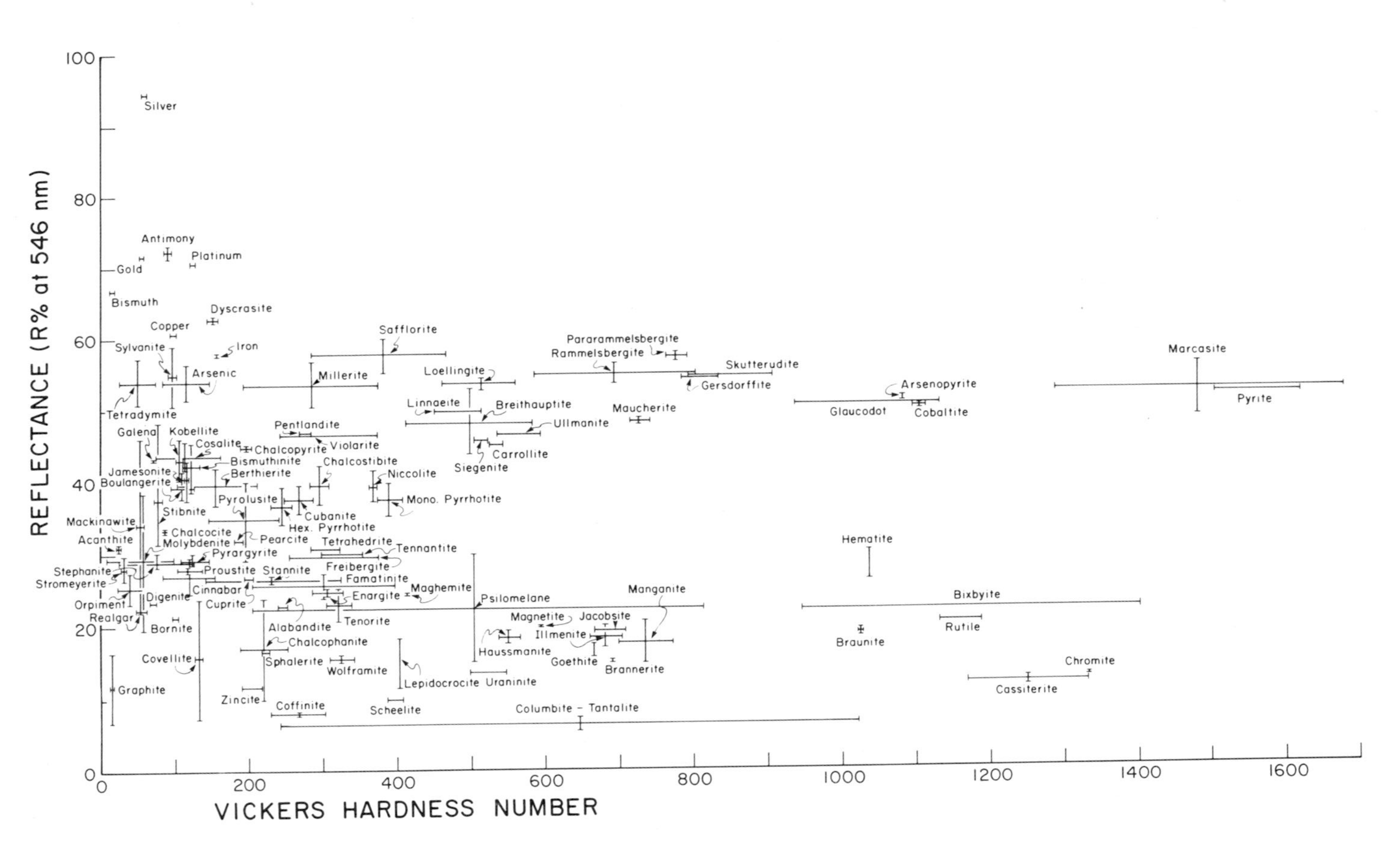
REFLECTANCE (R% at 546 nm)
VICKERS HARDNESS NUMBER
0
20
40
60
80
100
200
400
600
800
1000
1200
1400
1600
Silver
Antimony
Platinum
Gold
Bismuth
Dyscrasite
Copper
Iron
Sylvanite
Arsenic
Tetradymite
Galena
Kobellite
Cosalite
Chalcopyrite
Bismuthinite
Berthierite
Jamesonite
Boulangerite
Pyrolusite
Stibnite
Mackinawite
Chalcocite
Molybdenite
Acanthite
Pyrargyrite
Proustite
Stephanite
Stromeyerite
Digenite
Cinnabar
Orpiment
Realgar
Cuprite
Bornite
Covellite
Graphite
Zincite
Coffinite
Alabandite
Chalcophanite
Sphalerite
Wolframite
Scheelite
Safflorite
Millerite
Pentlandite
Violarite
Chalcostibite
Cubanite
Hex. Pyrrhotite
Pearcite
Tetrahedrite
Tennantite
Stannite
Freibergite
Famatinite
Enargite
Maghemite
Tenorite
Niccolite
Mono. Pyrrhotite
Loellingite
Linnaeite
Breithauptite
Ullmanite
Carrollite
Siegenite
Lepidocrocite
Uraninite
Psilomelane
Magnetite
Jacobsite
Ilmenite
Haussmanite
Goethite
Brannerite
Manganite
Columbite - Tantalite
Pararammelsbergite
Rammelsbergite
Skutterudite
Gersdorffite
Maucherite
Arsenopyrite
Glaucodot
Cobaltite
Hematite
Bixbyite
Rutile
Braunite
Chromite
Cassiterite
Marcasite
Pyrite

Author Index

Subject Index